Stability Theory

Stability Theory

2nd Edition

An Introduction to the Stability of Dynamic Systems and Rigid Bodies

Horst Leipholz

University of Waterloo
Waterloo, Ontario, Canada

B. G. TEUBNER
Stuttgart

JOHN WILEY & SONS
Chichester · New York · Brisbane · Toronto · Singapore

Copyright © 1987 by John Wiley & Sons Ltd. and
B. G. Teubner, Stuttgart.

CIP-Kurztitelaufnahme der Deutschen Bibliothek:
  Leipholz, Horst:
  Stability theory : an introd. to the
  stability of dynam systems and rigid H.
  bodies/Horst Leipholz. — 2nd ed. —
  Stuttgart : Teubner ; Chichester ; New York;
  Brisbane; Toronto; Singapore : Wiley, 1987.
    ISBN 3 519 02105 6 (Teubner)
    ISBN 0 471 91181 X (Wiley)

Library of Congress Cataloging-in-Publication Data:

Leipholz, H. H. E. (Horst H. E.), 1919–
  Stability theory.

  Translation of: Stabilitätstheorie.
  Includes index.
  1. Stability. I. Title
QA871.L47313   1987   531   86-15695
ISBN 0 471 91181 X

British Library Cataloguing in Publication Data:

Leipholz, H.
  Stability theory : an introduction to the
  stability of dynamic systems and rigid
  bodies. — 2nd ed.
  1. System analysis 2. Stability
  I. Title
  003 QA402

  ISBN 0 471 91181 X

Leipholz, *Stability Theory 2nd Edition*

ISBN 0 471 91181 X (Wiley)
ISBN 3 519 02105 6 (Teubner)

ERRATA

p. 7   lines 9 and 12 *for C read C_r*

p. 9   line 3, *for $>$ read $<$*

p. 12   line 21, *for* state *read* stated

p. 21   eqn (1.2.5), *for $u_{k1}\beta_1$ read $u_{kl}\beta_l$*

p. 22   eqn (1.2.6), *for ξ_1 read ξ_l*

        eqn (1.2.8), *for u_{1j} read u_{lj}*

        eqn (1.2.10), *for $\dfrac{\partial F_i}{\partial q_i}(q_k^0, \alpha_r^0, t)\xi_1$ read $\dfrac{\partial F_i}{\partial q_l}(q_k^0, \alpha_r^0, t)\xi_l$*

p. 24   line 10, *for ∂q_i read ∂q_j*

        lines 12 and 14, *for $q_{i,s}^0$ read $q_{j,s}^0$*

p. 39   line 3, *for $\dot{\xi}_2 = -\xi_1 + \xi_2 - \xi_3$ read $\dot{\xi}_2 = -\xi_1 + \xi_2 - 3\xi_3$*

p. 49   line 13, *for $\alpha_{rk}\xi_i^{(r)}(0)$ read $a_{rk}\xi_i^{(r)}(0)$*

p. 57   3 lines above eqn (1.3.29), *for t read τ*

p. 58   line 6, *for $\xi_{ik}^{(1)}$ read $\xi_{1k}^{(1)}$*

p. 59   Ex. 1 (a), *for a_k read a_{ik}*

p. 60   Ex. 5, line 4, for y_3 *read $\dot{\xi}_3$*

        Ex. 7, line 2, *for $g(x) \equiv 0$ read $g(x) \not\equiv 0$*

p. 63   footnote, *for k read h*

p. 68   last line, *for* cases pending *read* corresponding

p. 82   line 10, *for ξ_1^2 read ξ_1^3*

p. 84   line above eqn (1.5.8), *for B_f read B_j*

        2 lines below eqn (1.5.8), *for* coefficients α_{ik} *read* coefficients a_{ik}

p. 86   Ex. 1, line 3, *for ξ_3^3 read ξ_3^2*

p. 89   2 lines above eqn (1.6.3) and also in eqn (1.6.3), *for x_i* read $\dot{x}_i$

p. 94   eqn (1.6.39), *for $\dfrac{\partial p_j}{\partial q_i}$ read $\dfrac{\mathrm{d}p_j}{\mathrm{d}q_i}$*

p. 95   eqn (1.6.41), *for p read p_j*

p. 96   9 lines up, *for (1.6.46) read (1.6.34)*

p. 98   last line, *for ψ^2 read ω^2*

p. 101   2 lines after eqn (1.6.85), *for* ω^2 *read* ω_0^2

p. 109   3 lines up, *for* $\cos^4 \phi$ *read* $\cos^3 \phi$

p. 115   line after eqn (1.7.32), *for* $\dot{\xi}_{ir}$ *read* ξ_{ir}

           3 lines after eqn (1.7.32), *for* $\dot{\xi}_{i1}$ *read* ξ_{i1}

p. 116   eqn (1.7.35), *for* $\dot{\xi}_1$ *read* $\dot{\xi}_i$

           eqn (1.7.36), *for* $\dot{\xi}_1$ *read* $\dot{\xi}_i$

p. 120   3 lines above Section 1.7.5, *for* t_0 *read* t

p. 126   last line, *for* ϕk *read* ϕ_k

p. 128   8 lines after eqn (1.7.62), *for* $\mu w\ \delta \dot{w}$ *read* $\mu \dot{w}\ \delta w$

p. 136   eqn (1.7.81), *for* $\displaystyle\int_0^T f[\chi(\xi)\ \mathrm{d}\xi = 0$ *read* $\displaystyle\int_0^T f[\chi(\xi)]\ \mathrm{d}\xi = 0$

p. 138   line 4, *for* b_2^* *read* b_1^*

p. 140   2 lines up, *for* $\sin \omega_1$ *read* $\sin \omega_1 t$

p. 141   Ex. 4, line 4, *for* w^N *read* w^{IV}

p. 145   eqn (2.1.2), *for* $f(r) = (mM\gamma/r^2)$ *read* $f(r) = -(mM\gamma/r^2)$

           2 lines above eqn (2.1.4), *for* $\partial\dot{\phi}L/\partial$ *read* $\partial\dot{\phi}L/\partial\dot{\phi}$

p. 147   eqn (2.1.13) *for* r_n *read* r^n

           2 lines below eqn (2.1.14), *for* the (2.1.13) are unstable for $n < 3$ *read* the form (2.1.13) are unstable for $n > 3$

p. 154   10 lines up, *for* $U^2 yy$ *read* U_{yy}^2

p. 158   Ex. 2, line 8, *for* k *read* h

p. 167   eqn (2.2.6), *for* x_{ik} *read* x_{1k}

p. 173   line after eqn (2.2.16), *for* $F(\sigma) > 0$ *read* $\sigma F(\sigma) > 0$

p. 176   line 17, *for*
$$V = B(A-B)\xi_2^2 + C(A-C)\xi_2^3 + [B\xi_2^2 + C\xi_3^2 + A(\xi^2 + 2\xi_1\omega_{x0})^2$$
read
$$V = -B(A-B)\xi_2^2 - C(A-C)\xi_2^3 + [B\xi_2^2 + C\xi_3^2 + A(\xi^2 + 2\xi_1\omega_{x0})]^2$$
line 23, *for* ξ_2^3 *read* ξ_3^3; *for* ω_{z0} *read* ω_{x0}
last line (above footnote), *for* (2.1.54) *read* (2.2.27)

p. 180   eqn (2.2.23), *for* $\cong$ *read* $=$

p. 182   line 8, *for* $\dot{\xi}_1 = -(1/\alpha v_0^2)$ *read* $\dot{\xi}_1 = -(1/\alpha v_0^2)\xi_2$

p. 183   line above eqn (2.2.41), *for* $A\dot{\psi} \cos \theta$ *read* $A\dot{\psi}\ \dot{\theta} \cos \theta$

p. 186   7 lines below eqn (2.2.51), *for* $J_F + m_R l^2)\lambda^2$ *read* $(J_F + m_R l^2)\lambda^2$

p. 187   eqn (2.2.56), *for* $J_R c_{21} c_{21}$ *read* $J_R c_{12} c_{21}$

p. 190   3 lines below eqn (2.2.68), *for* ζ_x *read* γ_x

Table of Contents

Preface to the First Edition

There have been great advances in theory of stability in recent decades due to the requirements of control theory and flight mechanics, for example. We need only mention the theory of A. M. Lyapunov. A number of specialists have given a very mathematical and abstract description of the Lyapunov stability theory which resulted in a 'stability theory of motion' applicable to the kinetics of rigid bodies and systems. The stability theory of elastomechanics was developed independently. However, there have been a number of important developments in recent years, also with respect to this theory, dealing with the following problems: The concept of the 'follower forces', non-conservative loads, respectively, has been introduced in aeroelasticity. A number of problems in elastic kinetics that involve pulsating loads or periodically varying parameters has led to new stability questions. So-called 'kinetic' methods have become necessary in elastomechnics in order to determine the stability boundaries. An evaluation of the stability criteria of elastostatics, which have been assumed to be generally valid, has shown that they can only be applied to a limited number of problems under special assumptions. The transition from stability to instability is a kinetic process in elastomechanics. Therefore, the most general and most certain method of determining stability is the *kinetic stability criterion* even if in special cases the classical stability criteria of elastostatics may remain valid. This will be discussed in detail in Section 2.3.

On this basis, it is possible to go beyond the original framework of the theory and treat stability problems of kinetics of rigid bodies and systems as well as of elastomechanics using a single, unified theory. One of the primary purposes of the author was to show the *common traits* of stability problems which occur in all branches of mechanics. The stability problems of hydromechanics and aeromechanics were omitted, because they are discussed in separate lectures, outside of technical mechanics lectures in our schools. However, the interested and knowledgeable reader may readily apply the methods for determining stability also to stability problems of continuum mechanics of liquid and gaseous media.

Stability theory is primarily expressed in mathematical terms. For purposes of brevity, it was assumed that the reader is familiar with certain basic mathematical methods. In order to ensure that the reader is able to pick up the requisite mathematical knowledge, several references are given in an appendix. For example, in the section on mathematical approximation methods used to solve non-self-adjoint or non-linear problems, the Galerkin method was discussed in detail.

The Einstein summation convention was used in most cases, unless this was not possible or not appropriate. The usual index notation was used in vector and tensor calculus.

I would like to express particular appreciation to Professor Dr. K. Magnus, Stuttgart, who inspired me to write the book and who aided me with many valuable suggestions. I am indebted to my former assistant, Mr. Dipl.-Ing. H. J. Hoffmann, for his critical remarks. Dipl. Math. W. Hauger made the drawings and performed a number of necessary calculations. Mrs. L. Döttinger and Miss G. Walter worked with great dedication in completing the manuscript.

Preface to the Second Edition

Over the past decades, interest in new stability problems in elastomechanics and a redesigned theory of stability for that area have made a second edition of the book desirable and even necessary. This opportunity has been used by the author to revise and expand certain parts of the book intensively, so as to take essential developments into account which have led to a remarkable reformulation of the stability theory of elastomechanics.

Thanks are due to B. G. Teubner-Verlag, Stuttgart, for making the new edition possible. Also, I am indebted to Mrs. G. Moseley for preparing the typescript of the book in its new form in a competent and dedicated manner.

PART 1

Fundamentals

1.1
Basic Concepts and Definitions

There is no absolute definition of stability. It has continuously evolved during the course of history, and has always been adjusted to the special requirements of particular problems. The multilateral meanings of the stability concept must be studied in detail, as well as the various methods of determining stability, before stability theory proper in mechanics may be examined. Previous studies on the subject are either monographs devoted to questions of stability in a more or less restricted region of mechanics (1), or they comprise chapters of special books dealing with particular problems in mechanics (2), so that in this case the concept of stability is studied with respect to the particular problem in question. The majority of the cited references contain only *one* definition of stability as well as more or less extensive generalizations of it. If the different definitions included in these works are compared, the features they have in common are not always readily apparent.

An attempt will be made in this book to present the features that the various stability definitions have in common. We shall trace the bifurcations of the stability concept and the extent to which the various nuances of the concept have penetrated into the different branches of mechanics. In this way we may clarify the relations between the various points of view, as well as the manner in which they are interwoven, and thus arrive at a useful synthesis.

1.1.1 Historical Development of the Stability Concept

Based on an article by Magnus[1], it may be established that stability studies have their beginnings in history in the works of Aristotle and Archimedes. Two different points of view were expressed by these two men. The former author investigated the motion occurring after a perturbation, and he determined the stability of the unperturbed state from the *course of the motion*. In the latter case, the *purely geometric situation* occurring after the perturbation of a system was used to determine the stability of the unperturbed system.

[1]Magnus, K., Development of the stability concept in mechanics, *Naturwissenschaften*, **46**, pp. 590–595, 1959.

There are therefore two methods. The first is called the kinematic method. Beginning with Aristotle, it was used up until the period of Galileo. Except for astronomy, which used it continuously to investigate the stability of motion, it has again increased in importance with respect to the stability of motion of technical systems. The second method is called the geometric method. It is connected with the names of Archimedes, Torricelli, Baldi, and Lagrange and dominated the field of mechanics for an extended period of time, from about the 17th to the 19th century greatly overshadowing the first method. As Magnus pointed out, this was of little benefit to the general storehouse of knowledge. The energy method represents a third method. Energy criteria are used here to determine the stability of an equilibrium position. This method, which is of great importance in the theory of elastic bodies (the numerous results obtained by S. Timoshenko when using this method need only be recalled), also contributed to the development of a stability theory which appears to be almost complete but which, in its classical form, turned out to have only limited regions of applicability. However, it must be pointed out that E. J. Routh made an attempt to generalize this method. He tried to extend the energy criteria to an investigation of the stability of states of motion. This was the first attempt to establish a unified stability theory for all branches of mechanics. However, it was not completely successful. Even today the various methods exist separately, and independent theories of stability have been formulated, depending on whether dynamics is being studied — in which the kinematic method is predominant — or the theory of elasticity and statics is being studied, where the geometric–static and energy methods are used in most cases. At the present time, the foundations of the latter are being subjected to a critical examination, which will be discussed at a later point.

Just as the methods have led to a division between the theories, the various stability definitions have contributed to the division. Even if the multiple meanings of the stability concept in everyday usage are disregarded (durability, constancy, steadiness, immobility, etc.), numerous possibilities of interpretation remain even if the precise definitions of technical physics are used. It must first be stressed that the concept is relative and depends on the reference system employed. Inexact formulations can lead to a great deal of confusion. It is also very important to determine which characteristics are used for determining the stability of a system. Some well-known examples of this are given by the definition of Lagrange, who states that the trajectory of a planet is stable if its major axis remains bounded in spite of perturbations. On the other hand, we have the definition of Poisson stipulating that a planet is stable when it repeatedly passes arbitrarily close to every point of its motion. These represent two different opinions regarding the same subject: stability of a planet! How many more different stability concepts will there be when different problems are considered! The stability concept of Lyapunov has found wide acceptance recently and will be discussed in detail later on. However, in spite of a desire for unification, this concept is not all-inclusive, as it is frequently too restricted. Features that these concepts have in common, which will be different for each case and must be justified in each individual case, must be sought below the surface. The manner in which this may be done will be discussed.

1.1.2   Common Characteristics of Modern Stability Definitions

Let us first present the common characteristics of all stability definitions. They consist of the following stipulations and concepts. An unperturbed state, whose stability is being studied, is specified. A perturbation is then applied to the unperturbed state, so that it is transformed into a perturbed state. Certain quantities are emphasized, which we shall call norms, which characterize the states at any desired time. The change in the norms during the transition from the unperturbed state to the perturbed state under the influence of the perturbation is determined. Based on this behaviour, a conclusion may be reached regarding the stability of the unperturbed state or its instability. In order to define the change in the norms, a measure for the perturbation is defined, as well as a measure for the norm. The following definition is used: If the perturbation does not exceed this defined measure, the unperturbed state is called stable when the change in the norm caused by the perturbation does not exceed its established measure. Otherwise, the unperturbed state is unstable.

The manner in which a great number of different, special stability concepts emerge from this small number of simple stipulations will now be illustrated:

A certain reference system is established when the unperturbed state is selected. Thus, the relative nature of the stability concept becomes apparent, which was already mentioned above. The perturbed state, as well as what is commonly designated as stability, has a direct relationship to the unperturbed state selected and its reference system.

In addition, the choice of the norm is an important factor in reaching a conclusion regarding stability. There are again differences in the stability concept, depending on the choice of the norm. As a specific example of this, we would like to point to the opinions of Lagrange and Poisson regarding stability of planet trajectories that do not agree, as was already mentioned in Section 1.1.1. As a more abstract example, we shall select the following as the norm: In a metric space $\mathscr{R}$ with elements f, g and the metric d, let the point set $\mathscr{M}\{f_k\}$, $= 1, 2, 3, \ldots$, with elements f_k from $\mathscr{R}$ be given. Let $\mathscr{M}\{f_k\}$ characterize a state that we shall assume is an unperturbed state. A sphere $\mathscr{K}_r(f_k) = \{f/d(f, g, f_k) > r\}$ corresponds to each f_k with the center at f_k and the radius r. The sphere $\mathscr{K}_r$ is the norm, and the radius r is the measure with which the characteristic of the perturbed state — i.e., its distance from the unperturbed state — is determined in terms of the metric of the space. By combining all $\mathscr{K}_r(f_k)$, we may define a subspace

$$\mathscr{U} = U_k(\mathscr{K}_r(f_k)) = \mathscr{K}_r(f_1) U \mathscr{K}_r(f_2) U \ldots$$

of $\mathscr{R}$. Let the perturbed state that is created due to the perturbation $\mathscr{S}$ be

specified by the point set

$$\mathcal{N}_{\mathcal{Y}} = \mathcal{N}_{\mathcal{Y}}\{g_k\} \qquad k = 1, 2, 3, \ldots$$

where the g_k are also elements of $\mathcal{R}$. It may be stated that $\mathcal{M}\{f_k\}$ is stable with respect to any perturbation $\mathcal{S}$, if — in spite of the influence of $\mathcal{S}$ — the set $\mathcal{N}_{\mathcal{Y}}$ always remains in the subspace $\mathcal{U}$. In this way, a purely geometric–topological stability concept has been obtained.

For example, we may assume that $\mathcal{R}$ is the three-dimensional Euclidean space R_3, d is the distance in the Euclidean sense, and f, g are points in R_3 with the coordinates x_i^0, $x_i^{\mathcal{Y}}$, $i = 1, 2, 3$. The unperturbed state of the ball B at the point $P(x_i^0)$ (see Fig. 1) is stable when the ball can only reach perturbed positions under the influence of perturbations of given magnitude, i.e. it can reach points $Q(x_i^{\mathcal{Y}})$ which are located inside the sphere $\mathcal{K}_r(P)$ which have a finite radius r around the point P. Such is the case, for example, when the ball is moved away from its equilibrium position P by a finite force along the funnel T, which is directed opposite to the force of gravity. If the disturbing force is not sufficient to displace the ball completely out of the funnel, then every perturbed position Q of the ball will lie within a sphere around P with a finite radius r, i.e. the position P is stable. In this way, we have become acquainted with the concept of equilibrium position stability (because P is the equilibrium position of the ball). This concept is of great importance in kinetics, as well as in statics, and was the primary subject of stability studies from the 17th to the 19th century. During this time, the geometric method prevailed.

Because of its importance, we shall mention another example of the geometric–topological point of view. Let $\mathcal{R}$ be a $2n$-dimensional phase space R_{2n}, and let f, g be the points

$$P_k = P(q_i^0, p_i^0)_k \qquad Q_k = Q(q_i^{\mathcal{Y}}, p_i^{\mathcal{Y}})_k \qquad k = 1, 2, 3, \ldots$$

of this phase space with the generalized coordinates $(q_i^0, p_i^0)_k$, $(q_i^{\mathcal{Y}}, p_i^{\mathcal{Y}})_k$,

Fig. 1. Stable position

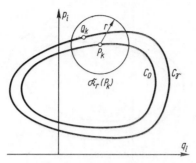

Fig. 2. Trajectories in phase space,
an example of trajectory stability

$i = 1, 2, \ldots, n$. Let point set $\mathcal{M} = \mathcal{M}\{(q_i^0, p_i^0)_k\}$, $k = 1, 2, 3, \ldots$, be represented in phase space by the trajectory C_0, and let another point set $\mathcal{N}_\mathcal{S} = \mathcal{N}_\mathcal{S}$ $\{(q_i^\mathcal{S}, p_i^\mathcal{S}_k\}$ be specified by the trajectory $C_\mathcal{S}$ (Fig. 2). Let the sequence of unperturbed states be specified by $\mathcal{M}$ or C_0, respectively, and let the sequence of states created by the perturbation $\mathcal{S}$ be described by $\mathcal{N}_\mathcal{S}$ or $C_\mathcal{S}$. By combining the spheres $\mathcal{K}_r(P_k)$, surrounding the points P_k of C_0, i.e., by means of $U_k\mathcal{K}(P_k)$, we may specify the subspace that is in the shape of a 'tube' with radius r and which surrounds the trajectory C_0. It may be stated that the trajectory C_0 is stable when the trajectories C resulting from the perturbation $\mathcal{S}$ lie entirely within the tube $\mathcal{U}$ that surrounds C_0, and this tube has a finite radius r when $\mathcal{S}$ does not exceed a certain amount. In other words, the trajectory C_0 possess orbital stability when the perturbed trajectories C are sufficiently close to the unperturbed trajectory C_0. A simple picture of this orbital stability concept is obtained when the Euclidean space R_3 is again selected for $\mathcal{R}$, which is a special case: it may be assumed that $q_i \equiv x_i$, $p_i \equiv 0$, $i = 1, 2, 3$, and that the trajectories are the orbits of point masses in motion. Thus, a rocket having an initial velocity which is very close to escape velocity of 11.3 km s^{-1} with respect to an earth-fixed reference system has orbital instability, because a very small perturbation could make the rocket escape the gravitational field of the earth and continue to increase its distance from the earth. In this case, a hyperbolic trajectory would result that would deviate from the unperturbed, elliptical trajectory by an arbitrary amount. Therefore, the condition of orbital stability would be violated.

We can readily see the extent to which the norm selection influences the stability concept by abandoning the choice of the sphere $\mathcal{K}_r(P_k)$ as the norm. We shall now assume that the points of the sets $\mathcal{M}$ and $\mathcal{M}_\mathcal{S}$ (they are represented as trajectories C_0 and $C_\mathcal{S}$ in phase space R_{2n} in Fig. 3) are identified by the value of a parameter t. Two image points may be visualized, running along C_0 and $C_\mathcal{S}$ when this parameter is continuously varied. The motion of these points indicates the change in the unperturbed and perturbed states. For example, if the parameter t is selected as the time, the changes in the states with time are obtained. This time we shall select the distance of the image points from each

8

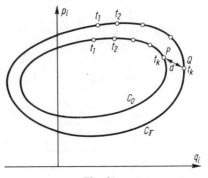

Fig. 3.

other as the norm, in the sense of the space metric, and this is the distance $d = d[\mathrm{P}(t_k), \mathrm{Q}(t_k)]$ of two points P and Q of the trajectories C_0 and $C_{\mathscr{S}}$, which correspond to the same parameter value, i.e., the norm is the distance of two geometric points which are occupied by the image points at the 'same time'. For stability it is required that $d[\mathrm{P}(t_k), \mathrm{Q}(t_k)] < r$ for each value t_k of the parameter t, i.e. it is required that 'at no time' can the image points on C_0 and $C_{\mathscr{S}}$ be too far from each other — that is, no farther than the measure r. Two features of this norm selection become readily apparent. In the first place, we have passed from the geometric–topological concept of stability to the kinematic stability concept. The kinematic nature of this idea becomes very apparent from the motion of the image points in phase space. In the second place, the new stability requirement is stricter because it is possible that a process having orbital stability may no longer be stable according to the new definition. For example, let us consider two point masses m_0 and $m_{\mathscr{S}}$ that move along concentric circles with a somewhat different period. If the differences in the circle radii are small enough, the geometric trajectories of the two points in motion will be close to each other and orbital stability is achieved. Owing to the difference in the period, which can be very small, the points will be out of phase repeatedly. In spite of the assumption that there should be a small distance between the points at the beginning of motion, there will be an unacceptably large distance between them repeatedly, so that stability does not exist in the kinematic sense (Fig. 4).

If this is to be visualized, we may substitute the Euclidean space R_3 for the abstract phase space R_{2n} (Fig. 5), which was done in the example given. The states are then motions, and the trajectories are the orbits. The image points moving in time coincide with the geometric points $\mathrm{P}(t)$ (along the trajectory C_0 of the unperturbed motion) and $\mathrm{Q}(t)$ (along the trajectory $C_{\mathscr{S}}$ of the perturbed motion) which also change with time. Since we are dealing with Euclidean space, we may also regard P and Q as the totality of their coordinates, and we may set

$$\mathrm{P}(t) = [x_i^0(t)], \mathrm{Q}(t) = [x_i(t)] \qquad i = 1, 2, 3.$$

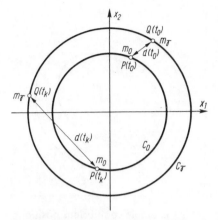

Fig. 4. Trajectories in phase space, an
example of trajectory stability and
instability according to Lyapunov

Instead of the stipulation $d[P(t), Q(t)] < r$, which is valid for any t, we may
formulate the coordinate condition

$$| x_i^0(t) - x_i^{\mathscr{G}}(t) | > r^*$$

for any i, every t, and a suitable r^*, in the sense of the Euclidean metric. We
therefore have a condition that is frequently encountered and is typical of the
stability of motion.

We would like to draw particular attention to the case in which the trajectory C_0 representing the unperturbed state may be reduced to a single point
P which coincides with one of the two image points at *every* time t. The point
then represents an *equilibrium position* (Fig. 6). If this equilibrium position is
to be stable, all perturbed states represented by $C_{\mathscr{G}}$ must satisfy the condition
$d[P, Q(t)] < r$ at all times, i.e., the other image point describing the perturbed

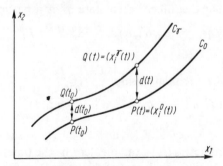

Fig. 5. Trajectories in Euclidean space,
an example of the stability definition of
Lyapunov

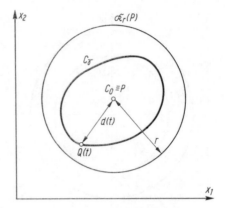

Fig. 6. Stable equilibrium position

states must move along points $Q(t)$ in such a way that C is always 'in the vicinity' of P. However, not only the kinematic condition but also the geometric condition are simultaneously satisfied because all points of $C_\mathscr{G}$ are located within the sphere $\mathscr{K}_r(P)$. This indicates that, when an equilibrium condition is stable, it is stable in the kinematic and geometric sense. Consequently, we shall regard an equilibrium position as a special case of motion, and shall apply the kinematic representation of stability, thus aiming at a unification of stability methods and concepts.

Further differences in the treatment of stability problems arise when various *types of perturbations* are considered. The perturbations can be instantaneous or permanent, and can differ in many ways with respect to their time dependence. This is most readily apparent if the mathematical representation is considered. Let us assume that a perturbed or unperturbed state can be described by a differential equation or a system of differential equations. It will customarily be assumed below that everything has been reduced to a system of first-order differential equations. The process is described in a multi-dimensional space R_n by means of the generalized coordinates q_i, $i = 1, 2, \ldots, n$, so that the unperturbed state q_i^0 is a particular solution of

$$\mathrm{d}q_i/\mathrm{d}t = \dot{q}_i = F_i(q_1, q_2, \ldots, q_n, \alpha_1^0, \alpha_2^0, \ldots, \alpha_m^0, t) \equiv F_i^0 \qquad i = 1, 2, \ldots, n$$

$$(1.1.1)$$

and therefore satisfies the relationship

$$\mathrm{d}q_i^0/\mathrm{d}t \equiv \dot{q}_i^0 = F_i(q_1^0, q_2^0, \ldots, q_n^0, \alpha_1^0, \alpha_2^0, \ldots, \alpha_m^0, t) \qquad i = 1, 2, \ldots, n.$$

$$(1.1.1')$$

The α_r^0 and t are parameters (for example, let t be the time), and the particular solutions q_i^0 of (1.1.1) are coordinates of the point $P_k = (q_i^0)_k$, which represent the unperturbed set $\mathscr{M}$ of the metric space R_n. Let the perturbed states be specified by the points $Q_k = (q_i)_k$ of the set $\mathscr{N}_\mathscr{G}$. The coordinates $q_i^\mathscr{G}$ of these points are either due to different initial conditions in solving (1.1.1) or due to changes from the parameters α_r^0 to $\alpha_r^{\mathscr{S}}$ of the functions F_i. They may also be due to the assumption of altered right-hand sides in (1.1.1). The latter case — i.e. a change in the right-hand sides of (1.1.1) — could, for example, be caused by the occurrence of perturbation functions $G_i(q_1, q_2, \ldots, q_n, \alpha_1^{\mathscr{S}}, \alpha_2^{\mathscr{S}}, \ldots, \alpha_m^{\mathscr{S}}, t)$.

We have thus presented three important types of pertubations: perturbations of the *initial conditions*, of the *parameters*, and of the *differential equations* (i.e. of the physical law that is described by these differential equations). We should consider parameter perturbations as a special case of perturbation of the differential equations. However, this case is of such great practical importance and has so many unique features that a separate study of it is justified. It frequently coincides with the case of perturbed differential equations when the parameter changes are so extensive that they lead to a qualitative change in the differential equations.

The most frequently studied case is that of *perturbations of the initial conditions*. Together with the kinematic representation of stability, this leads to the stability definition of Lyapunov, which has already been mentioned. This definition is as follows: Let the unperturbed state be specified by $q_i^0(t)$, the perturbed state by $q_i^{\mathscr{S}}(t)$, let $q_i^0(t_0)$, $q_i^{\mathscr{S}}(t_0)$ be their initial conditions, and let ε, $\eta(\varepsilon)$ be two arbitrarily small positive numbers. The sequence of unperturbed states is designated as stable when for each $\varepsilon > 0$ exists an $\eta > 0$ such that

$$d[q_i^0(t), q_i(t)] < \varepsilon \quad \text{for} \quad t > t_0$$

whenever the perturbation of the initial conditions is small enough so that

$$d[q_i^0(t_0), q_i(t_0)] \leqslant \eta.$$

The metric can be easily formulated for the Euclidean space, and the following definition may therefore be employed for this space: Stability as defined by Lyapunov occurs when, for ε there exists an $\eta(\varepsilon)$ such that

$$|q_i^0(t) - q_i(t)| < \varepsilon, t > t_0$$

holds whenever

$$|q_i^0(t_0) - q_i(t_0)| \leqslant \eta(\varepsilon) \tag{1.1.2}$$

The case of *parameter perturbation* is of great importance. *Structural stability* is referred to when stability with respect to the variation of parameters is being considered. Parameter values marking the beginning of a situation in which the perturbed states differ from the unperturbed states to a very great extent, and even qualitatively, are called *branch values* of the parameters. An entire theory for determining branch values has been developed that in essence is due to Poincaré.

Very frequently in practice it is necessary to consider perturbations caused by changes in initial conditions as well as parameter value changes. We shall discuss examples of this type at a later point.

12

The case of the continuous perturbation should also be mentioned, which mathematically corresponds to the addition of a *perturbation function G_i* to the right-hand side of (1.1.1). Recently, the special case in which G_i is a statistical function has been studied in great detail. This leads to the concept of *statistical stability*, which is quite important.

In addition to the three types of perturbations mentioned, there are many others which influence and change the stability concept. Special conditions can also be applied to the perturbations, which then leads to the concept of *conditional stability*. We shall not discuss this subject in detail. We shall simply indicate the train of thought: The conditions imposed on the stability can, for example, imply restrictions regarding the *magnitude of the perturbation* and the continuous *dependence* of the state characteristics on the size of the perturbation. In the case of the Lyapunov stability definition, the latter statement would mean that η is a continuous function of ε. The continuous dependence condition was studied in particular by Klein.[1]

Let us consider Fig. 7 in which this is illustrated visually by considering the motion of a small ball. As can be seen from the figures, it is possible that the position P′ in Fig. 7(a) is unstable with respect to small perturbations, because the ball passes immediately from the perturbed position Q′ to the position P. Larger perturbations cause the ball to pass from the perturbed position Q′ to the stable position P. It may then be state that the position P′ is unstable *in the small*, and is stable *in the large*. The reverse of the situation is shown in Fig. 7(b). There the position P is stable in the small, but is unstable in the large. In addition, it may be readily seen that in the case of Fig. 7(a) there is no continuous dependence of the perturbed stated characteristics on the perturbation size. This is due to the fact that, no matter how small the perturbation may be, the ball will always jump from position Q′ to position P. If the perturbation making the ball assume position Q is reduced and approaches

Fig. 7. (a) Stability in the large, instability in the small — 'Stability for practical purposes'. (b) Stability in the small, instability in the large — 'Instability for practical purposes'

[1]Klein, F., and Sommerfeld, A., *The Theory of the Gyroscope*, pp. 342–374. Johnson Reprint Corp. (Teubner), New York and Stuttgart, 1965.

zero, the ball will not return to the initial position P' in a continuous fashion but will always assume position P. Therefore, no stability exists in terms of Klein's definition. However, in the case of Fig. 7(a), it is usually assumed that the position P' is stable for *practical purposes*, if P is sufficiently close to P'. The reverse is true in the case of Fig. 7(b). In this case, position P is stable in terms of Klein's definition only for sufficiently small perturbations. It must be regarded as unstable for practical purposes if P' is close to P. The concepts of stability in the small and in the large may be clearly visualized when the corresponding limit cycles of an oscillator phase diagram are studied. This will be discussed in Section 1.4.2. Other definitions of stability for practical purposes lead to other differences. One of these is stability in a *finite time interval*. Even though stability over an infinite time interval is usually of interest, the requirement of an open time interval can frequently be disregarded. In order to surmount mathematical difficulties, stability for a finite time interval only may be considered. The Lyapunov definition modified in this way would result in the following stability condition:

$$|q_i^0(t) - q_i^{\mathscr{S}}(t)| < k\eta \quad \text{for} \quad t_0 \leqslant t \leqslant t_0 + \tau$$

whenever

$$|q_i^0(t_0) - q_i^{\mathscr{S}}(t_0)| \leqslant \eta.$$

The quantity k is a constant and τ is the time interval selected in such a way that stability, in terms of the selected definition, prevails over the entire interval τ. By way of an example, we would like to mention the studies of de Bra[1] regarding the position control of satellites: The motion of satellites is considered to be stable when certain motion parameters remain within certain bounds during 20 revolutions of a satellite. This condition is sufficient to guarantee the desired stable performance over the predictable lifetime of the satellite.

Finally, we would like to mention the fact that a special stability concept results when the norm measure depends on time. In this case, reference is made to *asymptotic* stability, when the measure tends to zero as the time is increased. For example, for a purely geometric representation, let us assume that the condition $r \to 0$ for $t \to \infty$ holds for the sphere radius $\mathscr{K}_r(P_k)$. Asymptotic orbital stability then prevails, because the trajectories $C^{\mathscr{S}}$ of the perturbed states come as close as desired in time to the trajectory C_0 of the unperturbed states. For the kinetic representation, $d(t) < \varepsilon$ as well as $d(t) \to 0$ for $t \to \infty$ must be satisfied for asymptotic stability, or in the case of the Lyapunov definition we must have

$$|q_i^0(t) - q_i^{\mathscr{S}}(t)| < \varepsilon$$

[1]De Bra, D. B., *The large attitude motions and stability due to gravity of a satellite*, etc. Stanford University, California, Dept. of Aeronautics and Astronautics, SUDAER No. 126, 1962.

as well as

$$|q_i^0(t) - q_i^{\mathscr{S}}(t)| \to 0 \quad \text{for} \quad t \to \infty. \tag{1.1.3}$$

We have thus presented a general description of the common foundations and variations of the stability concept. We should like to mention the fact that we shall restrict ourselves below to initial value perturbations and parameter perturbations, because they are the most important cases in practice. In addition, we shall adopt the stability definition of Lyapunov, unless we indicate the opposite.

We shall now discuss the question of obtaining *criteria* which make it possible to estimate whether and when the stability conditions are satisfied. For this purpose, let us assume that the *unperturbed state* q_i^0 is specified as a particular solution of (1.1.1) by means of the condition

$$\dot{q}_i^0 = F_i(q_1^0, q_2^0, \ldots, q_n^0, \alpha_1^0, \alpha_2^0, \ldots, \alpha_m^0, t) \equiv F_i^0(q_k^0, \alpha_r^0, t).$$

The *perturbed state* is correspondingly specified by

$$\dot{q}_i = F_i(q_1^{\mathscr{S}}, q_2^{\mathscr{S}}, \ldots, q_n^{\mathscr{S}}, \alpha_1^{\mathscr{S}}, \alpha_2^{\mathscr{S}}, \ldots, \alpha_m^{\mathscr{S}}, t) \equiv F_i^{\mathscr{S}}(q_k^{\mathscr{S}}, \alpha_r^{\mathscr{S}}, t) \tag{1.1.4}$$

We may set $q_i^{\mathscr{S}}(t) = q_i^0(t) + \xi_i(t)$ and designate $\xi_i(t)$ as the *variation* of $q_i^0(t)$. The following system of differential equations holds for the variations

$$\dot{\xi}_i(t) = F_i^{\mathscr{S}}(q_k^{\mathscr{S}}, \alpha_r^{\mathscr{S}}, t) - F_i^0(q_k^0, \alpha_r^0, t)$$

or

$$\dot{\xi}_i(t) = \Phi_i(\xi_1, \xi_2, \ldots, \xi_n, \beta_1, \beta_2, \ldots, \beta_m, t) \tag{1.1.5}$$

where $\xi_i = q_i^{\mathscr{S}} - q_i^0$, $\beta_k = \alpha_k^{\mathscr{S}} - \alpha_k^0$. The *variational equations* (1.1.5) possess the trivial solution $\xi_1 = \xi_2 = \ldots = \xi_n = 0$ for $\beta_i = 0$, which represents the origin of the n-dimensional space. Since they simultaneously satisfy $\dot{\xi}_i(t) = 0$, the origin is a stationary solution, an equilibrium position. The stability definition of Lyapunov can be expressed as follows by employing the variations

$$\left.\begin{array}{l} |\xi_i(t)| < \varepsilon \quad \text{for} \quad t > t_0 \\[2mm] |\xi_i(t_0)| < \eta(\varepsilon). \end{array}\right\} \tag{1.1.6}$$

whenever

Thus, everything has been reduced to investigating stability of the equilibrium position $\xi_i = 0$ for the case $\beta_i = 0$. In geometric terms this means that for stability the trajectory $C_{\mathscr{S}}$ described by $\xi_i(t)$ must remain in the vicinity of the origin, or even approaches the origin if the condition $|\xi_i(t)| \to 0$ for $t \to \infty$ is satisfied, i.e., in the case of asymptotic stability.

If a description of the *unperturbed state stability* is desired, the *time dependence of the variation* must be known or must be estimable. It is most natural to integrate the system (1.1.5) for the initial conditions $\xi_i(t_0) = q_i^{\mathscr{S}}(t_0) - q_i^0(t_0)$. The data necessary for determining the existence of stability are then immediately available.

In general, however, the integration of a system of differential equations will be avoided. The following possibilities then remain:

If a system of fundamental solutions is known, as is the case for linear differential equations, and if it is known which characteristic parameters influence the stability behavior of the solution, it is then possible to restrict the stability investigation to determining the course of these *characteristic parameters* and to obtain stability criteria in this way.

In the other cases, *intermediate integrals* are used. A system of n differential equations of the first order for the $n + 1$ variables ξ_i, $i = 1, 2, \ldots, n$, and t possesses n independent first intermediate integrals

$$\Psi_i(\xi_k, t) = C_i, \qquad i, k = 1, 2, \ldots, n.$$

All n first integrals must be known in order to obtain the complete solution of the system of differential equations. It is frequently simpler to specify one or more intermediate integrals instead of completely solving the system, i.e., instead of integrating it. These intermediate integrals represent surfaces in the space R_{n+1} of the variables ξ_i, t. These surfaces represent geometric loci in phase space along which the proper solution curve (ξ_i, t) must lie. The behaviour of the variations ξ_i and therefore the stability of the stationary, unperturbed solution follows from this more or less complete localization of the solution by means of the intermediate integral surfaces. For this purpose, the intermediate integral surfaces are often projected on to coordinate planes of the phase space, resulting in the well-known *phase curves* that — if they are multi-dimensional — can again be hypersurfaces. The use of energy integrals of conservative, dynamic systems represents a good example of employing intermediate integrals, from which the *energy criteria* of stability theory follow.

Example: Let the motion of a point mass be given by $\ddot{x} = -F(x)$. We set $x = \xi_1$, $\dot{x} = \xi_2$ and obtain the first-order system

$$\dot{\xi}_1 = \xi_2$$

$$\dot{\xi}_2 = -F(\xi_1)$$

which can also be represented as the simultaneous system

$$\frac{d\xi_1}{\xi_2} = \frac{d\xi_2}{-F(\xi_1)} = \frac{dt}{1}.$$

A first integral is obtained from

$$\frac{d\xi_1}{\xi_2} = \frac{d\xi_2}{-F(\xi_1)}$$

because by integration it follows that

$$\tfrac{1}{2}\xi_2^2 + \int F(\xi_1)\, d\xi_1 = C_1.$$

Using the abbreviation

$$U(\xi_1) = \int F(\xi_1)\, d\xi_1$$

we thus have

$$\Psi_1(\xi_1, \xi_2) = \tfrac{1}{2}\xi_2^2 + U(\xi_1) = C_1$$

which represents a cylinder with its generator parallel to the t axis of the phase space (ξ_1, ξ_2, t). By projecting this cylinder on to the ξ_1, ξ_2 plane, one obtains the phase curve

$$\tfrac{1}{2}\xi_2^2 + U(\xi_1) = C_1.$$

As can be readily seen, this is a geometric image of the energy integral because $\tfrac{1}{2}\xi_2^2$ is the kinetic and $U(\xi_1)$ is the potential energy of the point mass.

If the phase curve is simply closed and surrounds the equilibrium position $\xi_1 = \xi_2 = 0$ then the *stability of the equilibrium position* is assured, because the phase curve surrounds the region G of the ξ_1, ξ_2 plane in which the total energy of motion is smaller than C_1. Any motion of the point mass with a total energy $C_2 < C_1$ will progress in such a way that its phase curve will be inside G. The magnitudes ξ_1, ξ_2 remain bounded for all times so that stability prevails. Figure 8 shows the relationships for $F(x) = cx$ for a point mass oscillating about the position $x = 0$. The phase curves are the ellipses $\tfrac{1}{2}(\xi_2^2 + c\xi_1^2) = C_1$, and we have $U(\xi_1) = c/2 \cdot \xi_1^2$.

If we set $\tfrac{1}{2}\xi_2^2 + U(\xi_1) = E$, where E is the total energy, then region G in which $E < C_1$ holds is surrounded by the ellipse $\tfrac{1}{2}(\xi_2^2 + c\xi_1^2) = C_1$, along which $E = C_1$ holds.

A somewhat different question is raised when the *stability of the geometric trajectory* is examined.

Example: Let us consider the motion of a point mass in the x, y plane under the influence of conservation forces. The equations of motion are

$$\ddot{x} = -\partial U(x, y)/\partial x \qquad \ddot{y} = -\partial U(x, y)/\partial y.$$

By introducing new variables

$$\xi_1 = x \qquad \xi_2 = y \qquad \xi_3 = \dot{x} \qquad \xi_4 = \dot{y}$$

we obtain the system of simultaneous differential equations

$$\frac{d\xi_1}{\xi_3} = \frac{d\xi_2}{\xi_4} = \frac{d\xi_3}{-L} = \frac{d\xi_4}{-M} = \frac{dt}{1}$$

with

$$L = \frac{\partial U}{\partial x} \equiv \frac{\partial U(\xi_1, \xi_2)}{\partial \xi_1} \qquad \text{and} \qquad M = \frac{\partial U}{\partial y} \equiv \frac{\partial U(\xi_1, \xi_2)}{\partial \xi_2}.$$

From

$$\frac{d\xi_1}{\xi_3} = \frac{d\xi_3}{-L} \qquad \frac{d\xi_2}{\xi_4} = \frac{d\xi_4}{-M}$$

a quadrature results in

$$\tfrac{1}{2}\xi_3^2 + \int L \, d\xi_1 = C_1 \qquad \tfrac{1}{2}\xi_4^2 + \int M \, d\xi_2 = C_2.$$

The following result is obtained by addition and by taking into account the meaning of L and M:

$$\tfrac{1}{2}(\xi_3^2 + \xi_4^2) + \int\left[\frac{\partial U(\xi_1, \xi_2)}{\partial \xi_1}\,\mathrm{d}\xi_1 + \frac{\partial U(\xi_1, \xi_2)}{\partial \xi_2}\,\mathrm{d}\xi_2\right] = C_1 + C_2.$$

If we set

$$\int\left[\frac{\partial U}{\partial \xi_1}\,\mathrm{d}\xi_1 + \frac{\partial U}{\partial \xi_2}\,\mathrm{d}\xi_2\right] = \int \mathrm{d}U = U(\xi_1, \xi_2)$$

and $C_1 + C_2 = C$, we have the intermediate integral

$$\tfrac{1}{2}(\xi_3^2 + \xi_4^2) + U(\xi_1, \xi_2) = C$$

which is the energy theorem at the same time, because $\tfrac{1}{2}(\xi_3^2 + \xi_4^2)$ is the kinetic and $U(\xi_1, \xi_2)$ is the potential energy of motion.

Since $\tfrac{1}{2}(\xi_3^2 + \xi_4^2) \geqslant 0$, the relationship $U \leqslant C$ must always hold for motions having the total energy $E \leqslant C$. Let us assume that the simple closed, continuous curve $U(\xi_1, \xi_2) = C$ delimits a bounded region $U \leqslant C$ in the ξ_1, ξ_2 plane. Motions that originate in this region and whose total energies satisfy $E = C$ or $E < C$, also satisfy the relationship $U \leqslant C$. Consequently, their orbits must remain entirely in this region, because they could only cross its boundaries for $U > C$, which is not possible. Therefore, these motions are stable. The use of this 'energy criterion' for the determination of orbital stability is common practice in celestial mechanics.

If the criterion is applied to the oscillating point mass considered previously, we then have $U(\xi_1) = C_1$, from which it follows that $\xi_1 = \pm(2C_1/c)^{1/2}$. The points A and B along the ξ_1 axis in Fig. 8 are thus specified. The ξ_1, ξ_2 plane of the preceding example has now degenerated into the ξ_1 axis, and the region bounded by $U(\xi_1, \xi_2) = C$ has degenerated into the closed interval [A, B] of the ξ_1 axis. In any case, it follows that the oscillator motion is stable about the

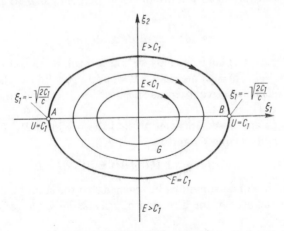

Fig. 8. Elliptical phase curves

origin, because for $E \leqslant C_1$ the oscillation amplitude will be smaller than $(2C_1/c)^{1/2}$.

The following holds for the intermediate integral $\Psi_i(\xi_k, t) = C_i$[1]:

$$\frac{d\Psi_i}{dt} = \frac{\partial \Psi_i}{\partial t} + \frac{\partial \Psi_i}{\partial \xi_k} \frac{d\xi_k}{dt} = 0.$$

This means that the solution curves corresponding to the integration constants C_i lie along the surface $\Psi_i = C_i$. We may also express this relationship in a geometric manner as follows:

$$d\Psi_i/dt = \text{grad } \Psi_i(ds/dt) = 0$$

using the vectors

$$\text{grad } \Psi_i = (\partial \Psi_i/\partial t, \partial \Psi_i/\partial \xi_k)$$

and

$$ds = (dt, d\xi_k).$$

Since the solution curve remains on the integral surface Ψ_i, the vector ds indicating the advance of an image point along the solution curve is perpendicular to the normal to Ψ_i, which is parallel to the vector grad Ψ_i. We therefore have grad $\Psi_i \cdot ds = 0$.

The purpose of identifying the solution curve in R_{n+1} space can be generalized in the following way. Instead of finding intermediate integrals and therefore integral surfaces of the corresponding differential equation, we may select arbitrary but suitable hypersurfaces $V_i(\xi_k, t) = C_i$, which divide the R_{n+1} space in such a way that the solution curves intersect them when they pass from one region to another. The course of the solution curves may also be obtained by considering the penetration points. If this generalized procedure is employed, i.e. if the solution curves are made to penetrate arbitrary families of surfaces $V_i = C_i$, then ds will no longer be perpendicular to grad V_i, so that we have

$$dV_i/dt = \text{grad } V_i(ds/dt) \neq 0.$$

The use of such test surfaces may be traced back to Lyapunov. Therefore, the functions $V_i(\xi_k, t) = C_i$ are called Lyapunov functions. The intermediate integrals $\Psi_i = C_i$ can be regarded as special cases of Lyapunov functions for which we have $dV_i/dt = 0$.

If the $V_i = C_i$ are closed surfaces around the origin of R_{n+1} which enclose themselves for increasing C_i and if

$$dV/dt = \text{grad } V_i(ds/dt) \equiv dC_i/dt < 0$$

always holds, the conclusion must be reached that the solution curves intersect the family of surfaces in the direction of *decreasing* values of the parameter

---

[1] We shall employ the *Einstein summation convention* from this point on.

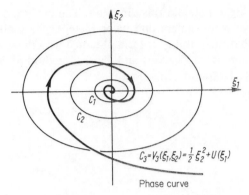

Fig. 9. Phase curve of a damped oscillator

C_i. They always are closer to the origin as t increases. Asymptotic stability may then exist.

Example: Let us consider the damped motion of an oscillating point mass, for which the system of differential equations is $\dot{x} = v$, $\dot{v} = -cx + D(x, v)$. We shall again employ the notations $x = \xi_1$, $v \cong \xi_2$, and we obtain the system

$$\dot{\xi}_1 = \dot{\xi}_2$$
$$\dot{\xi}_2 = -c\xi_1 + D(\xi_1, \xi_2).$$

Using the ideas of Lyapunov, we neither integrate this system nor do we determine its intermediate integrals. Instead, we select as suitable functions $V_i(\xi_1, \xi_2) = C_i$, the energy integrals of the corresponding conservative system that is obtained for $D \equiv 0$ (Fig. 9). They may be written as follows:

$$V_i = \tfrac{1}{2}(\xi_2^2 + c\xi^2) = C_i$$

and represent closed curves (ellipses) in the ξ_1, ξ_2 plane that envelop each other. They satisfy the following equation:

$$dV_i/dt = c\xi_1\dot{\xi}_1 + \xi_2\dot{\xi}_2 = \xi_2 D(\xi_1, \xi_2)$$

as may readily be seen by using the right-hand sides of the differential equation system. The condition $dV_i/dt < 0$, the course of the phase curve shown in Fig. 9, and therefore asymptotic stability of motion is assured if $D(\xi_1, \xi_2)$ and ξ_2 always have opposite sign. The damping D just satisfies this requirement, so that the well-known fact that damped oscillations of point masses are asymptotically stable has been verified.

The preceding statements have shown that there is a tendency toward obtaining stability criteria while trying to avoid the integration of differential equations as much as possible. The use of the Lyapunov functions has represented the biggest step in this direction. The method of Lyapunov functions is somewhat inconvenient and incomplete, as we shall see later. This is due to the fact that the stability criteria obtained with these functions are only sufficient, and there is no rigorous and general method of obtaining and formulating suitable Lyapunov functions.

1.2
Sensitivity Equations and Variational Equations

We assume an undisturbed state q_i^0, which satisfies the relationship

$$\dot{q}_i^0 = F_i(q_k^0, \alpha_r^0, t) \qquad \left.\begin{array}{l} i, k = 1, 2, \ldots, n \\ r = 1, 2, \ldots, m \end{array}\right\} \qquad (1.2.1)$$

in a general phase space R_n. The system of first-order differential equations $\dot{q}_i = F_i$ which correspond to a mechanical system and are basic to Eqs. (1.2.1), can, for example, be given by the set of canonical Hamilton differential equations. The α_r^0 are parameters, and t is the time. Owing to perturbations, which as a special case should also consist of parameter changes, the perturbed state $q_i^{\mathscr{S}}$ is obtained. If it is assumed that parameter changes cause α_r^0 to pass into $\alpha_r^{\mathscr{S}} = \alpha_r^0 + \beta_r$ and that the structure of the differential Eqs. (1.2.1) is not changed, then the following holds if we restrict ourselves to terms which are linear in β_r:

$$q_i = q_i^0(\alpha_r^0 + \beta_r, t) + \xi_i = q_i^0(\alpha_r^0, t) + (\partial q_i^0/\partial\alpha_r^0)\beta_r + \xi_i. \qquad (1.2.2)$$

The expression $(\partial q_i^0/\partial\alpha_r^0)\beta_r$ is the change of q_i^0 caused by parameter changes. For purposes of notation abbreviation, we shall set

$$u_{ir} = \partial q_i^0/\partial\alpha_r^0. \qquad (1.2.3)$$

If follows from (1.2.2) that

$$\dot{q}_i^{\mathscr{S}} = \dot{q}_i^0(\alpha_r^0, t) + \dot{u}_{ij}\beta_j + \dot{\xi}_i \qquad (1.2.4)$$

where we have used (1.2.3). From

$$\dot{q}_i^{\mathscr{S}} = F_i(q_k^{\mathscr{S}}, \alpha_r^0 + \beta_r, t)$$

together with (1.2.2), it follows that

$$\dot{q}_i^{\mathscr{S}} = F_i[q_k^0(\alpha_r^0, t) + u_{k1}\beta_1 + \xi_k, \alpha_r^0 + \beta_r, t]. \qquad (1.2.5)$$

If this expression is expanded in a Taylor series, the following expression is obtained if all terms containing products of ξ_k, β_r and containing the terms

21

ξ_k, β_r themselves raised to higher powers are neglected:

$$\dot{q}_i = F_i[q_k^0(\alpha_r^0, t), \alpha_r^0, t]$$

$$+ [(\partial F_i/\partial q_l)(q_k^0, \alpha_r^0, t)](u_{lj}\beta_j + \xi_1) + (\partial F_i/\partial \alpha_j^0)(q_k^0, \alpha_r^0, t)\beta_j. \quad (1.2.6)$$

Owing to (1.2.4) and (1.2.1), we have, on the other hand,

$$\dot{q}_i^{\mathcal{S}} = F_i[q_k^0(\alpha_r^0, t), \alpha_r^0, t] + \dot{u}_{ij}\beta_j + \dot{\xi}_i. \quad (1.2.7)$$

If (1.2.7) and (1.2.6) are compared, it can then be seen that the relationships

$$\dot{u}_{ij} = [(\partial F_i/\partial q_l)(q_k^0, \alpha_r^0, t)]u_{lj} + (\partial F_i/\partial \alpha_j^0)(q_k^0, \alpha_r^0, t) \quad (1.2.8)$$

and

$$\dot{\xi}_i = (\partial F_i/\partial q_l)(q_k^0, \alpha_r^0, t)\xi_l \quad (1.2.9)$$

must hold.

It follows from this that, in this degree of approximation, the influence of parameter changes can be treated independently of changes in the initial conditions. The system of differential Eqs. (1.2.8) for the sensitivity coefficients u_{ir} can be used for this purpose. In general, the initial conditions $u_{ir}(0) = 0$ are appropriate for this system.

On the other hand, if the influence of changes in the initial conditions is of interest, the variations ξ_i must be employed, and the system of differential Eqs. (1.2.9) must be used. In both cases, it is assumed the q_k^0 are known, because at the beginning of the calculation a decision must be made as to which undisturbed state will be investigated.

System (1.2.9) is known as the system of variational equations. It was introduced in the literature by H. Poincaré. In general, it is non-linear, because, if terms of higher order are considered in the expansion of (1.2.6) and if changes in the parameters α_k are not considered, the following system is obtained for the variation ξ_i:

$$\dot{\xi}_i = \frac{\partial F_i}{\partial q_i}(q_k^0, \alpha_r^0, t)\xi_1 + \dots + \frac{1}{n!}\frac{\partial^n F_i}{\partial q_l \dots \partial q_m}\xi_l \dots \xi_m + \dots \quad (1.2.10)$$

In the following, we shall first consider the linear Eqs. (1.2.9), and shall deal with the non-linear Eqs. (1.2.10) later on. In this connection, it might be of interest to note that there is a certain relationship between (1.2.8) and (1.2.9): as can be seen, (1.2.8) is composed of equation subgroups, which formally coincide with (1.2.9) with the exception of the additional terms $\partial F_i/\partial \alpha_j^0$.

Exercises 1.1 and 1.2.

1. The motion of a point mass is given by $\ddot{x} = -cx$, where c is a constant.
   a) Let $x = \xi_1$, $\dot{x} = \xi_2$ and transform the given differential equation into a first-order differential equation system in ξ_1, ξ_2.
   b) Find a first integral to the system in (a).

   c) Use the first integral to evaluate stability and to determine an upper bound for the magnitude of x.

2. Given the differential equation $\ddot{x} = 2(2x + 1)$

   a) Set $x = q_1$, $\dot{x} = q_2$ and transform the second-order differential equation into a simultaneous first-order system in q_1 and q_2.

   b) Using the procedure outlined in section 1.2, and assuming that the undisturbed state is given by $q_1^0 = -\frac{1}{2}$, $q_2^0 = 0$, find the system of variational equations.

   c) Determine a first integral to the variational equations in (b).

   d) Let the constant of integration in the first integral be C. Plot the phase curves corresponding to the first integral for $C = 0$ and $C = \frac{1}{2}$.

   e) Using the graph of these phase curves, determine whether the unperturbed solution $q_1^0 = -\frac{1}{2}$, $q_2^0 = 0$ is stable.

1.3
Linear Variational Equations

1.3.1 Linear Variational Equations with Constant Coefficients

We shall assume that the time t does not appear explicitly in the functions F_i in the right-hand side of (1.2.1). In this case, the process described by $q_i^0(t)$ is autonomous. In addition, we select the stationary state as the undisturbed state, for which $\dot{q}_i^0 = 0$ holds. We obtain it by solving the algebraic or transcendental system of equations $F_i(q_k^0, \alpha_r^0) = 0$. It can have one or more solutions $q_{i,s}^0$, which do not depend on the time, but rather represent fixed points in the phase space. If we formulate the system of variational equations with respect to this stationary solution, which represents the undisturbed state, according to (1.2.9), we then obtain

$$\dot{\xi}_i = (\partial F_i / \partial q_i)(q_{k,s}^0, \alpha_r^0)\xi_j.$$

As an abbreviation, we set

$$(\partial F_i / \partial q_k)(q_{i,s}^0, \alpha_r^0) = a_{ik,s}$$

or simply

$$(\partial F_i / \partial q_k)(q_{i,s}^0, \alpha_r^0) = \alpha_{ik}$$

because we carried out the calculation for a fixed s, so that the subscript s in the a_{ik} can be suppressed.

We then arrive at the system of first-order differential equations

$$\dot{\xi}_i = a_{ik}\xi_k \qquad i,k = 1,2,\ldots,n \qquad (1.3.1)$$

which has constant coefficients a_{ik}.

The structure of the matrix $\mathbf{A} = (a_{ik})$ is decisive for the type of solutions. We shall discuss this, and shall assume that matrix theory and the theory of systems of first-order differential equations with constant coefficients are known to the reader. The reader is referred to a book by Zurmühl (3) for a discussion of these topics.

By means of the so-called elementary transformations, the characteristic

24

matrix

$$\mathbf{B} = (a_{ik} - \lambda\delta_{ik})^1$$

which corresponds to any nth-order matrix $\mathbf{A} = (a_{ik})$, can be converted into the Smith canonical form

$$N = \begin{bmatrix} E_1 & & 0 \\ & \ddots & \\ 0 & & E_n \end{bmatrix}.$$

Except for the sign, or a constant factor, we obtain the relationship

$$\det \mathbf{B} \cong E_1 E_2 \ldots E_n$$

where E_ν are the elementary polynomials of the matrix $\mathbf{B}$. Let $\mathbf{A}$ have the $m \leqslant n$ different eigenvalues λ_σ, $\sigma = 1,2,\ldots,m$, which are obtained as the roots of the characteristic equation

$$\det(a_{ik} - \delta_{ik}\lambda) = 0,$$

as is well known. The following representation of the elementary polynomials then holds:

$$E_\nu = (\lambda - \lambda_1)^{e_{\nu 1}}(\lambda - \lambda_2)^{e_{\nu 2}} \ldots (\lambda - \lambda_m)^{e_{\nu m}}.$$

The $(\lambda - \lambda_\sigma)^{e_{\nu\sigma}}$ are called the elementary divisors of the matrix $\mathbf{A}$. The relationship

$$\sum_{\nu=1}^{n} e_{\nu\sigma} = p_\sigma$$

holds for the elementary divisor exponents $e_{\nu\sigma}$ that correspond to an eigenvalue λ_σ, where p_σ is the multiplicity of the eigenvalue λ_σ.

If the Smith canonical form of $\mathbf{B}$ or the Jordan standard form of $\mathbf{A}$ are known, it may then be used to determine the characteristics of $\mathbf{A}$, which can be represented in the following way:

	λ_1	λ_2	$\ldots$	λ_σ	$\ldots$	λ_m	
E_n	$e_{n,1}$	$e_{n,2}$	$\ldots$	$e_{n,\sigma}$	$\ldots$	$e_{n,m}$	
E_{n-1}	$e_{n-1,1}$	$e_{n-1,2}$	$\ldots$	$e_{n-1,\sigma}$	$\ldots$	$e_{n-1,m}$	
	$\ldots$	$\ldots$	$\ldots$	$\ldots$	$\ldots$		
	p_1	p_2	$\ldots$	p_σ	$\ldots$	p_m	$\sum p_\sigma = n$
	d_1	d_2	$\ldots$	d_σ	$\ldots$	d_m	$\sum d_\sigma = \varkappa$

---

[1] $\delta_{ik} = \begin{cases} 1 \text{ for } i = k \\ 0 \text{ for } i \neq k \end{cases}$ is the Kronecker delta.

In addition to known quantities, this table also contains d_σ, the rank decrease or defect, which the characteristic matrix

$$\mathbf{B} = (a_{ik} - \delta_{ik}\lambda)$$

undergoes when the eigenvalue λ_σ is substituted for λ, and $\varkappa$, which is the number of linearly independent eigenvectors of the matrix $\mathbf{A}$. The d_σ are equal to the number of elementary divisor exponents appearing in the column under λ_σ. The following relationship is of decisive importance for this discussion. If $d_\sigma = p_\sigma$, then all elementary divisor exponents $e_{\nu\sigma} = 1$ that correspond to λ_σ, and it may be stated that then the elementary divisors corresponding to λ_σ are linear. On the other hand, if $d_\sigma < p_\sigma$, then at least one of the elementary exponents $e_{\nu\sigma} > 1$. These facts will be employed shortly.

As is well known, the general solution of (1.3.1) is comprised of particular solutions having the form

$$\xi_{i,\sigma} = KP(t)\, e^{\lambda_\sigma t} \tag{1.3.2}$$

where K is a constant and $P(t)$ is a polynomial in t of order $e_{\nu\sigma} - 1$ or of a smaller order. As can be seen, the structure of these particular solutions is determined by λ_σ and $e_{\nu\sigma}$, which are quantities appearing in the characteristics of $\mathbf{A}$. *If the characteristic of $\mathbf{A}$ is known, all the possible particular solutions of* (1.3.1) *may be determined without having integrated the system at all.*

In order that the undisturbed solution q_i^0 be stable, we must require that the $\xi_i(t)$ remain sufficiently bounded for all times $t > t_0$. As Eq. (1.3.2) shows, the bounded nature of the particular solution and thus the general solution is essentially determined by the exponential term $e^{\lambda_\sigma t}$, and therefore by the behavior of λ_σ. H. Poincaré called the quantities λ_σ the characteristic exponents because they are decisive for stability.

It is obvious that the possible particular solutions (1.3.2) remain bounded or even decrease monotonically with time, when the real parts of all characteristic exponents are negative. We therefore obtain the following.

Theorem 1: *When all characteristic exponents λ_σ [roots of the characteristic equation of $\mathbf{A} = (a_{ik})$] have negative real parts, then the stationary solution is asymptotically stable.*

It may also be readily seen that, if the exponent λ_σ in only one of the particular solutions (1.3.2) has a positive real part, the general solution of (1.3.1) will contain at least one term which constantly increases with time. In this case, ξ_i does not remain bounded, and we have the following.

Theorem 2: *If only one of the characteristic exponents λ_σ has a positive real part, then the stationary solution q_i^0 is unstable.*

We have a special case when a characteristic λ_σ has a zero real part. Let the multiplicity of λ_σ be p_σ. Everything now depends on the relationship between the decrease in rank d_σ and the multiplicity p_σ. If we first assume $d_\sigma = p_\sigma$, then all the $e_{\nu\sigma}$ are equal to 1, as was shown previously. Then the polynomial $P(t)$ that appears in (1.3.2), which is of order $e_{\nu\sigma} - 1$, degenerates to a constant, which can be absorbed in K, and the particular solutions corresponding to $\lambda_\sigma = \genfrac{}{}{0pt}{}{+}{(-)}i\beta$ are of the form

$$\xi_{i,\sigma} = K e_{(-)}^{+ \, i\beta t}$$

which are therefore bounded. The stability has therefore not been impaired. If, however, we consider the second case $d_\sigma < p_\sigma$, then at least one of the $e_{\nu\sigma}$ will be larger than one, and $P(t)$ appearing in (1.3.2) is a true function of time t, which will be at least linear. For $\lambda_\sigma = \genfrac{}{}{0pt}{}{+}{(-)}i\beta$ there will be at least one particular solution in form

$$\xi_{i,\sigma} = K P(t) \, e_{(-)}^{+ \, i\beta t}$$

that apparently does not remain bounded, so that the bounded nature of the general solution ξ_i is cancelled. In this case, then stability disappears. This can be stated by means of the the the following.

Theorem 3: *When one of the characteristic exponents λ_σ has a zero real part, then the stationary solution q_i^0 is stable for $d_\sigma = p_\sigma$ (however, no longer asymptotically) and is unstable for $d_\sigma < p_\sigma$.*

Based on the preceding conclusions, we may outline the following procedure for investigating stability:

1. The characteristic equation $\det(a_{ik} - \delta_{ik}\lambda) = 0$ is solved, and the characteristic exponents λ_σ are obtained as the roots of this equation, which have the multiplicity p_σ.

2. If the real parts of λ_σ are all negative, then stability prevails. No questions remain, and the investigation is concluded.

3. If at least one of the real parts of λ_σ is positive, then instability prevails. Again, no questions remain, and the investigation is concluded.

4. If there is a λ_σ with zero real part, and if $p_\sigma = 1$, then we also have $d_\sigma = 1$, which means that the relationship $p_\sigma = d_\sigma$ is satisfied. Since the stability is not influenced by this, it is not necessary to make an additional investigation. If $p_\sigma > 1$, the root $\lambda_\sigma = \genfrac{}{}{0pt}{}{+}{(-)}i\beta$ must be substituted for λ in the characteristic matrix

$$\mathbf{B} = (a_{ik} - \delta_{ik}\lambda)$$

and the decrease in rank d_σ obtained in this way must be determined. If $d_\sigma = p_\sigma$, the stability is not influenced. On the other hand, if $d\sigma < p_\sigma$, this characteristic exponent $\lambda_\sigma = \genfrac{}{}{0pt}{}{+}{(-)}i\beta$ will cause instability. In this way, everything has been clarified.

Example: Let

$$\dot{\xi}_1 = -3\xi_1$$
$$\dot{\xi}_2 = -3\xi_2$$
$$\dot{\xi}_3 = -3\xi_1 + \xi_4 - \xi_5$$
$$\dot{\xi}_4 = 6\xi_2$$
$$\dot{\xi}_5 = -\xi_5$$

be the system of variational equations. Then the matrix **A** is of the form

$$\mathbf{A} = \begin{bmatrix} -3 & 0 & 0 & 0 & 0 \\ 0 & -3 & 0 & 0 & 0 \\ -3 & 0 & 0 & 1 & -1 \\ 0 & 6 & 0 & 0 & 0 \\ 0 & 0 & 0 & 0 & -1 \end{bmatrix}$$

and from its characteristic equation

$$\begin{vmatrix} -(3+\lambda) & 0 & 0 & 0 & 0 \\ 0 & -(3+\lambda) & 0 & 0 & 0 \\ -3 & 0 & -\lambda & 1 & -1 \\ 0 & 6 & 0 & -\lambda & 0 \\ 0 & 0 & 0 & 0 & -(1+\lambda) \end{vmatrix} = 0$$

it follows that $(3 + \lambda)^2 \lambda^2 (1 + \lambda) = 0$, from which the three characteristic exponents are obtained

$$\lambda_1 = -3 \qquad \lambda_2 = 0 \qquad \lambda_3 = -1$$

which the multiplicities

$$p_1 = 2 \qquad p_2 = 2 \qquad p_3 = 1.$$

We shall not expect the exponents λ_1, λ_3 to impair the stability, because they are negative and real. On the other hand, the zero exponent λ_2 must be investigated more closely. In order to do this, we substitute the value $\lambda_2 = 0$ for λ in

$$\mathbf{B} = \begin{bmatrix} -(3+\lambda) & 0 & 0 & 0 & 0 \\ 0 & -(3+\lambda) & 0 & 0 & 0 \\ -3 & 0 & -\lambda & 1 & -1 \\ 0 & 6 & 0 & -\lambda & 0 \\ 0 & 0 & 0 & 0 & -(1+\lambda) \end{bmatrix}$$

We then obtain the matrix

$$
\mathbf{B}_{(\lambda=0)} =
\begin{bmatrix}
-3 & 0 & 0 & 0 & 0 \\
0 & -3 & 0 & 0 & 0 \\
-3 & 0 & 0 & 1 & -1 \\
0 & 6 & 0 & 0 & 0 \\
0 & 0 & 0 & 0 & -1
\end{bmatrix}
$$

which in this special case must naturally coincide with **A**. By repeatedly performing the elementary operation of adding a row multiplied by an arbitrary
number with another row, in the sense of the Gaussian algorithm, and by
interchanging rows, we transform $\mathbf{B}(\lambda = 0)$ into

$$
\left.
\begin{bmatrix}
-3 & 0 & 0 & 0 & 0 \\
0 & -3 & 0 & 0 & 0 \\
0 & 0 & 0 & 1 & -1 \\
0 & 0 & 0 & 0 & -1 \\
\hline
0 & 0 & 0 & 0 & 0
\end{bmatrix}
\right\}
\begin{matrix} r \\ \\ d \end{matrix}
$$

from which it can be seen that $\mathbf{B}_{(\lambda=0)}$ is of order $n = 5$, and of rank $r = 4$.
Therefore, the decrease in rank is $d = 1$.

In this case we have $d_2 = 1 < p_2 = 2$ for $\lambda_2 = 0$, and therefore the zero root
$\lambda_2 = 0$ destroys the stability. It may therefore be concluded that the undisturbed and stationary state corresponding to the variational equations is
unstable.

As a matter of interest, we should also like to mention the fact that, by
substituting $\lambda_1 = -3$ in **B**, we obtain

$$
\mathbf{B}_{(\lambda=-3)} =
\left.
\begin{bmatrix}
0 & 0 & 0 & 0 & 0 \\
0 & 0 & 0 & 0 & 0 \\
-3 & 0 & +3 & +1 & -1 \\
0 & 6 & 0 & 3 & 0 \\
0 & 0 & 0 & 0 & +2
\end{bmatrix}
\right\}
\begin{matrix} d \\ \\ r \end{matrix}
$$

from which it follows that $d_1 = 2 = p_1$. In this simple case, it is therefore
possible to write the characteristic of the matrix **A**. It is given by

	$\lambda_1 = -3$	$\lambda_2 = 0$	$\lambda_3 = -1$	
E_5	1	2	1	
E_4	1	0	0	
p	2	2	1	$\Sigma = 5$
d	2	1	1	$\Sigma = 4$

We therefore have

$$E_5 = (\lambda + 3)\lambda^2(\lambda + 1) \qquad E_4 = \lambda + 3 \qquad E_3 = E_2 = E_1 = 1$$

and in fact

$$\det \mathbf{B} = E_1 E_2 \ldots E_5 = (\lambda + 3)^2 \lambda^2 (\lambda + 1)$$

as was already stated previously. In other situations, which are not so clear, it is necessary to go a step further to determine the various elementary divisor exponents by making the transformation to the Jordan canonical form, from which the characteristic can be read off immediately.

It is always desirable to minimize the work entailed in treating the formulated differential equations when investigating stability. Since we can decisively settle the question of stability by means of the first theorem mentioned in this chapter, according to which all roots of the characteristic equation must have a negative real part for stability to prevail, an attempt will be made to establish it without first completely solving the characteristic equation. The criterion used to decide whether an algebraic equation of nth order in λ has only roots with negative real parts is then simultaneously the decisive stability criterion for the case of linear variational equations with constant coefficients postulated here. If it is satisfied, then stability is definitely present.

There are many such criteria, and the two best known of these criteria will now be derived.

First Criterion: *This results in a graphical method, which dates back to Nyquist–Michailov (1932/1938), and is very extensively used in control theory.*

The characteristic equation is obtained by solving

$$\det(a_{ik} - \lambda\delta_{ik}) = 0$$

in the form

$$\lambda^n + a_1\lambda^{n-1} + a_2\lambda^{n-2} + \ldots + a_{n-1}\lambda + a_n = 0$$

where the coefficients a_m of this nth-order algebraic equation are certain functions of the elements a_{ik} of the matrix $\mathbf{A}$. The n roots $\lambda_\nu = x_\nu + iy_\nu$, $\nu = 1, 2, \ldots, n$ of the polynomial

$$f(\lambda) = \lambda^n + \alpha_1\lambda^{n-1} + \ldots + a_{n-1}\lambda + \alpha_n$$

are assumed to satisfy the condition of only negative real parts (Fig. 10).

In the following such a $f(\lambda)$ will be called a 'Hurwitz polynomial'. Under these conditions the roots represent points in a complex λ plane, which all lie to the left of the imaginary axis. We shall regard $f(\lambda)$ as the image of the λ plane. Then the origin of the complex f plane will be the image of the n roots λ_ν, and the imaginary iy axis of the λ plane will become a curve $f(iy)$ in the f plane (Fig. 11). This curve is called the *characteristic curve* or *locus curve* in control theory. If all roots λ_ν are on the left of the imaginary axis, then the

Fig. 10. Position of the roots of the
polynomial $f(\lambda) = 0$ in the λ plane

origin in the f plane will always be on the left of the characteristic curve, when
we proceed from $y = -\infty$ to $y = +\infty$.

The method of Nyquist–Michailov consists of deriving the variational
equations for a stability problem, forming the matrix **A** and formulating the
characteristic equation for **A**. Thus the polynomial $f(\lambda)$ is obtained. The
characteristic curve $f(iy)$ is then plotted in a f plane point by point. The
manner in which this curve passes around the origin of the f plane may be
observed. If $f = 0$ always remains on the left of the characteristic curve, when
the latter is traversed in the direction of increasing y, it may then be concluded
that $f(\lambda)$ is a Hurwitz polynomial. Stability then prevails.

There are a number of variations of the above method that we shall mention
only briefly. Owing to symmetry of the characteristic curve with respect to the
real axis, it is only necessary to plot one-half of the curve for $0 \leqslant y \leqslant +\infty$.
There is then another method of establishing the fact that the curve passes
around the origin in such a way that stability occurs. The radius vector in the
f plane drawn from the origin to the points of $f(iy)$ must rotate in the
mathematically positive direction by the angle $(n/2)\pi$ when ξ increases from 0
to $+\infty$, and when the polynomial $f(\lambda)$ is of degree n. Finally, separate curves
can be drawn for the real and imaginary parts of $f(iy)$, from which the

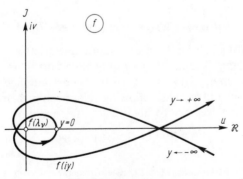

Fig. 11. Locus curve

Leonhardt and *Cremer criteria* follow. The reader is referred to the literature for more details. [1]

Up to the present we have discussed a graphic method. We shall now present an *algebraic method* advanced by Hurwitz (1895).

Second Criterion: *Let* $f(\lambda) = \lambda^n + a_1 \lambda^{n-1} + \ldots + a_n$ *be a nth-degree Hurwitz polynomial, a polynomial whose n roots* $\lambda_\nu = x_\nu + i y_\nu$ *are all on the left of the imaginary axis* $i y$.

The polynomial

$$f^*(\lambda) = \lambda^n - a_1 \lambda^{n-1} + \ldots + (-1)^n a_n$$

is formed with the Hurwitz polynomial, which has the roots $-\bar{\lambda}_\nu$, which are the mirror images of λ_ν with respect to the imaginary axis. It may be postulated that *it is always possible to formulate a Hurwitz polynomial* $F(\lambda)$ *of degree* $n+1$ *as well as another* $\Phi(\lambda)$ *of degree* $n-1$ *by using* $f(\lambda)$ *and* $f^*(\lambda)$ under the assumption that the coefficient a_1 in $f(\lambda)$ is positive and non-zero.

For this purpose, we assume

$$F(\lambda) = (\lambda + 2c)f(\lambda) + \lambda f^*(\lambda) \qquad c > 0$$

and

$$\Phi(\lambda) = (2a_1 - \lambda)f(\lambda) + \lambda f^*(\lambda).$$

It is obvious that $F(\lambda)$ has degree $n+1$. It is only necessary to determine whether $\Phi(\lambda)$ has degree $n-1$. After multiplication and a simple calculation, we find

$$\Phi(\lambda) = 2[\,a_1^2 \lambda^{n-1} + (a_1 a_2 - a_3)\lambda^{n-2} \ldots]$$

from which it follows that the degree of $\Phi(\lambda)$ is indeed $n-1$ due to the assumption $a_1^2 \neq 0$.

We shall now show that $F(\lambda)$ and $\Phi(\lambda)$ are truly Hurwitz polynomials. We may do this by using the same ideas as those employed to establish the Nyquist–Michailov criterion. We shall attempt to find the course of the characteristic curves $F(iy)$ and $\Phi(iy)$.

We have

$$F(iy) = 2cf(iy) + iy[\,f(iy) + f^*(iy)]$$

and we must distinguish between two cases. Let $f(iy) = u + iv$ be a point in the complex f plane. Then $f^*(iy) = u - iv$ is its mirror image with respect to the real axis for even n. $f^*(iy) = -u + iv$ is its mirror image with respect to the imaginary axis for odd n. The following relationship then results:

If n is *even* it is necessary to investigate the course of

$$F(iy) = 2cf(iy) + 2iyu$$

[1]Leonhardt, A., *Arch. Elektrotech.* **38**, p 17, 1944; Cremer, L., *Z. Angew. Math. Mech.* **25/27**, 161–163, 1947.

in the complex (u, v) plane. Fig. 12a,1 shows the relationships for $0 \leqslant y \leqslant +\infty$. The curve $F(iy)$ is obtained by first elongating the given curve $f(iy)$ by the amount $2c$, which leads to the intermediate curve $F'(iy)$. The points of this curve are then displaced along the perpendicular vector $2iyu$ whose magnitude varies. Its magnitude equals zero when $f(iy)$ crosses the imaginary axis corresponding to the intersection of $F'(iy)$ with this axis. Consequently, these points remain fixed when they are displaced in the perpendicular direction by an amount $2iyu$, which is exactly zero for these points. The curve $F(iy)$ is thus obtained, which is a deformed image of $F'(iy)$. The intersection points of $F'(iy)$ and $F(iy)$ with the y axis coincide. Therefore, the fundamental condition that the origin of the complex plane always is on the left of $F(iy)$ is never violated when the curve is traversed in the direction of increasing y. It follows from this that $F(\lambda)$ is a Hurwitz polynomial.

For *odd n*, we have the characteristic curve $F(iy) = 2cf(iy) - 2yv$, which is shown in Fig. 12a,2 for $0 \leqslant y \leqslant +\infty$. It is obtained from the given curve $f(iy)$ similarly to the method described above. The curve $F'(iy) = 2cf(iy)$ obtained by the elongation is displaced along the horizontal vector $2yv$. The magnitude of this vector changes in such a way that the points of $F'(iy)$ located along the real axis remain fixed. $F(iy)$ is again obtained by deformation of $F'(iy)$ so that the origin always is to the left of $F(iy)$ when the curve is traversed in the direction of increasing y. The Hurwitz property of $F(\lambda)$ remains intact.

It may also be seen that in both cases $F(iy)$ passes through one more quadrant of the complex plane than does $f(iy)$. This is a sign that the degree of $F(\lambda)$ is one greater than the degree of $f(\lambda)$.

The corresponding studies must be made for

$$\Phi(iy) = 2a_1 f(iy) + iy[f^*(iy) - f(iy)].$$

for *even n,* the curve $\Phi(iy) = 2a_1 f(iy) + 2yv$ must be considered, and is shown in Fig. 12b,1. The quantity $\Phi'(iy) = 2a_1 f(iy)$ is first obtained by an elongation. The desired $\Phi(iy)$ is then obtained by a horizontal displacement by $2yv$. The points of $\Phi'(iy)$ along the real axis remain fixed. The quantity $\Phi(\lambda)$ is a Hurwtiz polynomial due to the course of $\Phi(iy)$ around the origin. Finally, for *odd n*, the curve $\Phi(iy) = 2a_1 f(iy) - 2iyu$ shown in Fig. 12b,2 is investigated. It is obtained by the vertical displacement by $-2iyu$ of the curve $\Phi'(iy)$, which was obtained by elongation. The points of $\Phi'(iy)$ remain fixed along the imaginary axis. In spite of the deformation, $\Phi(iy)$ will of necessity again pass around the origin on the left side. $\Phi(\lambda)$ remains a Hurwitz polynomial. Figs. 12b,1 and 2 show that $\Phi(iy)$ passes through one quadrant less than does $f(iy)$ for even as well as odd n. This corresponds to the fact that the degree of $\Phi(\lambda)$ is one smaller than the degree of $f(\lambda)$.

We thus have proved the postulate that it is always possible to form two Hurwitz polynomials $F(\lambda)$ and $\Phi(\lambda)$ having the degree $n + 1$ and $n - 1$, respectively, from a given Hurwitz polynomial $f(\lambda)$.

After this presentation, two different procedures may be followed. In one

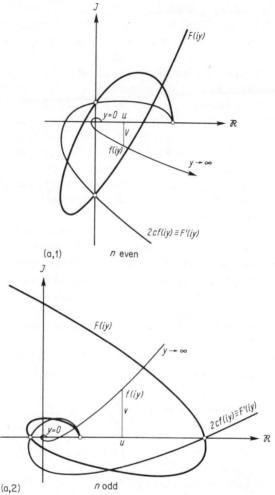

Fig. 12. Graphic representations

case, the expression $\Phi(\lambda)$ is used, which is a Hurwitz polynomial if $f(\lambda)$ is a Hurwitz polynomial. In order to illustrate the Hurwitz nature of $f(\lambda)$, a progression of $\Phi(\lambda)$ polynomials is formed beginning with $f(\lambda)$ until a second-degree polynomial is obtained. We may readily determine whether it satisfies the condition of two roots with negative real parts.

If this is the case, the preceding third-degree polynomial must be of the Hurwitz type. This means that this is also true for the polynomial of the next higher degree, etc. It is possible to show that for $c = a_1$, the relationship between $f(\lambda)$ and $\Phi(\lambda)$ equals the relationship between $F(\lambda)$ and $f(\lambda)$, except for a constant factor. This means that $f(\lambda)$ is also a Hurwitz polynomial if $\Phi(\lambda)$ is a Hurwitz polynomial. In this way it is possible to show that $f(\lambda)$ is of the Hurwitz type by carrying out the proof in the reverse direction.

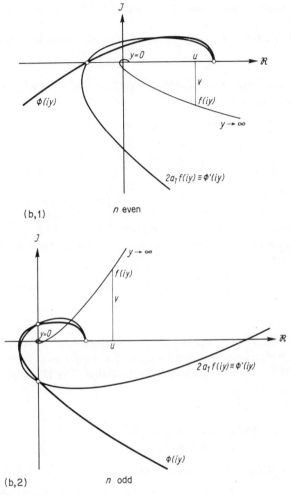

(b,1)   *n* even

(b,2)   *n* odd

proving the Hurwitz criterion

Example: Let us investigate $f(\lambda) \equiv \lambda^4 + 2\lambda^3 + 12\lambda^2 + 16\lambda + 16$. The following are formed:

$$\Phi_1(\lambda) = (4 - \lambda)(\lambda^4 + 2\lambda^3 + 12\lambda^2 + 16\lambda + 16)$$
$$+ \lambda(\lambda^4 - 2\lambda^3 + 12\lambda^2 - 16\lambda + 16)$$

$$\Phi_1(\lambda) = 8(\lambda^3 + 2\lambda^2 + 8\lambda + 8)$$

$$\Phi_2(\lambda) = 8(4 - \lambda)(\lambda^3 + 2\lambda^2 + 8\lambda + 8) + \lambda(\lambda^3 - 2\lambda^2 + 8\lambda - 8)$$

$$\Phi_2(\lambda) = 64(\lambda^2 + 2\lambda + 4) = 0.$$

The second-degree polynomial $\Phi_2(\lambda)$ has the roots $-1 \pm 3i$, and is therefore a Hurwitz polynomial. Thus, $\Phi_1(\lambda)$ and finally $f(\lambda)$ must also be Hurwitz polynomials.

In the other case, a *Hurwitz theorem* is used, which will now be described and proved.

Theorem of Hurwitz: *If the coefficients of the polynomial $f(\lambda)$ are used, we obtain the matrix*

$$\mathbf{H} = \begin{bmatrix} a_1 & 1 & 0 & 0 & 0 & 0 & \cdots \\ a_3 & a_2 & a_1 & 1 & 0 & 0 & \cdots \\ a_5 & a_4 & a_3 & a_2 & a_1 & 1 & \cdots \\ \multicolumn{7}{c}{\cdots\cdots\cdots\cdots\cdots\cdots\cdots\cdots\cdots} \end{bmatrix}$$

The diagonal terms of this matrix contain the $a_1, a_2, a_3, \ldots$ in succession. In addition, all $a_k = 0$ for $k > n$. The following is a necessary and sufficient condition for the polynomial to be a Hurwitz polynomial and have only roots with negative real parts. The principal minors Δ_ν of the Hurwtiz matrix $\mathbf{H}$, i.e., the

$$\Delta_1 = a_1, \Delta_2 = \begin{vmatrix} a_1 & 1 \\ a_3 & a_2 \end{vmatrix}, \Delta_3 = \begin{vmatrix} a_1 & 1 & 0 \\ a_3 & a_2 & a_1 \\ a_5 & a_4 & a_3 \end{vmatrix}, \ldots, \Delta_n = a_n \Delta_{n-1},$$

must all be positive.

Proof: Let $f(\lambda)$ be a Hurwitz polynomial. As was previously shown with the help of the characteristic curve $F(iy)$, this is a necessary and sufficient condition for $F(\lambda)$ to be a Hurwitz polynomial also. The Hurwitz matrix for the polynomial $\frac{1}{2} F(\lambda)$ is

$$\begin{bmatrix} c & 1 & 0 & 0 & \cdots \\ ca_2 & ca_1 + a_2 & c & 1 & \cdots \\ ca_4 & ca_3 + a_4 & ca_2 & ca_1 + a_2 & \cdots \end{bmatrix}.$$

The principal minors of this matrix are

$$D_1 = c, \qquad D_2 = c^2 \Delta_1, \qquad D_3 = c^3 \Delta_2, \ldots, \qquad D_n = c^n \Delta_{n-1},$$
$$D_{n+1} = c^{n+1} \Delta_n.$$

We assume that the Hurwitz theorem holds for polynomials of degree n. Then $\Delta_\nu > 0$ for $\nu = 1, 2, \ldots, n$. We have $D_\nu > 0$ for $\nu = 1, 2, \ldots, n+1$ due to the relationship between D_ν and Δ_ν, and because $c > 0$. Since $F(\lambda)$ is a Hurwitz polynomial, it follows that the Hurwitz condition of positive, principal minors is *necessary* for polynomials of degree $n + 1$.

This condition is also sufficient, because $\Delta_\nu > 0$, $\nu = 1, 2, \ldots, n$ follows from $D_\nu > 0$, $\nu = 1, 2, \ldots, n+1$. Since the Hurwitz conditions hold for polynomials having degree n, it follows that $f(\lambda)$ is a Hurwitz polynomial. If this is true for $f(\lambda)$, then $f(\lambda)$ must also be a Hurwitz polynomial due to its construction. Therefore, $D_\nu > 0$ was *sufficient* for this.

As we have just illustrated, if the Hurwitz condition holds for polynomials having degree n, then it holds very generally for polynomials having degree $(n + 1)$. This is because any arbitrary polynomial having degree $(n + 1)$ can be formed from one having degree n, just as $F(\lambda)$ was formed from $f(\lambda)$. The validity of the Hurwitz condition can be directly demonstrated for polynomials of the first and second degree. By mathematical induction from n to $n + 1$, it follows that the Hurwitz theorem holds for polynomials of arbitrary degree.

Example: Let us consider the polynomial

$$f(\lambda) = \lambda^4 + 2\lambda^3 + 12\lambda^2 + 16\lambda + 16$$

discussed previously. Its Hurwitz matrix is

$$\begin{bmatrix} 2 & 1 & 0 & 0 \\ 16 & 12 & 2 & 1 \\ 0 & 16 & 16 & 12 \\ 0 & 0 & 0 & 16 \end{bmatrix}$$

and the following principal minors are obtained:

$$\Delta_1 = 2, \qquad \Delta_2 = \begin{vmatrix} 2 & 1 \\ 16 & 12 \end{vmatrix} = 8, \qquad \Delta_3 = \begin{vmatrix} 2 & 1 & 0 \\ 16 & 12 & 2 \\ 0 & 16 & 16 \end{vmatrix} = 64$$

$$\Delta_4 = \begin{vmatrix} 2 & 1 & 0 & 0 \\ 16 & 12 & 2 & 1 \\ 0 & 16 & 16 & 12 \\ 0 & 0 & 0 & 16 \end{vmatrix} = 1024.$$

They are all positive, and therefore $f(\lambda)$ is a Hurwitz polynomial.

The condition of positive, principal minors in the Hurwitz theorem is called the *Hurwitz stability criterion*. This is due to the fact that it is the condition for negative, real parts of the characteristic equation roots, and therefore the condition for stability.

The criterion of Routh (1875) is another algebraic criterion. However, we shall not discuss it here. It is discussed in greater detail by Zurmühl (3).

We have seen that the position of the roots of the characteristic equation corresponding to a mechanical system in the complex plane determines whether the system is stable or unstable. If all the λ_v are on the left of the imaginary axis, the system is asymptotically stable. If a few of the roots lie along the imaginary axis, it is stated that the system lies along the *stability limit*. In the case of an asymptotically stable system, the variation $\xi_i(t)$ of the unperturbed solution $q_i^0(t)$ will damp out in time. This damping is caused by the negative real parts of the roots λ_n included in the exponents of exponential functions. The smallest absolute magnitude h of the real parts of all λ_v represents a certain measure of the damping rate, and therefore a measure of the stability, even though it is not completely conclusive. Geometrically, h is the distance from the imaginary axis to the closest root (Fig. 13). The quantity

38

Fig. 13. h is the degree of stability

h is designated as the *degree of stability*. It is obtained by displacing the imaginary axis by h in the λ plane, so that the system, which was originally stable, now lies along the stability limit. The λ plane becomes the μ plane owing to this displacement, with the coordinates $\eta = y$ and $\xi = x + h$. Because $\lambda = x + iy$, $\mu = \xi + i\eta$, we have $\lambda = \mu - h$. From the original equation

$$f(\lambda) = \lambda^n + a_1\lambda^{n-1} + \ldots + a_n = 0$$

we obtain the following new equation:

$$g(\mu) = \mu^n + A_1\mu^{n-1} + \ldots + A_n = 0$$

by substitution of $\lambda = \mu - h$. The coefficients A_ν are functions of h. We must set $h = h_0$ in such a way that at least one of the roots of the polynomial $g(\mu)$ has a vanishing real part but no root has a positive real part. The h_0 that satisfies this condition is the degree of stability.

Example: Let us consider $f(\lambda) = \lambda^2 + 2\lambda + 4 = 0$. Substituting $\lambda = \mu - h$, we obtain the equation

$$g(\mu) = \mu^2 + 2\mu(1 - h) + [(1 - h)^2 + 3] = 0$$

whose roots are $\mu_{1,2} = (h - 1) \pm 3i$. For $h = 1$, $\mu_{1,2}$ have vanishing real parts; therefore this h indicates the degree of stability, as may also be seen in Fig. 14.

It is of great importance to known the manner in which the parameters of a system must be selected so that a certain stability limit is obtained. These parameters appear in the coefficients of the variational equations as the elements of their matrix **A**, and finally as coefficients of the characteristic equation. This question leads us to the topic of structural analysis and optimization of mechanical systems. The corresponding theory is well formulated in control theory, for example, and we therefore refer to the literature (2, 12).

We have already discussed the question of how certain parameters α_r can influence the stability of a mechanical system. We cannot ignore this question. For purposes of simplicity, we shall restrict ourselves to two parameters α_1 and α_2, because in this way everything can be represented in a *parameter plane*.

Fig. 14. Example of calculating the
degree of stability

Example: Let the system of variational equations be specified by

$$\dot{\xi}_1 = (2 + \alpha_1)\xi_2$$
$$\dot{\xi}_2 = -\xi_1 + \xi_2 - \xi_3$$
$$\dot{\xi}_3 = (\alpha_2 - 1)\xi_1 + \xi_2 - 2\xi_3.$$

Its matrix is

$$\mathbf{A} = \begin{bmatrix} 0 & 2 + \alpha_1 & 0 \\ -1 & 1 & -3 \\ \alpha_2 - 1 & 1 & -2 \end{bmatrix}$$

and the characteristic equation is

$$\begin{vmatrix} \lambda & -(2 + a_2) & 0 \\ 1 & (\lambda - 1) & 3 \\ -(\alpha_2 - 1) & -1 & (2 + \lambda) \end{vmatrix} = \lambda^3 + \lambda^2 + (3 + \alpha_1)\lambda + 3(\alpha_2 - \tfrac{1}{3})(\alpha_1 + 2) = 0.$$

The coefficients of this equation are

$$a_1 = 1, \qquad a_2 = 3 + \alpha_1, \qquad \alpha_3 = 3(\alpha_2 - \tfrac{1}{3})(\alpha_1 + 2).$$

The Hurwitz matrix is

$$\mathbf{H} = \begin{bmatrix} a_1 & 1 & 0 \\ a_3 & a_2 & a_1 \\ a_5 & a_4 & a_3 \end{bmatrix}$$

and leads to the stability conditions

$$\Delta_1 = \alpha_1 > 0, \qquad \Delta_2 = a_1 a_2 - a_3 > 0, \qquad \Delta_3 = a_3 \Delta_2 > 0.$$

These may also be written as follows:

$$a_1 > 0, \qquad a_2 > 0, \qquad a_3 > 0, \qquad \Delta_2 = a_1 a_2 - a_3 > 0$$

which may be readily verified. Owing to the conditions imposed on the a_k, this is equivalent to the conditions

$$a_1 = 1 > 0$$

$$a_2 = (3 + \alpha_1) > 0 \qquad \text{i.e. } \alpha_1 > -3$$

$$a_3 = 3(\alpha_2 - \tfrac{1}{3})(\alpha_1 + 2) > 0 \qquad \text{i.e.} \begin{cases} \alpha_2 > \tfrac{1}{3}, \alpha_1 > -2 \\ \alpha_2 < \tfrac{1}{3}, \alpha_1 < -2 \end{cases}$$

$$\Delta_2 = a_1 a_2 - a_3 = 3 + \alpha_1 - 3(\alpha_2 - \tfrac{1}{3})(\alpha_1 + 2) > 0$$

$$\text{i.e. } -3\alpha_1\alpha_2 - 6\alpha_2 + 2\alpha_1 + 5 > 0.$$

The curves $\alpha_1 = -3$, $\alpha_2 = \tfrac{1}{3}$, $\alpha_1 = -2$, $3\alpha_1\alpha_2 + 6\alpha_2 - 2\alpha_1 - 5 = 0$ can be plotted in the α_1, α_2 plane. The boundaries of the region whose points correspond to the stability of the system are thus obtained (Fig. 15). For example, if the value $\alpha_1 = 1$ is held constant and if we proceed along a line parallel to the α_2 axis in a mathematically positive direction, we pass from an instability region to a stability region, and then again pass into an instability region. The *branching values* of the parameter α_2 are obtained when the stability limits are traversed. These values correspond to points at which the stability of the solution changes. For the example considered above, the branching value $\alpha_2 = 7/9$ is obtained at the limit $\Delta_2 = 0$ when the vertical line $\alpha_1 = 1$ is traversed.

If *several* parameters occur, the corresponding studies may be performed in n-dimensional parameter space, or the study is performed in various parameter planes. The *D decomposition* (2.5) of J. I. Nejmark which is frequently used in control theory represents another possibility for investigating

Fig. 15. Stability map in the parameter plane

the influence of parameters on stability. Let us again advance the basic concepts of the method using two parameters α_1, α_2.

Let us assume that a characteristic equation $f(\lambda, \alpha_1, \alpha_2) = 0$ is given. This equation represents an implicit equation for mapping the roots λ_ν of $f = 0$ in the *parameter plane* α_1, α_2. If $\lambda = iy$ is substituted in $f = 0$ and if we then solve for α_1, α_2, the α_1, α_2 values (and therefore points of the parameter plane) that correspond to purely imaginary roots of $f = 0$ are obtained.

A *limit curve* Γ is specified by $f(iy, \alpha_1, \alpha_2) = 0$. This curve divides the parameter plane into regions. Some of the regions contain points which correspond to the right half of the λ plane. The other regions contain points corresponding to roots in the left half of the λ plane. For purposes of stability, all roots of $f = 0$ must lie in the left λ half-plane. In the α_1, α_2 plane, we shall therefore determine the region delimited by the curve Γ corresponding to the mapping of points in the left λ plane. We thus obtain the stability region.

The method first calls for obtaining the curve Γ by using $f(iy, \alpha_1, \alpha_2) = 0$. At times it must be drawn point by point. The coordinates α_1^P, α_2^P of points P are then substituted in $f(\lambda, \alpha_1, \alpha_2) = 0$. These points are selected from the various regions of the parameter plane obtained by plotting Γ. Finally, the location of the roots λ in the λ plane obtained from $f(\lambda, \alpha_1^P, \alpha_2^P) = 0$ is determined. If $f = 0$ is an algebraic equation in λ having degree n, then the stability regions are defined by points P_{st} having the following property. The equation $f = 0$ has exactly n roots with negative real parts using the coordinates $(\alpha_1, \alpha_2)_{P_{st}}$ of P_{st}. In this case, all n roots lie in the left half of the λ plane. Let us describe such a calculation.

Example: We assume the characteristic equation

$$f(\lambda, \alpha_1, \alpha_2) = \lambda^3 + \lambda^2 + (3 + \alpha_1)\lambda + 3(\lambda_2 - \tfrac{1}{3})(\alpha_1 + 2).$$

Substituting $\lambda = iy$, we obtain the following:

$$i[-y^3 + y(3 + \alpha_1)] - y^2 + 3(\alpha_2 - \tfrac{1}{3})(\alpha_1 + 2) = 0.$$

The imaginary and real parts must vanish separately, so we have

$$-y^3 + y(3 + \alpha_1) = 0 \qquad -y^2 + 3(\alpha_2 - \tfrac{1}{3})(\alpha_1 + 2) = 0.$$

These two conditions are satisfied by $y = 0$ and $3(\alpha_2 - \tfrac{1}{3})(\alpha_1 + 2) = 0$. This leads to the branches $\alpha_2 = \tfrac{1}{3}, \alpha_1 = -2$ of the limit curve Γ. They appear in Fig. 15 as a result of the condition $a_3 = 0$. On the other hand, if $y \neq 0$ holds, then we must have $y^2 = 3 + \alpha_1$ due to the left-hand equation. If this is substituted into the right-hand equation

$$-(3 + \alpha_1) + 3(\alpha_2 - \tfrac{1}{3})(\alpha_1 + 2) = 0$$

is obtained, which corresponds to the equation

$$3\alpha_1\alpha_2 + 6\alpha_2 - 2\alpha_1 - 5 = 0$$

and therefore $\Delta_2 = 0$. This branch of the limit curve Γ is shown in Fig. 15.

In order to determine which region of the parameter plane divided by the

42

limit curve Γ is the stability region, we substitute the coordinates of the points $P_1(\alpha_1 = -1,\ \alpha_2 = 3)$ and $P_2(\alpha_1 = -1,\ \alpha_2 = 0.5)$ into $f(\lambda, \alpha_1, \alpha_2) = 0$ (see Fig. 15).

The coordinates of P_1 result in the equation $\lambda^3 + \lambda^2 + 2\lambda + 8 = 0$ having the roots $\lambda_1 = -1.98$, $\lambda_{2/3} = 0.53 \pm 1.93i$. Since only one root of the cubic equation has a negative real part, the region of the α_1, α_2 plane containing the point P_1 cannot be a stability region.

The coordinates of P_2 results in the equation

$$\lambda^3 + \lambda^2 + 2\lambda + 0.5 = 0$$

having the roots $\lambda_1 = -0.278$, $\lambda_{2,3} = -0.361 \pm 1.29i$. This time all three roots have a negative real part. We therefore have $P_2 \equiv P_{st}$, which defines a stability region of the α_1, α_2 plane. It is located below the right branch of the hyperbola $\Delta_2 = 0$ above the line $\alpha_2 = \frac{1}{3}$ and to the right of the vertical line $\alpha_1 = -2$. This result coincides with the previous calculation using the Hurwitz criterion.

It is interesting to note that the Nyquist-Michailov stability criterion may be regarded as a *special case* of a *D decomposition*. The limiting curve $f(iy) = \alpha$ which divides the α plane for the equation $f(\lambda) - \alpha = 0$ containing only one complex parameter α is identical to the characteristic curve of the Nyquist–Michailov stability criterion.

1.3.2 Linear Variational Equations with Variable Coefficients

The literature contains a great number of important general theorems (4) corresponding to a system of linear differential equations with variable coefficients. We assume that these theorems are known to the reader. We shall only consider the stability theory of such a system.

A system with variable coefficients is obtained if the right-hand sides of the system of equation $\dot{q}_i^0 = F_i(q_k^0, \alpha_r^0, t)$ contains time t explicitly. The mechanical process is then called *non-autonomous*. The linear or linearized variational equations can be obtained by the method outlined in Sections 1.2 and 1.3.1 as follows:

$$\dot{\xi}_i = a_{ik}(t)\xi_k. \qquad (1.3.3)$$

The terms

$$a_{ik}(t) = (\partial F_i/\partial q_k)(q_j^0, \alpha_r^0, t) \qquad (1.3.4)$$

are variable coefficients. It is assumed that they are continuous and bounded.

If we have a nth-order system, then (1.3.3) has n linearly independent particular *solution vectors* $\xi_i^{(k)}$, $k = 1, 2, \ldots, n$ representing a *fundamental system*. The *general solution* $\xi_i = C_k \xi_i^{(k)}$ is obtained from the fundamental system and the n integration constants C_k. The solution vectors $\xi_i^{(k)}$ (k_{fixed}) themselves are composed of n solution components $\xi_i^{(k)}(t)$, $i = 1, 2, \ldots, n$ (k_{fixed}). Therefore, a total of n^2 solution components $\xi_i^{(k)}(t)$, $i = 1, 2, \ldots, n$, $k = 1, 2, \ldots, n$ corresponds to (1.3.3). The *Wronskian deter-*

minant W(t) of the particular solutions at different times t_0 and t satisfies the relationship

$$W(t) = W(t_0) \exp\left(\int_{t_0}^{t} a_{ii}(t) \, \mathrm{d}t \right).$$ (1.3.5)

This relationship will be employed below.

An attempt will be made to estimate the time dependence of the general solution $\xi_i(t)$, which itself depends on the behaviour of the particular solution $\xi_i^{(k)}(t)$. If the *magnitudes* of all the $\xi_i^{(k)}(t)$ tend to zero with time, or their magnitude remains below a certain *limit*, it may be assumed that the unperturbed state q_i^0 underlying the variational equations is *stable*.

In the case of the linear differential equations with constant coefficients, it was possible to estimate their solutions (1.3.2) using the real part of the characteristic exponents.

This possibility no longer exists for solutions of equations with variable coefficients. An attempt will be made to estimate their growth pattern using similar methods. It is possible to use a method suggested by Lyapunov using the so-called *characteristic numbers* (abbreviation, CN).

Let us assume that the continuous function $x(t)$ defined for $t \geqslant 0$ is given. It is bounded if $|x(t)| < M$ is satisfied for $t > 0$ and $M > 0$, where M is finite. It is unbounded if it is possible to find values $t > 0$ for an arbitrarily large M for which $|x(t)| > M$ holds. The unbounded property of the function is defined by: $\lim |x(t)| = \infty$ for $t \to \infty$.

Let us assume that two numbers λ and μ exist for $x(t)$ so that the product $x(t) \, \mathrm{e}^{\lambda t}$ is unbounded and the product $x(t) \, \mathrm{e}^{\mu t}$ tends to zero asymptotically in time. In this case there exists a *characteristic number* $\varkappa$ for which $\mu \leqslant \varkappa \leqslant \lambda$ holds and which has the following properties: for arbitrarily small $\varepsilon > 0$, the function $x(t) \exp[(\varkappa + \varepsilon)t]$ is unbounded and $x(t) \exp[(\varkappa - \varepsilon)t]$ tends to zero asymptotically with time. The number $\varkappa$ is therefore a measure for the growth behaviour of $x(t)$. Let us assume $x(t) = t^2$, for example, and the characteristic number is $\varkappa = 0$. This is because $x \, \mathrm{e}^{\varkappa t} = t^2 \, \mathrm{e}^{kt}$ equals

$$t^2 \exp(o + \varepsilon t) = t^2 \, \mathrm{e}^{\varepsilon t}$$

for $\varkappa + \varepsilon$, which is unbounded. For $\varkappa - \varepsilon$, it tends to zero in the same way as

$$t^2 \exp(o - \varepsilon t) = t^2 \, \mathrm{e}^{-\varepsilon t}.$$

Let us consider the solutions (1.3.2) of the equations with *constant* coefficients. We must form $KP(t) \, \mathrm{e}^{x_\sigma t} \, \mathrm{e}^{\mathrm{i} y_\sigma t} \, \mathrm{e}^{\varkappa t}$, where $\lambda_\sigma = \chi_\sigma + \mathrm{i} y_\sigma$ has been used. It may be readily seen that we must have $\varkappa = -x_\sigma$ because $KP(t) \, \mathrm{e}^{\mathrm{i} y_\sigma t} \, \mathrm{e}^{\varepsilon t}$ is obtained for $\varkappa + \varepsilon$, which is unbounded. We obtain $KP(t) \, \mathrm{e}^{\mathrm{i} y_\sigma t} \, \mathrm{e}^{-\varepsilon t}$ for $\varkappa - \varepsilon$, which tends to zero in time. It is apparent that the *characteristic number equals the real part of the characteristic exponent λ_σ multiplied by* -1. This is not surprising, because in Section 1.3.1 we have already employed this real part of λ_σ in order to establish stability by estimating the growth pattern of the variations. Therefore, the theory of characteristic numbers presented here may be regarded as a logical generalization of the theory of characteristic exponents discussed in Section 1.3.1.

The following three *theorems* hold for the characteristic numbers.

Theorem 1: *The characteristic number of the sum of two functions x_1 and x_2 equals the smallest characteristic number of these functions.*

Proof: Let $\varkappa_1 < \varkappa_2$ be the characteristic numbers of x_1, x_2, so that the products $x_1 \exp[(\varkappa_1 + \varepsilon)t]$, $x_2 \exp[(\varkappa_2 + \varepsilon)t]$ are unbounded. On the other hand, we have $x_1 \exp[(\varkappa_1 - \varepsilon)t] \to 0$ and $x_2 \exp[(\varkappa_2 - \varepsilon)t] \to 0$ for $t \to \infty$. The sum $x = x_1 + x_2$ with $\varkappa = \mathrm{CN}(x)$ and $\varepsilon > 0$ is assumed to satisfy the following conditions: $x \exp[(\varkappa + \varepsilon)t]$ becomes unbounded, and $x \exp[(\varkappa - \varepsilon)t] \to 0$ for $t \to \infty$. We now have

$$x \exp[(\varkappa + \varepsilon)t] = x_1 \exp[(\varkappa + \varepsilon)t] + x_2 \exp[(\varkappa + \varepsilon)t].$$

If we set $\varkappa = \varkappa_1 = \varkappa_2 - \rho, \rho > 0$, we then have

$$x \exp[(\varkappa + \varepsilon)t] = x_1 \exp[(\varkappa_1 + \varepsilon)t] + x_2 \exp[(\varkappa_2 - \rho + \varepsilon)t]$$

which will definitely become infinitely large in time owing to the same property of the first term, regardless of the behaviour of the second term. On the other hand, we have

$$x \exp[(\varkappa - \varepsilon)t] = x_1 \exp[(\varkappa_1 - \varepsilon)t] + x_2 \exp[(\varkappa_2 - \rho - \varepsilon)t]$$

which tends to zero with time because both terms continuously decrease. Therefore, the number $\varkappa = \varkappa_1$ satisfies all the requirements of a characteristic number for the sum $x = x_1 + x_2$. It is therefore the characteristic number of x.

Theorem 2: *The characteristic number of the product of two functions x_1 and x_2 is equal to the sum of the characteristic numbers of x_1 and x_2 or is larger than this sum.*

Proof: Let us assume that $x = x_1 \cdot x_2$. We set $\mu = \varkappa_1 + \varkappa_2 - \varepsilon$, where $\varkappa_1$ and $\varkappa_2$ are the characteristic numbers of x_1 and x_2 and $\varepsilon > 0$, which is assumed to be arbitrarily small. According to the definition of μ, the condition $x e^{\mu t} \to 0$ is to be satisfied for $t \to \infty$. This is the case because according to our selection of μ, we have

$$xe^{\mu t} = x_1 x_2 \exp[(\varkappa_1 + \varkappa_2 - \varepsilon)t]$$

which can be transformed into the expression

$$x e^{\mu t} = x_1 \exp[(\varkappa_1 - \varepsilon/2)t] x_2 \exp[(\varkappa_2 - \varepsilon/2)t].$$

This expression indeed tends to zero, because there are two continuously decreasing factors on the right-hand side due to the definitions of $\varkappa_1$ and $\varkappa_2$. Therefore, the relationship $\varkappa_1 + \varkappa_2 \leqslant \varkappa$ holds for the characteristic number $\varkappa$ of the product x according to the definition of μ. This is what has been postulated. Using the functions $x_1 = e^{at \cos t}$, $x_2 = e^{-at \cos t}$, $a > 0$ as an example, we shall show that Theorem 2 is true not only for $x = x_1 + x_2$ but also for $x > x_1 + x_2$. It may be readily seen that $x_1 = CN(x_1) = x_2 = CN(x_2) = -a$. We have

$$x_{1,2} \exp[-(a+\varepsilon)t] = \exp\{[a(\pm\cos t - 1) - \varepsilon]t\} \to 0$$

because we always have $a(\pm\cos t - 1) - \varepsilon < 0$. On the other hand,

$$x_{1,2} \exp[-(a-\varepsilon)t] = \exp\{[a(\pm\cos t - 1) + \varepsilon]t\}$$

and we may always find t values such that $a(\pm\cos t - 1) + \varepsilon > 0$. Therefore, $x_{1,2} \exp[-(a-\varepsilon)t]$ becomes infinitely large with time. Therefore, $-a$ is the common characteristic number for x_1, x_2. Finally, the product $x = x_1 x_2 = 1$ = const, and it has the characteristic number zero. Hence the relationship

$$x = CN(x) = 0 > x_1 + x_2 = -2a$$

holds, which was postulated previously.

Theorem 3: *If the characteristic number x of a function x is positive, then* lim $x = 0$ *for* $t \to \infty$. *If it is negative, then* lim $|x| = \infty$ *holds for* $t \to \infty$.

Proof: We have $x \exp[(x-\varepsilon)t] \to 0$ for $t \to \infty$ according to the definition of x. We assume that $x > 0$ and ε is arbitrarily small. Therefore, ε can be selected so that $x - \varepsilon > 0$. Then the relationship $x \exp(x-\varepsilon)t] \to 0$ can only hold for $t \to \infty$ and positive x, if $x \to 0$ for $t \to \infty$. This is due to the fact that the factor $\exp[(x-\varepsilon)t]$ continuously increases because of $x - \varepsilon > 0$. On the other hand, $x < 0$ and ε is arbitrarily small. According to the definition of x, $x \exp[(x+\varepsilon)t]$ will become unbounded. We have

$$x \exp[(x+\varepsilon)t] = x \exp[(\varepsilon - |x|)t]$$

and ε may be selected small enough that $\varepsilon - |x| < 0$. Because of this, $\exp[(\varepsilon - |x|)t]$ will continuously decrease. In spite of this, the quantity $x \exp[(\varepsilon - |x|)t]$ will become unbounded for $t \to \infty$. This is only possible if x becomes unbounded, i.e., $\lim |x(t)| = \infty$ for $t = \infty$.

We shall now introduce the concept of the characteristic number x_k of a particular solution vector $\xi_i^{(k)}$, $i = 1,2,\ldots,n$ (k fixed). Let $x_{j,k} = CN(\xi_j^{(k)})$ (j fixed, k fixed) be the characteristic numbers of the *components* of the vector $\xi^{(k)}$. We shall then assume that

$$x_k = \min(x_{j,k}) \qquad j = 1,2,\ldots,n \ (k \text{ fixed})$$

holds, i.e., the characteristic number of the vector is assumed to be the *smallest* of the characteristic numbers of its components.

Using x_k, we may now formulate two theorems that are completely analogous to those given in Section 1.3.1.

Theorem 4: *If all characteristic numbers x_k of the solution vectors $\xi_i^{(k)}$ of the fundamental system of the system of differential equations (1.3.3) are positive, then the stationary solution q_i^0 is asymptotically stable.*

Theorem 5: *If only one of the characteristic numbers $\varkappa_k$ is negative, then the stationary solution q_i^0 is unstable.*

Proof: If $\varkappa_k > 0$ holds for all $\varkappa$, then the $\varkappa_{j,k}$ of the components $\xi_i^{(k)}$ of all solution vectors are also greater than zero for all j and k. Therefore, all these particular solution components strive asymptotically to zero in time. The general solution ξ_i of (1.3.3) is the sum of the particular solution components and therefore tends asymptotically to zero, which proves the stability of q_i^0.

On the other hand, if only one of the characteristic numbers $\varkappa_k$ is negative, then at least one of the characteristic numbers $\varkappa_{j,k}$ of the components of the solution vectors is negative. There is at least one particular solution component that becomes infinitely large with time. Because of this component, the general solution ξ_i of the variational equations must also become infinitely large, which brings about the instability of q_i^0.

The characteristic numbers $\varkappa_{j,k}$ of variational equations with *constant* coefficients are the real parts of the characteristic exponents λ_k multiplied by -1. If the two theorems given above are applied to variational equations with constant coefficients, it follows from $\varkappa_{j,k} > 0$ for all j,k that all characteristic exponents λ_k have negative real parts. If at least one of the $\varkappa_k$ and therefore at least one of the $\varkappa_{j,k}$ are negative, it follows that at least one of the characteristic exponents λ_k will have a positive real part. Asymptotic stability results in the first case and instability results in the second case. It can be seen that the corresponding theorems from Section 1.3.1 are only special cases of Theorems 4 and 5. Later on we will use the concept of a *regular system*. We will now introduce this concept. We first determine the *characteristic number of the Wronskian determinant* $W(t)$ of the solutions of the variational equations. On the one hand, we have

$$W(t) = \det[\xi_i^{(k)}(t)]$$

so that

$$W(t) = \sum (-1)^{J(\alpha,\beta,\gamma,\dots,\nu)} \xi_\alpha^{(1)} \xi_\beta^{(2)} \xi_\gamma^{(3)} \dots \xi_\nu^{(n)}.$$

The sum signs $\sum$ indicate that all permutations of the subscripts $1,2,3,\dots,n$ are to be used in the terms of the sum. $J(\alpha,\beta,\gamma,\dots,n)$ is the number of inversions of the corresponding permutation. On the other hand, we have the relationship

$$W(t) = W(t_0) \exp\left(\int_{t_0}^t a_{ii}\, dt\right)$$

where a_{ii} is the trace of the matrix $\mathbf{A} = [a_{ik}(t)]$ of the differential equations.

According to the theorem of the characteristic number of a product, we have

$$\mathrm{CN}(\xi_\alpha^{(1)} \xi_\beta^{(2)} \dots \xi_\nu^{(n)}) \geqslant \mathrm{CN}(\xi_\alpha^{(1)}) + \mathrm{CN}(\xi_\beta^{(2)}) + \dots + \mathrm{CN}(\xi_\nu^{(n)}).$$

The characteristic numbers of the solution components are always larger than the characteristic numbers x_k of the corresponding solution vectors corresponding to them, or they are equal to them. We also therefore have the relationship

$$\text{CN}(\xi_\alpha^{(1)} \xi_\beta^{(2)} \ldots \xi_n^{(n)}) \geqslant \sum_{k=1}^{n} x_k.$$

The characteristic number γ of the determinant $W(t)$ is the characteristic number of a sum, the terms of which have the characteristic numbers $\text{CN}(\xi_\alpha^{(1)} \xi_\beta^{(2)} \ldots \xi_\nu^{(n)})$. Consequently, γ is equal to the smallest of the characteristic numbers $\text{CN}(\xi_\alpha^{(1)} \xi_\beta^{(2)} \ldots \xi_\nu^{(n)})$, but always larger than

$$\sum_{k=1}^{n} x_k.$$

This is because all characteristic numbers $\text{CN}(\xi_\alpha^{(1)} \ldots \xi_\nu^{(n)})$, even the smallest one, are larger than $\sum_{k=1}^{n} x_k$. We therefore have

$$\gamma \geqslant \sum_{k=1}^{n} x_k.$$

We now form the product of the functions $W(t)$ and

$$e^{-p(t)} \left(p(t) = \int_{t_0}^{t} a_{ii}(t) \, dt \right)$$

and we will assume that they have the characteristic numbers γ and μ. The product

$$W(t) \, e^{-p(t)} = W(t_0) \, e^{p(t)} \, e^{-p(t)} = W(t_0) = \text{const}$$

has the characteristic number zero. According to the theorem of characteristic numbers of a product, we must therefore have

$$0 \geqslant \mu + \gamma \geqslant \mu + \sum_{k=1}^{n} x_k.$$

Lyapunov called the system of variational equations *regular*, if

$$\mu + \sum_{k=1}^{n} x_k = 0$$

is satisifed. We will show that the variational equations *having constant coefficients are always regular*.

Assume such a system is given by $\dot{\xi}_i = a_{ik}\xi_k$. Its coefficient matrix is $\mathbf{A} = (a_{ik})$ with the constant elements a_{ik}. The particular solutions of the system have the form $K_{kj}P_{kj} \exp(iy_k t) \exp(x_k t)$ and therefore have the characteristic numbers $x = -x_k$. The quantities x_k are the real parts of the roots λ_k of the characteristic equation

$$f(\lambda) = \lambda^n + a_1 \lambda^{n-1} + \ldots + a_n = 0.$$

As is well known, the coefficients a_1 of this equation satisfy $a_1 = -\text{tr}\mathbf{A}$, that is, $a_1 = -a_{ii}$. The real parts of its roots satisfy the relationship $\Sigma_k x_k = -a_1$, so that

$$\sum_k x_k = +a_{ii}$$

also holds. These relationships will be used.

The terms appearing in the sum representing $W(t)$ are equal to

$$(-1)^J K_{\alpha 1} P_{\alpha 1}(t)\, e^{\lambda_1 t} K_{\beta 2} P_{\beta 2}(t)\, e^{\lambda_2 t} \ldots K_{\nu n} P_{\nu n}(t)\, e^{\lambda_n t}$$

in this case, which leads to

$$(-1)^J \exp\left(\sum_k x_k t\right) \exp\left(\mathrm{i}\sum_k y_k t\right) K_{\alpha 1} P_{\alpha 1}(t) K_{\beta 2} P_{\beta 2}(t) \ldots K_{\nu n} P_{\nu n}(t)$$

due to $\lambda_k = x_k + \mathrm{i} y_k$. It is apparent that the characteristic number of such a sum is always equal to $-\Sigma_k x_k$, regardless of which term in the sum is under consideration. The characteristic number γ of $W(t)$ is the characteristic number of the sum of these terms. However, we have just seen that all terms of the sum have the same characteristic number $-\Sigma_k x_k$. Consequently, the characteristic number γ of the sum is also equal to $-\Sigma_k x_k$. Therefore, the characteristic number γ of the Wronskian determinant satisfies the relationship $\gamma = -\Sigma_k x_k \equiv -a_{ii}$.

Since a_{ii} is constant, we have

$$p(t) = \int_{t_0}^{t} a_{ii}\, \mathrm{d}t = a_{ii}(t - t_0)$$

Therefore, we have $e^{-p(t)} = \text{const}\, e^{-a_{ii}t}$, so that the characteristic number μ of $e^{-p(t)}$ must be equal to $+a_{ii}$.

If $\gamma = -a_{ii}$ and $\mu = +a_{ii}$, then we must have $\gamma + \mu = -a_{ii} + a_{ii} = 0$, which proves the regularity property of systems of variational equations with constant coefficients.

1.3.3 Linear Variational Equations with Periodic Coefficients

An important special case of linear variational equations with variable coefficients occurs when the coefficient functions are periodic. Owing to their great practical importance in celestial mechanics or the theory of vibrations, for example, a special theory has been developed for systems of differential equations with periodic coefficients. We assume that this theory is known to the reader. Detailed descriptions of this theory are given in the literature, especially under such topics as *Hill* and *Mathieu differential equations,* which are the most important types of equations with periodic coefficients (5).

$$\dot{\xi}_i = p_{ik}(t)\xi_k \qquad i = 1, 2, \ldots, n, \ k = 1, 2, \ldots, n \qquad (1.3.6)$$

in which the p_{ik} are real, bounded, and periodic functions of t with the common period ω. We shall assume that a fundamental system of solutions $\xi_i^{(k)}(t)$, $i = 1, 2, \ldots, n; k = 1, 2, \ldots, n$, is known, and we shall also assume that it satisfies the initial conditions $\xi_i^{(k)}(0) = \delta_{ik}$. Due to the periodic nature of the coefficients p_{ik}, the function $\xi_i^{(k)}(t + \omega)$ is also a system of (1.3.6). If the fundamental system $\xi_i^{(k)}(t)$ is substituted in (1.3.6), and if the transition from t to $t + \omega$ is made and the relationship $p_{ik}(t + \omega) = p_{ik}(t)$ is used, we have

$$\xi_i^{(k)}(t + \omega) = p_{ir}(t + \omega)\xi_r^{(k)}(t + \omega) = p_{ir}(t)\xi_r^{(k)}(t + \omega)$$

which illustrates the fact that $\xi_i^{(k)}(t + \omega)$ also satisfies the system (1.3.6). If this is true, then $\xi_i^{(k)}(t + \omega)$ must be a linear combination of the solutions of the fundamental system, so that we have

$$\xi_i^{(k)}(t + \omega) = a_{rk}\xi_i^{(r)}(t) \qquad i, r, k, = 1, 2, \ldots, n \qquad (1.3.7)$$

with the corresponding constants a_{rk}. These constants satisfy the relationship

$$\xi_i^{(k)}(\omega) = \alpha_{rk}\xi_i^{(r)}(0) = a_{rk}\,\delta_{ir}$$

because of the initial conditions that are valid for the fundamental system. It is obtained by substituting $t = 0$ in (1.3.7).

The relationship

$$a_{ik} = \xi_i^{(k)}(\omega) \qquad (1.3.8)$$

follows from this.

According to the theorem of *Floquet*, (1.3.6) has *multiplicative solutions* of the form

$$\xi_s(t + \omega) = \rho\xi_s(t) \qquad s = 1, 2, 3, \ldots, n \qquad (1.3.9)$$

and ρ is the *characteristic multiplier*. It must also be possible to obtain these solutions from the fundamental system solutions, and therefore the following relationships hold:

$$\xi_s(t) = b_i\xi_s^{(i)}(t) \qquad (1.3.10)$$

$$\xi_s(t + \omega) = b_i\xi_s^{(i)}(t + \omega) \qquad (1.3.11)$$

by introducing suitable constants b_i. By means of (1.3.7), we also obtain the following from (1.3.11):

$$\xi_s(t + \omega) = b_i a_{ri}\xi_s^{(r)}(t). \qquad (1.3.12)$$

It is thus possible to transform (1.3.9) into the following by using (1.3.10) and (1.3.12):

$$b_i a_{ri}\xi_s^{(r)} = \rho b_r\xi_s^{(r)}. \qquad (1.3.13)$$

The coefficients of the functions $\xi_s^{(r)}$ in the latter equation may be compared, leading to

$$(a_{ri} - \rho\delta_{ri})b_i = 0. \qquad (1.3.14)$$

This is a system of determining equations for the constants b_i. Because of the fact that it is a linear, homogeneous, algebraic system of equations, the requirement that b_i have non-trival solutions leads to the following:

$$\det(a_{ri} - \rho\delta_{ri}) = 0. \qquad (1.3.15)$$

It may be seen from this that the *characteristic multipliers ρ are the eigenvalues of the characteristic matrix* $\mathbf{A} = (a_{ri})$. This is analogous to the case of linear variational equations with constant coefficients: As we shall see shortly, the structure of the characteristic matrix $\mathbf{A}$ is decisive in establishing the type of solutions for the system of differential equations (1.3.6)

Let us first assume the simplest case, in which all elementary divisors of $\mathbf{A}$ are *linear*. Equation (1.3.15) then has n simple roots ρ_k, $k = 1, 2, \ldots, n$. The multiplicative solutions then comprise a fundamental system that satisfies

$$\xi_i^{(k)}(t + \omega) = \rho_k \xi_i^{(k)}(t) \qquad i,k = 1, 2, \ldots, n. \qquad (1.3.16)$$

This is a functional equation which is satisfied by

$$\xi_i^{(k)}(t) = e^{\lambda_k t}\phi_i^{(k)}(t) \qquad (1.3.17)$$

where

$$\lambda_k = (1/\omega) \ln \rho_k \qquad (1.3.18)$$

is the *characteristic exponent* and $\phi_i^{(k)}(t)$ is a function having the period ω. This may be demonstrated by substituting (1.3.17) and (1.3.18) in (1.3.16) because

$$\xi_i^{(k)}(t + \omega) = \exp(\ln \rho_k) \exp\left[(t/\omega) \ln \rho_k\right]\phi_i^{(k)}(t + \omega)$$

$$= \rho_k\, e^{\lambda_k t}\phi_i^{(k)}(t + \omega) = \rho_k\, e^{\lambda_k t}\phi_i^{(k)}(t) = \rho_k\xi_i^{(k)}(t)$$

is identically satisfied only when (1.3.18) holds and when $\phi_i^{(k)}(t)$ is periodic. If follows from this that (1.3.6) will have particular solutions of the form (1.3.17) if $\mathbf{A}$ only has linear elementary divisors.

On the other hand, if the elementary divisors are *not linear*, and if elementary divisor exponents of $\mathbf{A}$ exist which are larger than one, particular solutions are then obtained of (1.3.6) having the more complicated form

$$\xi_i^{(k)}(t) = P(t)\, e^{\lambda_k t}. \qquad (1.3.19)$$

The polynomials $P(t)$ are polynomials in t with periodic coefficients. The degree of these polynomials is one less than the corresponding elementary divisor exponent. Since this proof is very similar to that given in Section 1.3.1 for equations with constant coefficients, we shall not elaborate upon this further nor present the proof. The reader is referred to the literature (1.1) for a detailed discussion of this subject. A knowledge of the fundamental form of the (1.3.6) solutions is sufficient for performing stability studies; this is provided by (1.3.17) or (1.3.19). It may again be seen that the properties of the characteristic exponent λ_k are decisive for stability. The following theorems can thus be formulated, corresponding to those presented in Section 1.3.1:

Theorem 1: *When all the characteristic exponents $\lambda_k = 1/\omega \cdot \ln\rho_k$ have negative real parts, i.e., the magnitudes of the characteristic multipliers ρ_k are smaller than 1, the stationary solution basic to the variational equations is asymptotically stable.*

Proof: If the conditions presented in this theorem hold, all factors $e^{\lambda_k t}$, and therefore all particular solutions (1.3.17) or (1.3.19) and the general solution of (1.3.6), decrease continuously with time.

Theorem 2: *If only one of the characteristic exponents λ_k has a positive real part — i.e., the magnitude of one of the characteristic multipliers ρ_k is larger than 1 — the stationary solution is unstable.*

Proof: If the conditions presented in the theorem hold, at least one of these factors $e^{\lambda_k t}$, as well as one particular solution and the general solution of (1.3.6), increases continuously in time.

Theorem 3: *When one of the characteristic exponents λ_k has zero real parts —i.e., the magnitude of the corresponding multiplier ρ_k is 1 — the stationary solution is stable for $d_k = p_k$ (but no longer asymptotically stable), and is unstable for $d_k < p_k$.*

Proof: If the multiplicity p_k of the characteristic root ρ_k, whose magnitude is one, equals the decrease in rank d_k produced by it in the matrix $\mathbf{B} = (a_{ri} - \rho\delta_{ri})$, then the elementary divisors corresponding to ρ_k are linear and the corresponding polynomials $P(t)$ have degree zero with respect to t. Therefore, they degenerate to periodic functions. The solutions corresponding to this ρ_k therefore have the following form:

$$\xi_i^{(k)}(t) = \exp\left[(\pm)\mathrm{i}\arg\rho_k t/\omega\right]\phi_j^{(k)}(t)$$

and are also periodic and do not disturb the stability. On the other hand, if the multiplicity p_k of ρ_k is larger than d_k, then elementary divisor exponents $e_{\nu k} > 1$ are possible. At least one of the polynomials $P(t)$ will have the degree $e_{\nu k} - 1 = m \geqslant 1$, so that a particular solution having the following form:

$$\xi_j^{(k)}(t) = \exp\left[(\pm)\mathrm{i}\arg\rho_k t/\omega\right]t^m\phi_j^{(k)}(t)$$

exists. It represents a *secular term* of the (1.3.6) general solution and causes a continuous increase in time, thus destroying the stability.

The behaviour of the roots ρ_k of the characteristic equation

$$\det(a_{ri} - \rho\delta_{ri}) = 0$$

depending on the structure of the matrix $\mathbf{A} = (a_{ri})$, is decisive in establishing stability or instability. This equation may also be written in the following form:

$$\rho^n + a_1\rho^{n-1} + \ldots + a_{n-1}\rho + a_n = 0 \qquad (1.3.20)$$

where the coefficients a_i are functions of the elements of a_{ri} of $\mathbf{A}$. The unit circle of the complex ρ plane is transformed into the left half of the complex ω plane by means of the transformation $\rho = 1 + \omega/1 - \omega$. The characteristic equation (1.3.20) becomes

$$\left(\frac{1+\omega}{1-\omega}\right)^n + a_1\left(\frac{1+\omega}{1-\omega}\right)^{n-1} + \ldots + a_{n-1}\left(\frac{1+\omega}{1-\omega}\right) + a_n = 0. \qquad (1.3.21)$$

According to Theorem 1, the absolute magnitudes of the roots ρ_k of (1.3.20) must be smaller than one for stability to prevail. Points inside the unit circle of the ρ plane are therefore specified by the ρ_k. The ρ_k are transformed into the ω_k by means of the transformation $\rho_k = (1 + \omega_k)/(1 - \omega_k)$, and the ω_k are located in the left half of the ω plane. It may therefore be assumed that stability is assured when all roots ω_k of (1.3.21) have negative real parts. It is thus possible to apply the stability criteria already introduced in Section 1.3.1 — for example, the Hurwitz criterion — to Eq. (1.3.21).

We shall now show that *variational equations with periodic coefficients are always regular*. This means we must postulate the following: If γ is the characteristic number of the Wronskian determinant and if μ is the characteristic number of the function $\exp(- \int_{t_0}^t p_{ii}\, dt)$(where p_{ii} is the trace of the (1.3.6) coefficient matrix) we must then have $\gamma + \mu = 0$.

The following equation holds for the coefficient a_n of the characteristic Eq. (1.3.20):

$$(-1)^n a_n = \det \mathbf{A} = \det(a_{ik})$$

and therefore according to (1.3.8) we also have

$$(-1)^n a_n = \det[\xi_i^{(k)}(\omega)].$$

The following holds for the Wronskian determinant $W(t)$:

$$W(t) = W(t_0) \exp\left(\int_{t_0}^t p_{ii}\, dt\right) = \det[\xi_i^{(k)}(t)].$$

It we substitute $t_0 = 0$, $t = \omega$, it follows that

$$W(\omega) = W(0) \exp\left(\int_0^\omega p_{ii}\, dt\right) = \det[\xi_i^{(k)}(\omega)] = (-1)^n a_n.$$

However, we have $W(0) = \det[\xi_i^{(k)}(0)]$, and therefore, owing to $\xi_i^{(k)}(0) = \delta_{ik}$, we also have $W(0) = 1$, so that we finally obtain

$$(-1)^n a_n = \exp\left(\int_0 p_{ii}\, dt\right). \qquad (1.3.22)$$

Since $W(t) = \det[\xi_i^{(k)}(t)]$ and since in the most general case for $\xi_i^{(k)}(t)$, we have

$$\xi_i^{(k)}(t) = e^{\lambda_k t} P(t) = \exp[(1/\omega) \ln |\rho_k| \, t] \, \exp[i(1/\omega)\arg \rho_k t] \, P(t)$$

just as in Section 1.3.2, it follows that the characteristic number γ of $W(t)$ is given by

$$\gamma = -\frac{1}{\omega} \sum_k \ln |\rho_k| = -\frac{1}{\omega} \ln \left(\prod_k |\rho_k| \right).$$

According the Vieta root theorems and based on (1.3.22), we have

$$\prod_k |\rho_k| = (-1)^n a_n = \exp\left(\int_0^\omega p_{ii} \, dt \right)$$

so that the result is

$$\gamma = -\frac{1}{\omega} \int_0^\omega p_{ii} \, dt. \tag{1.3.23}$$

Let us now attempt to find the characteristic number μ of

$$\exp\left(- \int_{t_0}^t p_{ii} \, dt \right)$$

We may find it form the requirement that $e^{\mu t} \exp(-\int_0^t p_{ii} \, dt)$ must remain bounded for $t \to \infty$. This the case if we set

$$\mu = (1/\omega) \int_0^\omega p_{ii} \, dt \tag{1.3.24}$$

because

$$\mu t - \int_0^t p_{ii} \, dt = (t/\omega) \int_0^\omega p_{ii} \, dt - \int_0^t p_{ii} \, dt$$

is a function having the period ω and is therefore bounded for any t. Therefore, $e^{\mu t} \exp(-\int_0^t p_{ii} \, dt)$ always remains bounded. Consequently, $\mu = (1/\omega)\int_0^\omega p_{ii} \, dt$ is the characteristic number of $\exp(-\int_0^t p_{ii} \, dt)$.

With reference to (1.3.23) and (1.3.24), the regularity condition $\mu + \gamma = 0$ is satisfied.

Without discussing this further, we would like to mention that the variational equations with periodic coefficients are according to Lyapunov reducible, i.e., they may be transformed into equations with constant coefficients. We refer the reader to the literature (1.1) for a more detailed discussion.

As was already mentioned in Section 1.3.1, an attempt will always be made to determine stability or instability by using the simplest method possible. The characteristic exponents λ_k may be used for this purpose in the case of variational equations with constant coefficients. These exponents are obtained as the roots of the characteristic equation. We wished to avoid solving the characteristic equation, or even the variational equations, by applying stability criteria — for example, the criteria of Nyquist–Michailov or Hurwitz.

The same situation apparently exists in the case of variational equations with periodic coefficients when λ_k or ρ_k or ω_k are studied. The Hurwitz criterion could be applied to (1.3.21) in order to determine the behaviour of ω_k. Unfortunately, this is not quite true. According to (1.3.22) only the coefficient a_n of the characteristic equation is known. All other coefficients are functions of the elements a_{ik} of the characteristic matrix $\mathbf{A}$. Equation (1.3.8) holds for the a_{ik}. Therefore, it would be necessary to already know the particular solutions of the variational equations as well as specific values $\xi_i^{(k)}(\omega)$ in order to formulate the characteristic equation and to investigate the behaviour of the characteristic multipliers ρ_k. Therefore, one faces serious difficulties, and an attempt will be made to avoid them by applying approximation methods (perturbation method, series development method, infinite determinants method, etc.). This will now be discussed.

Problems encountered in practice often lead to differential equations with the following form:

$$\ddot{y} + [\lambda - hf(t)]\, y = 0$$

in which λ and h are parameters, and $f(t)$ is a function having the period ω. This is a *Hill differential equation*. By substituting $y = \xi_1$, $\dot{y} = \xi_2$, we obtain the following system from it:

$$\dot{\xi}_1 = \xi_2, \qquad \dot{\xi}_2 = (hf - \lambda)\xi_1$$

i.e., we have

$$p_{11} = 0 \qquad p_{12} = 1 \qquad p_{21} = hf - \lambda \qquad p_{22} = 0$$

so that $p_{ii} = p_{11} + p_{22} = 0$, and the Wronskian determinant satisfies $W(t) = W(0)\exp(\int_0^\omega p_{ii}\,dt) = W(0) = \text{const.}$ We have selected $\xi_i^{(k)}(t)$ as the fundamental system with the initial conditions $\xi_i^{(k)}(0) = \delta_{ik}$, $i, k = 1, 2$ so that $W(0) = \det[\xi_i^{(k)}(0)]$ holds. In this case, we have $W(t) = \text{const} = 1$ for any value of t and also for $t = \omega$, so that we obtain

$$W(\omega) = \xi_1^{(1)}(\omega)\xi_2^{(2)}(\omega) - \xi_1^{(2)}(\omega)\xi_2^{(1)}(\omega) = 1. \tag{1.3.25}$$

The characteristic equation follows from

$$\begin{vmatrix} a_{11} - \rho & a_{12} \\ a_{21} & a_{22} - \rho \end{vmatrix} = 0$$

or from the following due to $a_{ik} = \xi_i^{(k)}(\omega)$

$$\begin{vmatrix} \xi_1^{(1)}(\omega) - \rho & \xi_1^{(2)}(\omega) \\ \xi_2^{(1)}(\omega) & \xi_2^{(2)}(\omega) - \rho \end{vmatrix} = 0$$

and is given by

$$\rho^2 - [\xi_1^{(1)}(\omega) + \xi_2^{(2)}(\omega)]\rho + [\xi_1^{(1)}(\omega)\xi_2^{(2)}(\omega) - \xi_1^{(2)}(\omega)\xi_2^{(1)}(\omega)] = 0.$$

We have $a_1 = \xi_1^{(1)}(\omega) + \xi_2^{(2)}(\omega)$ and, because of (1.3.25), we have

$a_2 = \xi_1^{(1)}(\omega)\xi_2^{(2)}(\omega) - \xi_1^{(2)}(\omega)\xi_2^{(1)}(\omega) = 1$. It follows from $\rho^2 - a_1\rho + 1 = 0$ that

$$\rho_1 = \tfrac{1}{2}a_1 + \tfrac{1}{2}(a_1^2 - 4)^{\frac{1}{2}} \qquad \rho_2 = \tfrac{1}{2}a_1 - \tfrac{1}{2}(a_1^2 - 4)^{\frac{1}{2}}.$$

When $a_1^2 < 4$, the ρ_1, ρ_2 are complex, and the magnitude of both is 1. This is due to the fact that we must have $|\rho_1| |\rho_2| = a_2 = 1$ according to the Vieta root theorems. Stability is thus guaranteed.

When $a_1^2 > 4$, the ρ_1, ρ_2 are real and have different magnitudes. Due to $|\rho_1| |\rho_2| = 1$, we must have $|\rho_1| > 1$ and $|\rho_2| < 1$. Instability consequently prevails. In the case $a_1^2 = 4$ — that is, $a_1 = \pm 2$ — we have

$$\rho_1 = \rho_2 = \rho = \tfrac{1}{2} a_1 = \pm 1$$

i.e., we have $|\rho| = 1$, and this case corresponds to the stability limit.

We have the following conditions:

$$\tfrac{1}{2} |a_1| < 1 \qquad \text{for stability}$$

$$\tfrac{1}{2} |a_1| = 1 \qquad \text{for the stability limit.}$$

Using the definition of a_1, we may also write these as follows:

$$\tfrac{1}{2} |\xi_1^{(1)}(\omega) + \xi_2^{(2)}(\omega)| < 1 \qquad \text{for stability}$$

$$\tfrac{1}{2} |\xi_1^{(1)}(\omega) + \xi_2^{(2)}(\omega)| = 1 \qquad \text{for the stability limit.}$$

It is now apparent that in order to evaluate these stability criteria, solutions $\xi_1^{(1)}(t)$, $\xi_2^{(2)}(t)$ must be obtained, which also satisfy the initial conditions $\xi_1^{(1)}(0) = 1$, $\xi_2^{(2)}(0) = 1$. Once these have been determined, they may be written in the following form:

$$\xi^{(1)} = \xi_1^{(1)}(\lambda, h, t) \qquad \xi^{(2)} = \xi_2^{(2)}(\lambda, h, t)$$

i.e., as functions of the parameters λ and h included in the differential equation. The stability criterion

$$\tfrac{1}{2} |\xi_1^{(1)}(\lambda, h, \omega) + \xi_2^{(2)}(\lambda, h, \omega)| = 1$$

may also be regarded as an implicit equation connecting λ and h. The solution of this equation results in the limiting curves of the stability regions in the λ, h plane, which is the parameter plane.

Example: If we set $f(t) = 2 \cos 2t$, we obtain the *Mathieu differential equation*. The *Strutt map* (Fig. 16) specifies the division of the λ, h plane.

The small oscillations of a vertical, heavy pendulum having length l, whose point of suspension harmonically oscillates in the vertical direction according to $u = A \cos \omega t$, satisfy the differential equation

$$\phi''(x) + (\lambda - 2h \cos 2x)\phi(x) = 0$$

where $x = \tfrac{1}{2}\omega t$, $\lambda = -4g/\omega^2 l$, $2h = -4A/l$. This is a Mathieu differential equation. The system

$$\dot{\xi}_1 = \xi_2 \qquad \dot{\xi}_2 = (2h \cos 2\tau - \lambda)\xi_1 \qquad (1.3.26)$$

is obtained by using the relationships $\phi = \xi_1$, $\phi' = \xi_2$, $x = \tau$. Differention with respect to the argument again is indicated by a point.

56

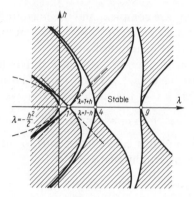

Fig. 16. Stability map of Strutt

We have $\omega = \pi$, and the fundamental system

$$\xi_1^{(1)}(\tau) \qquad \xi_1^{(2)}(\tau)$$
$$\xi_2^{(1)}(\tau) \qquad \xi_2^{(2)}(\tau)$$

is employed.

Let us assume that τ is replaced by $-\tau$, and let $\xi_1^{(1)}(-\tau) = \xi_1^{(1)}(\tau)$ be a solution of the system. Substitution in (1.3.26) indicates that $\xi_1^{(1)}(-\tau) = -\xi_1^{(1)}(\tau)$ must hold. In the same way, we may assume that $\xi_1^{(2)}(-\tau) = -\xi_1^{(2)}(\tau)$ is a solution by transition from τ to $-\tau$.

Substitution in (1.3.26) results in the relationship $\xi_2^{(2)}(-\tau) = \xi_2^{(2)}(\tau)$. We therefore have

$$\left. \begin{array}{ll} \xi_1^{(1)}(\tau) = \xi_1^{(1)}(-\tau) & \xi_1^{(2)}(\tau) = -\xi_1^{(2)}(-\tau) \\ \xi_2^{(1)}(\tau) = -\xi_2^{(1)}(-\tau) & \xi_2^{(2)}(\tau) = \xi_2^{(2)}(-\tau). \end{array} \right\} \qquad (1.3.27)$$

According to (1.3.7), we have

$$\xi_1^{(1)}(\tau + \pi) = a_{11}\xi_1^{(1)}(\tau) + a_{21}\xi_1^{(2)}(\tau)$$
$$\xi_1^{(2)}(\tau + \pi) = a_{12}\xi_1^{(1)}(\tau) + a_{22}\xi_1^{(2)}(\tau).$$

If we substitute $\tau = -\pi$, if follows that

$$\xi_1^{(1)}(0) = 1 = a_{11}\xi_1^{(1)}(-\pi) + a_{21}\xi_1^{(2)}(-\pi)$$
$$\xi_1^{(2)}(0) = 0 = a_{12}\xi_1^{(1)}(-\pi) + a_{22}\xi_1^{(2)}(-\pi).$$

According to (1.3.8) and (1.3.27), for $\omega = \pi$, we have

$$\xi_1^{(1)}(-\pi) = \xi_1^{(1)}(\pi) = a_{11}$$
$$\xi_1^{(2)}(-\pi) = -\xi_1^{(2)}(\pi) = -a_{12}$$

so that we finally obtain

$$1 = a_{11}^2 - a_{21}a_{12}$$

$$0 = a_{12}a_{11} - a_{22}a_{12}.$$

The relationship

$$1 = a_{11}a_{22} - a_{12}a_{21}$$

also follows from (1.3.8) and (1.3.25).

If the first equation is subtracted from this one, it follows that

$$a_{11}^2 - a_{11}a_{22} = 0.$$

Therefore, the relationships

$$a_{11}(a_{11} - a_{22}) = 0 \qquad a_{12}(a_{11} - a_{22}) = 0 \qquad a_{11}a_{22} - a_{12}a_{21} = 1$$

must hold simultaneously. Only in the case $a_{11} \neq 0$, $a_{12} \neq 0$, $a_{11} = a_{22}$ is there no contradiction. The following relationships therefore hold for the Mathieu differential equation:

$$a_{11} = a_{22} \qquad \text{or} \qquad \xi_1^{(1)}(\pi) = \xi_2^{(2)}(\pi).$$

Therefore,

$$\xi_1^{(1)}(\pi) + \xi_2^{(2)}(\pi) = 2\xi_1^{(1)}(\pi)$$

and the stability conditions becomes $|\xi_1^{(1)}(\pi)| < 1$ for stability, $|\xi_1^{(1)}(\pi)| = 1$ for the stability limit.

We must now find the solution $\xi_1^{(1)}(\tau)$. This will be done according to a method formulated by Poincaré, which is known as the *perturbation method*. The trial solution

$$\left.\begin{array}{l} \xi_1^{(1)} = \xi_{10}^{(1)} + h\xi_{11}^{(1)} + h^2\xi_{12}^{(1)} + \cdots \\ \xi_2^{(1)} = \xi_{20}^{(1)} + h\xi_{21}^{(1)} + h^2\xi_{22}^{(1)} + \cdots \end{array}\right\} \tag{1.3.28}$$

is used, assuming that h is a sufficiently small parameter for the series to converge.

If (1.3.28) is substituted into the system (1.3.26), we obtain the following new systems, arranged according to powers of h:

$$\dot{\xi}_{10}^{(1)} = \xi_{20}^{(1)}, \ldots, \qquad \dot{\xi}_{11}^{(1)} = \xi_{21}^{(1)}, \ldots, \qquad \dot{\xi}_{12}^{(1)} = \xi_{22}^{(1)}, \ldots,$$

$$\dot{\xi}_{20}^{(1)} = -\lambda\xi_{10}^{(1)}, \qquad \dot{\xi}_{21}^{(1)} = -\lambda\xi_{11}^{(1)} + 2\cos 2\tau\xi_{10}^{(1)}, \qquad 22$$

$$\ddot{\xi}^{(1)} = -\lambda\xi_{12}^{(1)} + 2\cos 2t\xi_{11}^{(1)}, \ldots.$$

Instead of this we may also write

$$\ddot{\xi}_{10}^{(1)} = -\lambda\xi_{10}^{(1)}, \qquad \ddot{\xi}_{11}^{(1)} = -\lambda\xi_{11}^{(1)} + 2\cos 2\tau\xi_{10}^{(1)}$$

$$\ddot{\xi}_{12}^{(1)} = -\lambda\xi_{12}^{(1)} + 2\cos 2\tau\xi_{11}^{(1)}, \tag{1.3.29}$$

$$\vdots$$

$$\ddot{\xi}_{1k}^{(1)} = -\lambda\xi_{1k}^{(1)} + 2\cos 2\tau\xi_{1(k-1)}^{(1)},$$

as can easily be seen. The corresponding initial conditions are $\xi_{10}^{(1)}{}'(0) = 1$, $\xi_{1k}^{(1)}(0) = 0$ for $k \geqslant 1$, which follows from the fact that $\xi_1^{(1)}{}'(0) = 1$ should hold for every h.

Taking into account the initial conditions, the solutions of (1.3.29) are

$$\xi_{10}^{(1)} = \cos \lambda^{\frac{1}{2}} \tau$$

$$\xi_{ik}^{(1)} = \frac{2}{\lambda^{\frac{1}{2}}} \int_0^{\tau} \xi_{1(k-1)}^{(1)}(x) \cos 2x \sin \lambda^{\frac{1}{2}}(\tau - x) \, dx, \qquad k \geqslant 1.$$

We thus obtain

$$\xi^{(1)}(\pi) = \sum_{k=0}^{\infty} h^k \xi_{1k}^{(1)}(\pi)$$

with

$$\xi_{10}^{(1)}(\pi) = \cos \lambda^{\frac{1}{2}} \pi \qquad \xi_{11}^{(1)}(\pi) = 0$$

$$\xi_{12}^{(1)}(\pi) = \frac{\pi \lambda^{\frac{1}{2}} \sin \pi \lambda^{\frac{1}{2}}}{4\lambda(\lambda - 1)} \qquad \xi_{13}^{(1)}(\pi) = 0.$$

If this series is already truncated at the first term, the following condition is obtained for the stability limit

$$\xi_{10}^{(1)}(\pi) = \cos \lambda^{\frac{1}{2}} \pi = 1 \qquad \text{or} \qquad \lambda = n^2.$$

As can be seen, we have obtained the limiting points for $h = 0$ in the Strutt map along the λ axis. If the third term in the series is considered, then the following condition holds for the stability limit

$$\cos \lambda^{\frac{1}{2}} \pi + h^2 \frac{\pi \lambda^{\frac{1}{2}} \sin \pi \lambda^{\frac{1}{2}}}{4\lambda(\lambda - 1)} = 1.$$

However, if we postulate that not only h but also λ is small, we may stipulate the following:

$$\cos \lambda^{\frac{1}{2}} \pi \approx 1 - \frac{\lambda \pi^2}{2}$$

$$\frac{\pi \lambda^{\frac{1}{2}} \sin \pi \lambda^{\frac{1}{2}}}{4\lambda(\lambda - 1)} \approx -\frac{\pi^2}{4}.$$

The condition of the stability boundary becomes

$$1 - \frac{\lambda \pi^2}{2} - \frac{h^2 \pi^2}{4} = 1, \qquad \lambda = \frac{h^2}{2}.$$

As can be seen from Fig. 16, we have indeed obtained the approximate course of the limiting curve of the Strutt map in the vicinity of the origin of the λ, h plane. The course of the remaining limiting curves of this map can be obtained to any degree of accuracy by applying other methods, for example, by a Mathieu[1] series development. If stability is to prevail, we must require that

$$1 - \frac{\lambda \pi^2}{2} - \frac{h^2 \pi^2}{4} < 1 \qquad \lambda + \frac{h^2}{2} > 0.$$

For the vertical pendulum mentioned previously, this means that

$$4g/\omega^2 l < 2A^2/l^2$$

so that its point of suspension must be displaced with a frequency $\omega^2 > 2gl/A^2$ in order that the vertical position of the pendulum be stable.

Exercises 1.3

1. Given the simultaneous first-order system of variational equations.

$$\dot{\xi}_1 = 3\xi_1 - \xi_2 + \xi_3$$
$$\dot{\xi}_2 = 2\xi_1 + \xi_3$$
$$\dot{\xi}_3 = \xi_1 - \xi_2 + 2\xi_3$$

   a) Setting $\dot{\xi}_i = a_k \xi_k$ identify the matrix $\mathbf{A} = (a_{ik})$ and the characteristic matrix $\mathbf{B}$
   b) Using $\mathbf{B}$ determine the Smith normal form $N(\lambda)$.
   c) Using $N(\lambda)$, find the characteristic of $\mathbf{A}$.
   d) Using the characteristic of $\mathbf{A}$ determine the set of fundamental solutions $\xi_{i,o}, i = 1, 2, 3$.
   e) Using the fundamental solutions, find, the general solution to the system of variational equations above.

2. Given the simultaneous first-order system of variational equations

$$\dot{\xi}_1 = \xi_2$$
$$\dot{\xi}_2 = -\xi_1 - 2\xi_2$$
$$\dot{\xi}_3 = \xi_4,$$
$$\dot{\xi}_4 = -\xi_3 - 2\xi_4 + \xi_1 + \xi_2,$$
$$\dot{\xi}_i = a_{ik}\xi_k$$

   a) Determine the eigenvalues and the characteristic of $\mathbf{A} = (a_{ik})$.
   b) Determine the fundamental solutions of the system of variational equations.
   c) Determine the general solution of the system.

3. Given the simultaneous first-order system of variational equations

$$\dot{\xi}_1 = -2\xi_1 + 2\xi_2 - 3\xi_3,$$
$$\dot{\xi}_2 = 2\xi_1 + \xi_2 - 6\xi_3,$$
$$\dot{\xi}_3 = -\xi_1 - 2\xi_2,$$
$$\dot{\xi}_i = a_{ik}\xi_k.$$

---

[1]See, for example, Rothe, R., and Szabo, I., *Higher Mathematics*, Vol. VI, p. 195. Teubner, Stuttgart, 1965. Also, W. J. Cunningham, *Introduction to Nonlinear Analysis*, pp. 259–274, McGraw-Hill Inc., New York, 1958.

Decide on the stability of the trivial solution using the appropriate theorem of the three given in section 1.3.1.

4. Given the following system of first-order differential equations:

$$\dot{q}_1 = q_1 + q_2 \qquad \dot{q}_2 = q_1^2 + q_2.$$

a) Find the stationary solutions of the systems.

b) For each stationary solution derive the variational equations and use these to determine stability or instability of the respective stationary solution.

5. Given the simultaneous first-order system of variational equations

$$\dot{\xi}_1 = -2\xi_2,$$
$$\dot{\xi}_2 = -\xi_1 + \xi_2 - 5\xi_3,$$
$$\dot{y}_3 = (1 - \mathbb{C})\xi_1 + \xi_2 - 2\xi_3 \qquad \mathbb{C} = \text{constant.}$$

Use Hurwitz' stability criterion in order to determine how $\mathbb{C}$ has to be chosen so that the trival solution of the variational equations turns out to be stable.

6. Prove that any linear first-order system of variational equations with periodic coefficients is *regular* in the sense of Lyapunov.

7. Given the differential equation $y'' + g(x)y = 0$, where

$$g(x) \leqslant 0 \qquad g(x) \equiv 0, \qquad g(x) \text{ periodic with } \omega.$$

Show that the solutions of this differential equation are unstable for

$$y_1(0) = 1 \qquad y_1'(0) = 0 \qquad y_2(0) = 0 \qquad y_2'(0) = 1$$

where the y_i, $i = 1, 2$, are the fundamental solutions of the differential equations.

8. Given the differential equation $\ddot{y} + (1 + \cos t)y = 0$, investigate the stability of the solution of this equation when the fundamental solutions satisfy the conditions $y_1(0) = 1$, $\dot{y}_1(0) = 0$, $y_2(0)$, $\dot{y}_2(0) = 1$.

1.4
Investigations in Phase Space

Let us assume that a mechanical process, similar to that which we have already discussed, is described by the system of differential equations $\dot{q}_i = F_i(q, \alpha^0)$. We will restrict ourselves to autonomous systems. In order to investigate the stability of a particular solution $\dot{q}_i^0 = F_i(q^0, \alpha^0)$ we will attempt to obtain our answer as easily as possible, as described in previous chapters. In other words, we will attempt to reach conclusions regarding stability *without the complete integration* of the system of differential equations whenever possible. In addition to the possibilities discussed in Section 1.3, we may use the *intermediate integrals* for this purpose, which geometrically represent *phase surfaces* and which have already been mentioned. The course of the solutions lying along the phase surfaces can be obtained from the properties of the phase surfaces. We may at least determine whether *orbital stability* exists or not.

The simplest and most easily understandable condition is obtained for a system with two variables, which is described by

$$\dot{q}_1 = F_1(q_1, q_2, \alpha^0) \qquad \dot{q}_2 = F_2(q_1, q_2, \alpha^0). \qquad (1.4.1)$$

The phase space then becomes the *phase plane*, and the phase surfaces become *phase curves*. This is shown in Fig. 17 for the case of orbital stability: the undisturbed solution q_i^0 corresponds to the phase curve C_0, and the adjacent, disturbed solutions correspond to the phase curves $C_{\mathscr{S},1}$, $C_{\mathscr{S},2}$, which remain entirely in the 'tube' (strip, in this case) $U_k \mathscr{K}_r(P_k)$ (Section 1.1.2). Therefore, we have orbital stability.

We should mention that the representaton of an entire family of phase curves is called a *phase diagram*. With a phase diagram it is possible to obtain the stability behavior of not only one, but an arbitrary number of solutions.

We have restricted ourselves to *autonomous* processes for purposes of simplicity. Naturally, it is possible to treat *non-autonomous* systems in this same way. In this case it is no longer possible to use the q_1, q_2 *plane*, but it is necessary to employ the q_1, q_2, t *space*. The phase curves become curves in space which often have a very complicated course, which makes the problem difficult to follow. The method of phase curves cannot be recommended in most cases, and will therefore not be treated here. On the other hand, the

61

62

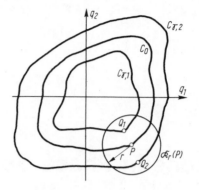

Fig. 17. Phase curves with orbital stability

use of a *phase cylinder* is to be recommended when the phase curves of autonomous processes are periodic in at least one of the variables, for example in q_1 (Fig. 18). In this case, there is only an apparent spatial representation. In reality, the phase curve is cut off along the length of period of q_1, and this piece is rolled up into a folder.

On the other hand, it is not difficult to study *non-linear* processes if it is possible to find intermediate integrals for them. This is especially true for *conservative* processes where the energy theorem represents such an intermediate integral. This method is primarily used when dealing with the theory of non-linear oscillations of a system having one degree of freedom.

Example: Let us assume that the vibration process is given by $\ddot{x} + f(x) = 0$, where $f(x)$ is a non-linear restoring force. By means of the substitutions $x = q_1$, $\dot{x} = q_2$ we obtain the system of differential equations.

$$\dot{q}_1 = q_2, \qquad \dot{q}_2 = -f(q_1).$$

As can easily be seen, the intermediate integral is

$$\tfrac{1}{2} q_2^2 + V(q_1) = h \qquad (1.4.2)$$

Fig. 18. Phase plane (development of the phase cylinder), phase cylinder

Fig. 19. Energy surface and phase curves

with $V(q_1) = \int_0^{q_1} f(q_1)\, dq_1$ and the parameter h. Since it is possible to interpret $\frac{1}{2}q_2^2$ as kinetic energy, $V(q_1)$ as the potential energy, and h as an energy constant, we have indeed found the energy integral of the process.

The relationship (1.4.2) can be represented as a phase curve in the q_1, q_2 plane for a certain value of the parameter h (Fig. 19). A complete family of phase curves $C_{h_i} i = 1, 2, \ldots,$[1] is obtained — that is, the phase diagram — when the contour curves on the energy surface $h = \frac{1}{2}q_2^2 + V(q_1) = H(q_1, q_2)$ resulting from the selection of the parameter h_i, $i = 1,, 2, \ldots$, are projected onto the q_1, q_2 plane.

In the most general case, the phase curves corresponding to a particular solution of interest are first plotted. Perturbations (we first only consider perturbations in the initial conditions) are then carried out. The phase curves corresponding to the perturbed solutions are then plotted in the phase diagram. Their course with respect to the curve for the unperturbed solution makes it possible to conclude whether or not orbital stability exists. In most cases, the stability of two types of solutions — that is, of stationary and periodic solutions — will be of most interest. Stationary solutions are characterized by $\dot{q}_i = 0$. They represent *equilibrium positions* and are represented by points $q_{i,\,s}$ in the phase space. Periodic solutions, i.e., solutions

---

[1] k is the number of the intermediate integral.

that are periodic with respect to the parameter t (which usually represents time), lead to closed curves in the phase space. These curves are called *limit cycles* if the process corresponding to them is the limiting state of *self-excited vibrations*.

1.4.1 Stationary Points

The course of the phase curves in the vicinity of stationary points is obtained as follows.

We set $q_i \equiv q_{i,s}$, $\dot{q}_i \equiv \dot{q}_{i,s} = 0$ in the system of differential equations $\dot{q}_i = F_i(q, \alpha^0)$. A system of determining equations for the stationary points $q_{i,s}$ is thus obtained with $F_i(q_s, \alpha^0) = 0$. Then $q_i = q_{i,s} + \xi_i$ is substituted into the original system of differential equations, and one obtains the following system of differential equations for the variations ξ_i considering that $\dot{q}_{i,s} = 0$, $F_i(q_s, \alpha^0) = 0$ holds:

$$\dot{\xi}_i = (\partial F_i / \partial q_k)_s \xi_k + \Phi_i [(\partial^n F_i / \partial q_k^n)_s, \xi_k] \qquad (1.4.3)$$

The terms of the Taylor series development, represented by $\Phi_i[(\partial^n F_i / \partial q_k^n)_s, \xi_k]$, contain the higher derivatives of F_i and the ξ_k raised to higher powers.

If the coefficients $(\partial F_i / \partial q_k)_s$ are not zero, the point is a *simple* stationary point. On the other hand, if they are all equal to zero for a given i, then we have $\dot{\xi}_i = \Phi_i(\xi_k)$ for this i. This point is then a *non-simple* stationary point.

The course of the phase surfaces in the vicinity of the stationary point is described in Eq. (1.4.3). If the point is a simple stationary point and if we restrict ourselves to its immediate vicinity, then only the linear terms will be important. We may then use the *abbreviated, linear* variational equation

$$\dot{\xi}_i = (\partial F_i / \partial q_k)_s \xi_k = a_{ik} \xi_k.$$

The results presented in Section 1.3 may be used when discussing the solutions of these equations.

For purposes of simplicity, we shall again assume below that we are dealing with variational equations and two variables. The abbreviated variational equations are then the following:

$$\left. \begin{array}{l} \dot{\xi}_1 = a_{11}\xi_1 + a_{12}\xi_2, \\ \dot{\xi}_2 = a_{21}\xi_1 + a_{22}\xi_2. \end{array} \right\} \qquad (1.4.4)$$

The form of the solutions depends on the roots $\lambda_{1,2}$ of the characteristic equation $\det(a_{ik} - \delta_{ik}\lambda) = 0$. If particular solutions of (1.4.4) have been based on certain initial conditions and on the $\lambda_{1,2}$, the phase curves may be plotted in the ξ_1, ξ_2 plane in the vicinity of the stationary point. The stationary point itself is transformed to the origin of the ξ_1, ξ_2 plane by variable transformation. Once the phase diagram surrounding it is known, conclusions may be reached regarding its stability in the small.

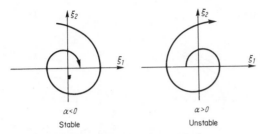

Fig. 20. Unstable and stable focus

For example, if the roots $\lambda_{1,2} = \alpha \pm i\beta$ are conjugate complex roots, a stable focus is obtained for $\alpha < 0$, and an unstable focus is obtained for $\alpha > 0$ (Fig. 20). In the first case, we have (1.1.6), i.e.,

$$|\xi_i(t)| < \varepsilon \qquad \text{for } t > t_0, i = 1, 2, \ldots,$$

if $|\xi_i(t_0)| < \eta(\varepsilon)$ holds for the chosen ε. Orbital stability as well as stability satisfying the stricter requirements of Lyapunov then prevail. In the second case, instability is proved in a corresponding way.

We shall now discuss the *types of stationary points* in general. The roots of the characteristic equation satisfy

$$\lambda_{1,2} = \tfrac{1}{2}\{(a_{11} + a_{22}) \pm [(a_{11} + a_{22})^2 + 4(a_{12}a_{21} - a_{11}a_{22})^{1/2}\}.$$

If both roots are real and have the same sign, a *node* is obtained. It is stable for $a_{11} + a_{22} < 0$ and unstable for $a_{11} + a_{22} > 0$ (Fig. 21).

If both roots are real and have opposite sign, a *saddle* (Fig. 22) is obtained corresponding to instability.

If both roots are purely imaginary, a *center* (Fig. 23) is obtained which is stable.

If the roots are conjugate, complex roots, a *focus* is obtained, which is stable for $\alpha = a_{11} + a_{22} < 0$ and unstable for $a_{11} + a_{22} > 0$ (Fig. 20).

Let us again assume only two variable q_1, q_2 or ξ_1, ξ_2, respectively, and we shall restrict ourselves to the immediate vicinity of a *non-simple*, stationary

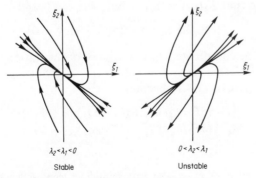

Fig. 21. Stable and unstable node

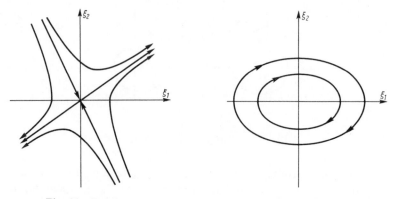

Fig. 22. Saddle   Fig. 23. Center

point. For a given i, we have $(\partial F_i/\partial q_k)_s = 0$. The abbreviated, non-linear, variational equation

$$\dot{\xi}_i = \tfrac{1}{2}(\partial^2 F_i/\partial q_j\partial q_k)_s\xi_j\xi_k \qquad j, k = 1, 2, \ldots$$

is obtained from

$$\dot{\xi}_i = \Phi_i[(\partial^n F_i/\partial_k^n), \xi_k] \tag{1.4.5}$$

This is due to the fact that second-order terms primarily determine the course of the phase curves in the ξ_1, ξ_2 plane[1] in the vicinity of the stationary point. A system of non-linear, differential equations is obtained. An attempt will first be made to find an intermediate integral, so that approximate phase curves can be plotted in the vicinity of the stationary point. These curves may then be employed to determine stability. If this is not possible, the course of the phase curves must be determined indirectly from (1.4.5) using the *isocline method*, which is a graphic method.

Example: Let us consider the following system:

$$\dot{q}_1 = F_1(q_1, q_2) = q_2, \qquad \dot{q}_2 = F_2(q_1, q_2) = aq_1 + bq_1^2 + cq_1^3. \tag{1.4.6}$$

We shall assume that $F_2(q_s) \equiv aq_{1s} + bq_{1s}^2 + cq_{1s}^3 = 0$ and

$$(\partial F_2/\partial q_1)_s = a + 2bq_{1s} + 3cq_{1s}^2 = 0$$

are satisfied for the coordinates $q_{1s} = \alpha$, $q_{2s} = 0$ of the stationary point. (These two conditions are satisfied by $b = -2a/\alpha$, $c = a/\alpha^2$.)

According to (1.4.3), (1.4.6) becomes

$$\left.\begin{aligned}\dot{\xi}_1 &= (\partial F_1/\partial q_1)_s\xi_1 + (\partial F_1/\partial q_2)_s\xi_2 + \Phi_1\\ \dot{\xi}_2 &= (\partial F_2/\partial q_1)_s\xi_1 + (\partial F_2/\partial q_2)_s\xi_2 + \Phi_2.\end{aligned}\right\} \tag{1.4.7}$$

---

[1] If the second derivatives of F_i are also zero, third-order terms with respect to ξ_k must be considered, etc.

We have

$$\partial F_1/\partial q_1 \equiv 0, \qquad \partial F_1/\partial q_2 \equiv 1, \qquad \partial F_2/\partial q_2 \equiv 0,$$

and, according to our assumption, $(\partial F_2/\partial q_1)_s = 0$. The following linear equation results if Φ_1 is neglected in the first line of (1.4.7):

$$\dot{\xi}_1 = \xi_2.$$

All the linear terms vanish in the second line of (1.4.7). The quadratic terms in Φ_2 must then be considered. The approximate equation

$$\dot{\xi}_2 = \frac{1}{2}\left[\left(\frac{\partial^2 F_2}{\partial q_1^2}\right)_s \xi_1^2 + 2\left(\frac{\partial^2 F_2}{\partial q_1\,\partial q_2}\right)_s \xi_1\xi_2 + \left(\frac{\partial^2 F_2}{\partial q_2^2}\right)\xi_2^2\right]$$

is obtained in that way. The equations

$$\frac{\partial^2 F_2}{\partial q_1\,\partial q_2} \equiv 0, \qquad \frac{\partial^2 F_2}{\partial q_2^2} \equiv 0, \qquad \left(\frac{\partial^2 F_2}{\partial q_1^2}\right)_s = 2b + 6cq_{1s} = \frac{2a}{\alpha}$$

must also be taken into account. The special values of $(\partial^2 F_2/\partial q_1^2)_s$ for q_{1s}, b, and c have been substituted. The following non-linear system of variational equations is finally obtained for the non-simple stationary point $(\alpha, 0)$:

$$\dot{\xi}_1 = \xi_2, \qquad \dot{\xi}_2 = (a/\alpha)\xi_1^2.$$

The following relationship may be derived from these variational equations:

$$\frac{d\xi_2}{d\xi_1} = \frac{a}{\alpha}\frac{\xi_1^2}{\xi_2}$$

so that along the isoclines

$$\xi_1 = 0, \qquad \xi_2 = 0, \qquad \xi_2 = \pm(a/\alpha)\xi_1^2$$

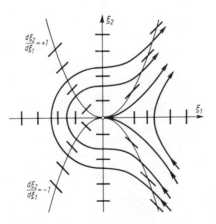

Fig. 24. Non-simple stationary point

68

the tangent inclinations of the phase curves of the ξ_1, ξ_2 plane are

$$d\xi_2/d\xi_1 = 0, \qquad d\xi_2/d\xi_1 = \infty, \qquad d\xi_2/d\xi_1 = \pm 1.$$

This small amount of information is sufficient to plot the phase curve in the vicinity of the stationary point. This is found to be an unstable point (Fig. 24).

1.4.2 Limit Cycles

Particular solutions of the periodic type are of as much interest as the stationary solutions. They correspond to closed phase curves. If a vibration problem is given by the system

$$\dot{q}_1 = q_2, \qquad \dot{q}_2 = - \omega^2 q_1,$$

the intermediate integral

$$(q_2^2/2h) + (\omega^2/2h)q_1^2 = 1$$

is usually obtained. It leads to a family of ellipses in the phase plane (q_1, q_2) whose semi-axes are $(1/\omega)(2h)^{1/2}$ and $(2h)^{1/2}$ depending on the parameter h (Fig. 25). The origin becomes a center. All solutions are periodic and pass around the stationary point (origin) and are closed curves. This represents a simple illustration of the fact that closed phase curves correspond to the periodic solutions of the vibration equation.

Isolated, periodic solutions may represent a special case in the general case of a non-linear system of differential equations. The phase diagram then reveals isolated, closed phase curves. The course of the phase curves in the vicinity of these curves indicates whether there are stable or unstable periodic solutions (Fig. 26) in the sense of orbital stability.

For example, if the trajectories of the perturbed solution (for example, $C\gamma_1, C\gamma_2$) come arbitrarily close to the closed trajectory C_a for $t \to \infty$, as is the case for C_a, the cases pending periodic solution is stable. If the adjacent

Fig. 25. Elliptical phase curves

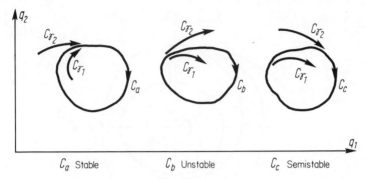

Fig. 26. Limit cycles

trajectories continue to move away for $t \to \infty$, as is the case for C_b, then the periodic solutions is unstable. If both cases occur simultaneously, the periodic solution is semistable. The case C_c represents an example of this. A semistable solution is unstable for practical purposes.

Isolated periodic solutions are of special interest in the fields of self-sustained and self-excited vibrations. There are many mechanical and electrical examples, for example, clock movements, electric oscillation circuits with grid tubes, etc. Several closed curves may prevail in the phase diagram (Fig. 27). The phase curve C_a corresponds to a stable, periodic solution with *soft self-excitation*, because its *region of self-oscillation* G_1 surrounds the origin of the phase plane which is an unstable focus. At the same time, instability in the small with respect to the origin may occur concurrently with stability in the large, due to C_a. The phase curve C_b is unstable in a way such that trajectories within its *damping region* G_2 asymptotically approach the stable curve C_a. It may be readily seen that the closed phase curves may be regarded as asymptotic limiting curves of the adjacent trajectories. It is for this reason that they are also designated as *limit cycles*.

Fig. 27. Limit cycles

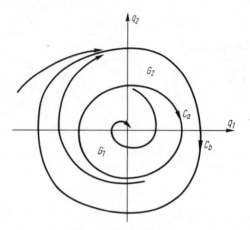

Fig. 28. Limit cycles

Fig. 28 illustrates the opposite conditions. The cycle C_a is unstable, and C_b is stable. Stability in the small and instability in the large prevail. G_1 is the damping region, and G_2 is the self-excitation region. This is the case of *hard self-excitation*, because a perturbation of sufficient magnitude to make the perturbed trajectory enter the region of self-excitation G_2 is necessary to produce an asymptotic transition to the vibration represented by C_b.

There are several theorems predicting closed phase curves, particularly those advanced by Poincaré and Bendixon.

Let us first consider the criterion of Poincaré stipulating the requisite condition for closed phase curves. We shall restrict ourselves to a system of two variables. The gradient field in the phase plane is specified by

$$\dot{q}_1 = F_1(q_1, q_2), \quad \dot{q}_2 = F_2(q_1, q_2).$$

The inclination of every line element of the field with respect to the q_1 axis of the phase is given by the angle $\Theta = \arctan F_1/F_2$. If a closed curve K not passing through any stationary point is traversed in the mathematically positive direction, line elements on K will be rotated by an angle $2\pi J$ if the image point on the curve K rotates by an angle 2π. According to Poincaré, the rational number J is called the *index* of the curve. It is calculated from

$$J = \frac{1}{2\pi} \int_K d\left(\arctan \frac{F_2}{F_1}\right) = \frac{1}{2\pi} \int_K \frac{F_1 \, dF_2 - F_2 \, dF_1}{F_1^2 + F_2^2}.$$

Criterion of Poincaré: The index J of a curve K is zero if K encloses only regular, but no stationary points. It equals $+1$ if K encloses a focus, center, or node. It equals -1 if K encloses a saddle. These values may also be regarded as the indices of the points surrounded by K. If the region of the phase plane

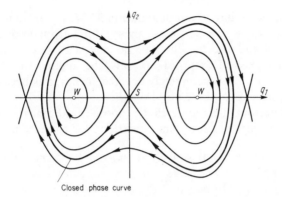

Fig. 29. Closed Phase curve contains a saddle S
and two centres W

delimited by K contains the n stationary points $S_1 \ldots, S_n$ we have

$$J(K) = \sum_{k=1}^{n} J(S_k). \qquad (1.4.8)$$

The index of a closed phase curve is $+1$. Since the index of a regular point is zero, it follows that a closed phase curve must enclose at least one stationary point. In addition, it follows from (1.4.8) that *a closed phase curve can never enclose only one saddle S*, but must enclose other stationary points, for example, two centers W. Only in this way do we have $J(S) + 2 J(W) = -1 + 2 = 1$, so that the index of the closed phase curve equals $+1$, as must be the case (Fig. 29).

The stipulation that the index of a closed phase curve equal $+1$ is necessary, but is not sufficient in itself. If $J = +1$, a limit cycle does not definitely exist. It is therefore advantageous to employ other theorems in addition. We shall first mention the theorem advanced by Bendixson. This theorem may be employed to demonstrate the existence of a limit cycle in a region B of the phase plane based on a sufficient condition.

Theorem of Bendixson: *If the sign of the expression $\partial F_1/\partial q_1 + \partial F_2/\partial q_2$ does not change in B, no closed phase curves can exist in B, provided this expression is not identically zero.*

Proof: If a closed phase curve C exists in B, the following equation must hold for the line integral taken along C:

$$\int_C (F_1 \, dq_2 - F_2 \, dq_1) = 0.$$

According to Green's theorem, we have

$$\int_C (F_1 \, dq_2 - F_2 \, dq_1) = \int\int_{B^*} \left(\frac{\partial F_1}{\partial q_1} + \frac{\partial F_2}{\partial q_2} \right) dq_1 \, dq_2$$

for $B^* \in B$, and according to our assumption $\partial F_1/\partial q_1 + \partial F_2/\partial q_2$ must have the same sign everywhere in B and must always differ from zero. Consequently, the area integral and also the line integral cannot be zero — i.e., the closed curve C cannot exist. This proof may also be applied to arbitrary partial regions B^* of B until B is completely covered. It has thus been demonstrated that there are no closed phase curves in B.

A third, important theorem is that proposed by Poincaré-Bendixson presenting the necessary and sufficient conditions for closed phase curves.

Theorem of Poincaré-Bendixson: *If a trajectory C remains within a finite region B of the phase plane without approaching any stationary points, C is itself a limit cycle or approaches one.*

The study by LaSalle[1] is an example of the application of this theorem in determining a limit cycle in the phase diagram of the Van der Pol equation. The various trajectories of the phase diagram correspond to different initial conditions. If one of these trajectories is called an 'unperturbed' solution, the other trajectories will be 'perturbed' solutions with respect to this trajectory and will correspond to different initial conditions. It may be concluded that they correspond to *'perturbations of the initial conditions'* in terms of the discussion presented above.

Let us now assume that a parameter α (or several of them) is included in the differential equations. A *perturbation of the parameter* may then also occur which can influence the phase diagram. The *branching values* α_{crit} of α, which are of particular interest, are characterized by the fact that the stationary points and limit cycles of the phase picture change qualitatively when the variable α passes through the value α_{crit}. For example, a stable point may become unstable, and at the same time a stable limit cycle may appear that justifies the name of 'branching value'. It may be assumed that the stable stationary solution has 'branched' into an unstable stationary solution and a stable periodic solution. More frequently, when α changes, two limit cycles — a stable cycle and an unstable cycle — approach each other, coinciding as a semistable limit cycle for α_{crit}. The latter disappears completely when α has passed through the branching value. The calculation of the branching values is a very important problem in stability theory, because this determines whether stability or instability prevails.

Example: Let us assume that a load P is applied to two elastic rods which are supported and connected by pin joints, as shown in Fig. 30. As long as the load P is small enough, there is a stable equilibrium position A that is maintained by the system. As the load P increases, a value P_{crit} is reached for which the framework *'snaps through'* into the new equilibrium position B. The load P plays the role of the parameter α, and P_{crit} is the parameter value for which the nature of the initial, stable equilibrium position A changes in such

---

[1]LaSalle, J., *Quart. Appl. Math.* 7, pp. 1–19, 1949.

Fig. 30. Snap-through problem   Fig. 31. Snap-through oscillator

a way that stability no longer exists. It may be shown that this value P_{crit} may be found by considering the phase diagram.

For this purpose, we shall assume that a single mass m is tied to the load P as the two rods of the framework are assumed to be massless, elastic 'springs.' We shall now assume that the mass m performs small oscillations around the equilibrium position A. If the deflections from position A are denoted by x, the small oscillations are described by the differential equation

$$m\ddot{x} + P + 2K \sin(\phi + \Delta\phi) = 0$$

(Fig. 31).

We have

$$K = EF\,\Delta l/l = c\,\Delta l$$

$$\Delta l = [a^2 + (x+b)^2]^{1/2} - l$$

$$\sin(\phi + \Delta\phi) = (x+b)/[a^2 + (x+b)^2]^{1/2}.$$

and the differential equation becomes

$$m\ddot{x} + P + 2c(x+b)\{1 - l/[a^2 + (x+b)^2]^{1/2}\} = 0.$$

Introducing the new notation,

$$(x+b)/a = \xi_1, \qquad \dot{\xi}_1 = \xi_2, \qquad l/a = \alpha,$$

and performing simple transformations, we obtain the system of differential equations

$$\dot{\xi}_1 = \xi_2,$$

$$\dot{\xi}_2 = -\left[\frac{P}{am} + \frac{2c}{m}\xi_1 - \frac{2c\alpha}{m}\frac{\xi_1}{(1+\xi_1^2)^{1/2}}\right] = -f(\xi_1).$$

One intermediate integral, the energy integral, is given by

$$\tfrac{1}{2}\xi_2^2 + \int_0^{\xi_1} f(\xi_1)\,\mathrm{d}\xi_1 = E^h.$$

The phase curves are then

$$\xi_2 = \pm 2[E^h - V(\xi_1)]^{1/2}$$

where

$$V(\xi_1) = \frac{P}{am}\,\xi_1 + \frac{c}{m}\,\xi_1^2 - \frac{2c\alpha}{m}\,[(1 + \xi_1^2)^{1/2} - 1].$$

As Fig. 32a illustrates, the phase diagram has the two stable equilibrium positions A and B as well as the unstable equilibrium position C for $P = P_0 < P_{crit}$.

If the load P is increased, the stable equilibrium position A approaches the unstable equilibrium position C (Fig. 32b). A and C coincide when the parameter P attains the 'branching value' P_{crit}. Only *one* stable equilibrium position B remains, and the framework snaps through into this position for P_{crit} (Fig. 32c). If P exceeds P_{crit}, C and A vanish, and the stable position B remains. The framework remains in this position after snapping through.

In addition, a change in the phase diagram may be determined from the change in $V(\xi_1)$, which results from varying P. The condition for which A and C coincide is given by $dV/d\xi_1 = d^2V/d\xi_1^2 = 0$, from which we may calculate

$$\xi_1 = [(l/a)^{2/3} - 1]^{1/2} \qquad P_{crit} = 2EF[1 - (a/l)^{2/3}]^{3/2}.$$

P_{crit} may be called the 'branching value.' If we pass through the values of P in reverse order, the phase curve is closed up to P_{crit}, where a separating curve (separatrix) occurs. In addition to B, the equilibrium positions A and C also occur.

Exercises 1.4

1. Given the simultaneous first-order system of differential equations

$$\dot{q}_1 = q_2, \qquad \dot{q}_2 = q_1 + q_1^2 + \tfrac{1}{4}q_1^3.$$

   a) Find the stationary points of this system in the phase space.
   b) Determine whether the stationary points are simple or not.
   c) If a stationary point should turn out to be non-simple, derive the corresponding non-linear variational equations.
   d) Use these variational equations and the isocline method in order to plot the phase curves in the vicinity of the nonsimple stationary point. Subsequently decide on stability or instability of the stationary point.
2. Investigate the stability of all stationary points of the system.

$$\dot{q}_1 = q_2, \qquad \dot{q}_2 = q_1 - \tfrac{2}{3}q_1^2 + \tfrac{1}{9}q_1^3$$

   by plotting the respective phase curve diagrams around the stationary points.

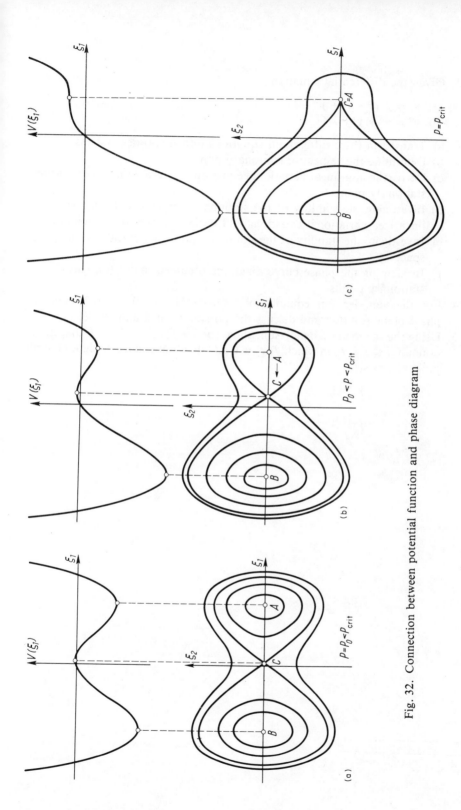

Fig. 32. Connection between potential function and phase diagram

3. Given the differential equation

   (i) $y'' = y'(y-1) + y^2 - 1$
   (ii) $y'' = y^2 + y'y + 4y$

   a) Transform these differential equations into first-order systems.
   b) Determine the respective stationary points.
   c) Derive the non-linear variational equations corresponding to the various stationary points.
   d) Investigate the stability of the stationary points analytically using the linearized variational equations and decide whether this is admissible.
   e) Investigate the stability of the stationary points graphically in the phase space.
   f) In view of the phase curve diagrams obtained under (e), classify the stationary points.

4. For the van der Pol equation $\ddot{x} - \varepsilon(1 - x^2)\dot{x} + x = 0$, $\varepsilon = 0.2$, find the phase-plane solution and discuss the possibility of a limit cycle.

5. Using the theorem of Bendixson, show for $\ddot{x} + f(x)\dot{x} + g(x) = 0$ that in the domain $a \leqslant x \leqslant b$, in which $f(x) \neq 0$ and does not change its sign, no limit cycle can exist.

1.5

The Direct Method of Lyapunov

1.5.1 The Stability Definition of Lyapunov

Let us consider a problem of mechanics defined by the system of ordinary differential equations

$$\dot{q}_i = F_i(t, q_1, q_2, \ldots, q_n), \qquad i = 1, 2, \ldots, n, \qquad (1.5.1)$$

and let us investigate the stability of a particular solution $q_i^0(t)$. For purposes of simplicity, we have omitted the influence of any parameters. The following variational equations are obtained by the well-known transformation (1.2) $q_i = q_i^0 + \xi_i$:

$$\dot{\xi}_i = (\partial F_i / \partial q_k)(q_j^0, t)\xi_k + \Phi_i(t, \xi_1, \ldots, \xi_n) \qquad (1.5.2)$$

$$i, k = 1, \ldots, n.$$

According to Lyapunov, $q_i^0(t)$ is stable when the variations satisfy $|\xi_i(t)| < \varepsilon$ for $t > t_0$ if $|\xi_i(t_0)| < \eta(\varepsilon)$ holds at the initial time t_0. The quantity ε is an arbitrarily small positive number, just like η, which is a function of ε. This is equivalent to the postulate that the stationary solution $\xi_i = 0$, $i = 1, \ldots, n$ of the variations is stable if the conditions mentioned previously are satisfied. We shall demonstrate this in the following discussion.

As was already mentioned in Section 1.1.2, the Lyapunov stability definition pertains to *perturbations of the initial conditions*. At this point, we will restrict ourselves to this case. However, the stability concept may be extended, so that perturbations in the parameters or in the differential equation can be taken into consideration. The reader is referred to the literature for a discussion of this subject, for example, Malkin (1.1).

The purpose of the direct *Lyapunov* method is to determine stability *without performing integration*. This goes one step further than previous methods, because solutions to the linear variational equations were known. We indirectly assumed that the integration of the differential equations have been carried out. The characteristics of the solutions (for example, characteristic exponents) were then used to determine stability. When the phase space method was used, intermediate integrals were found in order to establish phase surfaces along which the solutions of (1.5.2) were found. The direct method of *Lyapunov* uses instead *test surfaces* $V(t, \xi_1, \ldots, \xi_n) = \text{const}$, so-called *Lyapunov functions*. These functions make it possible to determine the course of the solution curves of (1.5.2) in the $t, \xi_1, \ldots, \xi_n$ space and to determine the stability of the stationary solution $\xi_i = 0$, $i = 1, \ldots, n$. For this purpose, let us first substitute the variable $t \equiv \xi_0$. We will use the relationships

$$\Omega_0 \equiv 1, \qquad \Omega_i = (\partial F_i/\partial q_k)(q_j^0, t)\xi_k + \Phi_i, \qquad i, k = 1, 2, \ldots, n.$$

Equation (1.5.2) can then be written in the form

$$\dot{\xi}_i = \Omega_i(\xi_0, \ldots, \xi_n), \qquad i = 0, 1, \ldots, n. \tag{1.5.3}$$

Let us assume that the path element $d\xi_i$, $i = 0, 1, \ldots, n$ describes a portion of the (1.5.3) solution curve. The vector grad $V = (\partial V/\partial \xi_i)$, $i = 0, \ldots, n$, is perpendicular to the test surface V in the $t, \xi_1, \ldots, \xi_n$ space. Let us form the scalar product $(\partial V/\partial \xi_i)\, d\xi_i$.

The course of the solution curve of (1.5.2) in the $\xi_1, \ldots, \xi_n$ space can be inferred from the time derivative of the scalar product.

$$Z = (\partial V/\partial \xi_i)\dot{\xi}_i = (\partial V/\partial \xi_i)\Omega_i, \qquad i = 0, 1, \ldots, n,$$

which can also be written as

$$Z \equiv \frac{dV}{dt} = \frac{\partial V}{\partial t} + \frac{\partial V}{\partial \xi_i}\, \Omega_i \qquad i = 1, 2, \ldots, n. \tag{1.5.4}$$

The decision regarding stability can be made on this basis. In order to do this, we must now deal with the special properties of Lyapunov functions.

1.5.2. Lyapunov Functions

Often it is necessary to select Lyapunov functions which are *definite* and under certain circumstances *have an arbitrarily small upper limit*.

Let us first consider the definiteness property. For this purpose, let us consider two functions $V(t, \xi_1, \ldots, \xi_n)$ and $W(\xi_1, \ldots, \xi_n)$, which vanish for $\xi_1 = \ldots = \xi_n = 0$. It is also assumed that they are single-valued within the region in which they are defined and that they have continuous partial derivatives in this region. The function V is called definite if its sign is prescribed. $V(t, \xi_k)$ is *positive definite* or *negative definite* when the relationship

$$V(t, \xi_1, \ldots, \xi_n) \geqslant W(\xi_1, \ldots, \xi_n) \qquad \text{or} \qquad V(t, \xi_1, \ldots, \xi_n) \leqslant -W(\xi_1, \ldots, \xi_n)$$

holds for $t \geqslant t_0 > 0$ and $|\xi_k| \leqslant h$ (where t_0 and h are sufficiently large or small constants, respectively), where $W(\xi_k)$ is assumed to be a positive definite function which is independent of t.

A function $W(\xi_k)$ is called *positive definite* if it only takes on positive values for $|\xi_k| \leqslant h$ and vanishes for $\xi_1 = \ldots = \xi_n = 0$.

Example: The functions $W = \xi_1^2 + \xi_2^2 + \xi_3^4$ and

$$W = \xi_1^2 + \xi_2^2 + \xi_3^2 - \xi_3^3$$

(for sufficiently small h) are positive definite functions W for $n = 3$. On the other hand, the functions $W = \xi_1^2 + \xi_3^2$ and $W = \xi_1^2 + \xi_2^2 - \xi_3^4$ are not positive definite for $n = 3$. In the first case, the function W can certainly be zero if ξ_1 and ξ_3 are both equal to zero, but at the same time ξ_2 can have an arbitrary value that can be different from zero. In the second case, the function W can change sign. A number of criteria for the property of being definite can be found in the work by Malkin (1.1).

The semidefinite property must also be mentioned in this connection. A function is called *semidefinite* if it has only one sign, but can become zero without all of its arguments being zero.

The function $W = \xi_1^2 + \xi_3^2$ mentioned above is therefore semidefinite.

Example: The function

$$V = t(\xi_1^2 + \xi_2^2) - 2\xi_1\xi_2 \cos t$$

is a positive definite Lyapunov function for $n = 2$ and $t > 2$, which depends on t. If we set $W = \xi_1^2 + \xi_2^2$, we have

$$V - W = (t - 1)(\xi_1^2 + \xi_2^2) - 2\xi_1\xi_2 \cos t > 0$$

for $t > 2$, which must hold for a positive definite V. On the other hand, the function $V = e^{-t}(\xi_1^2 + \ldots + \xi_n^2)$ is not positive definite because we have $V \to 0$ for $t \to \infty$, which means that no positive definite W can exist that will always satisfy the inequality $V \geqslant W$.

The function $V(t, \xi_1, \xi_2, \ldots, \xi_n)$ has an infinitely small upper limit or is, in other words, *uniformly small*, if the inequality

$$|V(t, \xi_1, \ldots, \xi_n)| \leqslant l \qquad \text{for } t \geqslant t_0, \, |\xi_k| \leqslant \mu,$$

is satisfied for the arbitrarily small positive numbers l and μ.

Example: This condition is satisfied by $V = \sin t(\xi_1 + \ldots + \xi_n)$ but not by $V = \sin[t(\xi_1 + \ldots + \xi_n)]$. In the latter case, V is bounded but there is no infinitely small upper limit.

1.5.3. The Stability and Instability Theorems of Lyapunov

We assume that the variational equations are given by

$$\dot{\xi}_i = \Omega_i(t, \xi_1, \ldots, \xi_n), \qquad i = 1, \ldots, n,$$

where the Ω_i are continuous for $t \geqslant t_0$, $|\xi_k| \leqslant h$ in such a way that the variational equations have unique solutions for prescribed initial conditions.

The following theorem holds for stability.

Theorem 1: *If a continuous, positive, or negative definite function $V(t, \xi_k)$ having an infinitely small upper limit can be prescribed for the variational* *equations for which the expression* $dV/dt = \partial V/\partial t + (\partial V/\partial \xi_i)\Omega_i$ *is semidefinite with the opposite sign of V or is identically zero, the unperturbed solution is stable.*

Proof: We assume that V is positive definite, so that

$$V \geqslant W \geqslant 0$$

for $t \geqslant t_0$ and $\xi = \max(|\xi_1|, \ldots, |\xi_n|) \leqslant h$. According to Theorem 1, $dV/dt \leqslant 0$. Owing to the positive definite property, $W = 0$ can only hold for $\xi_1 = \ldots = \xi_n = 0$.

Let $\xi = \varepsilon$ where ε is arbitrarily small, and assume the condition $0 < \varepsilon < h$ is satisfied. The continuous function W takes on its lower limit $l > 0$ under the assumptions $\xi = \varepsilon$, $|\xi_k| \leqslant \varepsilon$. The condition $W \geqslant l$ is just satisfied for $\xi = \varepsilon$. Therefore, $\xi < \varepsilon$ must be satisfied for $W < l$ and $|\xi_{k0}| < \xi$, where the ξ_{k0} are the initial values of ξ_k.

$V(t_0, \xi_k)$ is independent of t for $t = t_0$. If we select $|\xi_k| \leqslant \eta < \varepsilon$, it is possible to have $V(t_0, \xi_k) < l$. We also have $V(t_0, \xi_{k0}) = V_0 < l$ for the initial conditions $|\xi_{k0}| \leqslant \eta$.

According to the assumptions made for dV/dt, we have

$$V - V_0 = \int_{t_0}^{t} (dV/dt)\, dt \leqslant 0$$

so that $V \leqslant V_0$ results. This means $V \leqslant V_0 < l$ according to the statements made above. Owing to the positive definite property, we also have $W \leqslant V$. Both inequalities together result in $W \leqslant V \leqslant V_0 \leqslant l$, or $W < l$. This means that $\xi < \varepsilon$ or $|\xi_k| < \varepsilon$, respectively, are just satisfied for $k = 1, \ldots, n$ and $t \geqslant t_0$. This is true for arbitrarily small ε. This means that the stability condition is satisfied for the stationary solution $\xi_1 = 0, i, = 1, \ldots, n$.

The following represents another stability theorem.

Theorem 2: *If the assumptions of Theorem 1 are satisfied and if dV/dt is definite with the opposite sign of V and if V has an infinitely small upper limit, the unperturbed solution is asymptotically stable.*

It is possible to give a *geometric representation* for Theorems 1 and 2 if Ω_i does not depend on t and if $n \leqslant 3$. In this case, the positive definite function $V(\xi_1, \xi_2, \xi_3) = C$ specifies a family of surfaces that surround the coordinate origin. In addition, the surface $V = C_1$ lies entirely within the surface $V = C_2$ if $C_1 < C_2$.

For Theorem 1, consider the case where $dV/dt = (\partial V/\partial \xi_i)\dot{\xi}_i$ equals zero. This means that the corresponding perturbed solution is completely located on $V = C_k$. If the appropriate initial conditions were selected for the solution, then its 'distance' $d[\xi_i, \xi_i^0 = 0]$ from the stationary solution $\xi_i^0 = 0$ will remain sufficiently small. Consequently $\xi_i^0 = 0$ is stable. As may be readily deduced, a definite intermediate integral of the variational equations can be used as a

Lyapunov function, because the solutions of the variational equations must lie on the family of surfaces specified by the intermediate integral.

For Theorem 2, let us consider the case where dV/dt is always negative. The condition of an infinitely small upper bound for V means that the family of surfaces specified by $V = C$ is contracted to the origin. The relationship

$$dV/dt = (\partial V/\partial \xi_i)\dot{\xi}_i < 0$$

means that an image point corresponding to the perturbed solution moving along the space curve will aways penetrate the family of surfaces $V = C$ from the outside to the inside. Since, in addition, $V = C$ passes arbitrarily close to the origin for $C \to 0$, the image point must finally reach $\xi_i = 0$ for $t \to \infty$. This means that the stationary solution is asymptotically stable (Fig. 33).

Without presenting the proof, let us mention the instability theorems of A. M. Lyapunov and N. B. Chetayev. The proofs can be found in the book by Malkin (1.1).

Theorem 3: *Let us assume that a function $V(t, \xi_1 \ldots, \xi_n)$ exists which has an infinitely small upper bound and that dV/dt is definite. Also, let us assume that V can have the same sign as dV/dt for arbitrarily large t when the ξ_k are arbitrarily small. The unperturbed solution is then unstable.*

Theorem 4: *Let us assume that a region $V > 0$ exists in any arbitrarily small vicinity of the origin for arbitrarily large values of t, in which V is bounded and dV/dt is positive. In this case the unperturbed solution is unstable for autonomous problems.*

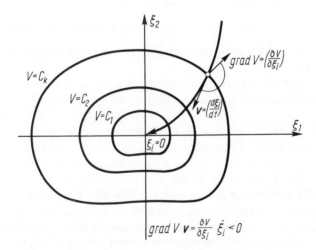

Fig. 33. Representation of the second stability theorem of Lyapunov

Example (for Theorem 1):

$$\dot{\xi}_1 = -(\xi_1 - 3\xi_2)(1 - 2\xi_1^2 - 4\xi_2^2), \qquad \dot{\xi}_2 = -(\xi_1 + \xi_2)(1 - 2\xi_1^2 - 4\xi_2^2),$$

where $V = \xi_1^2 + 3\xi_2^2$. We have $dV/dt = (\partial V/\partial \xi_1)\dot{\xi}_1 + (\partial V/\partial \xi_2)\dot{\xi}_2$, which results in

$$dV/dt = [2\xi_1(3\xi_2 - \xi_1) - 6\xi_2(\xi_1 + \xi_2)](1 - 2\xi_1^2 - 4\xi_2^2)$$
$$= -2(\xi_1^2 + 3\xi_2^2)(1 - 2\xi_1^2 - 4\xi_2^2).$$

V is positive definite and $dV/dt \leqslant 0$ for sufficiently small values of ξ_1, ξ_2, which means it is semidefinite with the opposite sign to V. Therefore, the unperturbed solution $\xi_1 = \xi_2 = 0$ is stable.

Example (for Theorem 2):

$$\dot{\xi}_1 = -\xi_1^2 - 3\xi_2, \qquad \dot{\xi}_2 = 3\xi_1 - 5\xi_2^3$$

with the positive definite function $V = \xi_1^2 + \xi_2^2$. We have therefore

$$dV/dt = -2\xi_1(\xi_1^3 + 3\xi_2) + 2\xi_2(3\xi_1 - 5\xi_2^3) = -2(\xi_1^4 + 5\xi_2^4).$$

$dV/dt < 0$ is negative definite. Because V also has an infinitely small upper limit (the family of curves $V = C$ contracts to the origin for $C \to 0$), the unperturbed solution, $\xi_1 = \xi_2 = 0$ is asymptotically stable.

Example (for Theorem 4):

$$\dot{\xi}_1 = \xi_1^2 + \xi_2, \qquad \dot{\xi}_2 = \xi_1 + \xi_2^2,$$

where

$$V = (\xi_1^3/3) + 2\xi_1\xi_2 + (\xi_2^3/3).$$

Apparently a region surrounding the origin of the ξ_1, ξ_2 plane can be specified in which $V > 0$. In addition, V is bounded in it and

$$dV/dt = (\xi_1^2 + 2\xi_2)(\xi_1^2 + \xi_2) + (2\xi_1 + \xi_2^2)(\xi_1 + \xi_2^2) > 0.$$

The conditions of Theorem 4 are satisfied, and the unperturbed solution $\xi_1 = \xi_2 = 0$ is unstable.

As an important *example* from mechanics, let us derive the theorem of Lagrange-Dirichlet regarding the *stability of an equilibrium position*, which is as follows.

When the potential energy U has a minimum for an equilibrium position, the equilibrium position is stable.

Proof: Let us go from the Cartesian coordinates χ_i to the generalized Lagrangian coordinates q_i by means of a coordinate transformation in such a way that the kinetic energy T is a positive definite function of the velocities $\dot{q}_i$, and the equilibrium position $q_{i,s}$ corresponds to the origin. The second condition can always be satisfied. The first condition can be satisfied when the coordinate transformaton is independent of time, which means that the system is time independent. In addition, we assume that the potential energy U is only a function of the coordinates.

Let us now consider the problem, in the $2n$-dimensional phase space $\dot{q}_i$, q_i. The function U should have a minimum equal to zero. Since U is a homogeneous function of the q_i, it takes on the value zero for $q_{i,s} = 0$. U is therefore positive for sufficiently small q_i, which means that U is a positive definite function. T is known to be positive definite, and therefore the Hamilton function $H = T + U$ is positive definite. The energy of the mechanical system is specified by H, and H represents an intermediate integral of the canonical system of differential equations. H can therefore be used as a Lyapunov function which also satisfies $dH/dt = 0$. Therefore, all the conditions of Theorem 1 are satisfied for the 'Lyapunov function' H. Therefore, the equilibrium position is stable.

We started with the variational Eqs. (1.5.2). Up to the present, we have considered linear as well as non-linear terms in the ξ_k appearing on the right-hand side of the system. One very important question is whether or not stability can be determined if the non-linear terms Φ_i are dropped. We would like to establish a criterion stipulating when so-called *stability in the first approximation* can be analyzed to a sufficient accuracy using the abbreviated, *linear equations of the first approximation*

$$\dot{\xi}_i = a_{ik}\xi_k, \qquad a_{ik} = (\partial F_i/\partial q_k)(q_j^0, t). \qquad (1.5.5)$$

In this case, the methods described in Section 1.3 would be sufficient.

The answer to this question is given in the following theorem of Lyapunov.

Theorem 5: *If a system of the first approximation is regular and if all its characteristic numbers are positive, the unperturbed solution is stable. However, if only one of the characteristic numbers is negative, then the unperturbed solution is unstable. Whether stability or instability prevails under these assumptions can be decided without considering the non-linear terms in* (1.5.2).

Two important special cases are contained in this theorem. Assume all the a_{ik} are *constant*. Equation (1.5.5) represents a system with constant coefficients which is regular. The characteristic numbers are all positive if the real parts of the characteristic exponents are all negative. If only one of the real parts of the characteristic exponents is positive, there will be one negative characteristic number. In this case, where the coefficients are constant, it is only necessary to analyze the real parts of the characteristic exponents.

Assume the a_{ik} are *periodic* functions ($a_{ik} \equiv p_{ik}$). System (1.5.5) is again a regular system. If the magnitudes of the characteristic multipliers are all smaller than one, all the characteristic numbers are positive. If only one characteristic multiplier is present with a magnitude larger than one, a negative characteristic number will exist. In this case, it is possible to analyze the situation with only the properties of the characteristic multipliers.

In Theorem 5 it was also assumed that the Φ_i are analytic, which means that they can be represented as a power series in the ξ_k that begins with terms of the second or higher order. This assumption can also be somewhat relaxed by requiring only that the Φ_i are continuous and satisfy the inequality.

$$| \Phi_i(t, \xi_1, \ldots, n) | \leqslant M(| \xi_1 |^{1+\alpha} + \ldots + | \xi_n |^{1+\alpha}).$$

M and α are positive constants. The book by Petrovski contains an example of Theorem 5 under these assumptions for constant a_{ik}, and when the characteristic exponents all have negative real parts. [1]

Let us consider one case in more detail and make the following assumptions. In the equations

$$\dot{\xi}_i = a_{ik}\xi_k + \Phi_i, \qquad i, k = 1, 2, \ldots, n, \tag{1.5.6}$$

the a_{ik} are constant coefficients. The abbreviated equations (1.5.5), i.e. $\dot{\xi}_i = a_{ik}\xi_k$, have only characteristic exponents with negative real parts. The Φ_i are analytic, independent of t, and can be developed in power series. In this case the characteristic equation $\det(a_{ik} - \lambda\,\delta_{ik}) = 0$ corresponding to (1.5.5) has only roots λ_σ with negative real parts, and the coefficients of the characteristic equation satisfy the Hurwitz criterion.

Assuming asymptotic stability, let us construct a positive definite Lyapunov function V for the abbreviated Eqs. (1.5.5). It can be obtained from the condition that dV/dt is a negative definite form U. This is because dV/dt must be always negative for positive definite V if the assumption of asymptotic stability is to be satisfied. We therefore have

$$\frac{dV}{dt} = \frac{\partial V}{\partial \xi_i} \dot{\xi}_i = \frac{\partial V}{\partial \xi_i} a_{ik}\xi_k = U. \tag{1.5.7}$$

Let $A_k, k = 1, 2, \ldots,$ and $B_j, j = 1, 2, \ldots,$ be the coefficients of the positive or negative, definite forms V and U. If the forms are expanded in (1.5.7) and if the coefficients of equal powers of ξ_i are compared, one obtains a system of algebraic equations that determine the A_k if the B_f are specified:

$$\alpha_{ik}A_k = B_i \tag{1.5.8}$$

The desired positive definite form V can indeed be produced in the form of a Lyapunov function if $\det(\alpha_{ik})$ is not zero. The coefficients α_{ik} of (1.5.5) are contained in the elements α_{ik} of this determinant. These elements have special properties, because the characteristic equation formed with them must satisfy the Hurwitz criterion. Based on this, it can be shown that $\det(\alpha_{ik}) \neq 0$ is always satisfied and therefore (1.5.8) always has a solution. Consequently, a form V can always be obtained based on (1.5.8) if the form U is specified.

Example: Let us show this for quadratic forms. Equations (1.5.5) then become

$$\left.\begin{array}{l} \dot{\xi}_1 = a_{11}\xi_1 + a_{12}\xi_2 \\ \dot{\xi}_2 = a_{21}\xi_1 + a_{22}\xi_2. \end{array}\right\} \tag{1.5.9}$$

---

[1]Petrovski, J. G., *Lectures on the Theory of Ordinary Differential Equations*, p. 162. Teubner, Leipzig, 1954.

The characteristic equation is

$$\lambda^2 - (a_{11} + a_{22})\lambda + a_{11}a_{22} - a_{12}a_{21} = 0$$

and the Hurwitz criterion requires that

$$\left.\begin{array}{l} a_{11} + a_{22} < 0 \\ a_{11}a_{22} - a_{12}a_{21} > 0. \end{array}\right\} \qquad (1.5.10)$$

Let the quadratic forms be

$$V = A_1\xi_1^2 + A_2\xi_1\xi_2 + A_3\xi_2^2 \qquad U = B_1\xi_1^2 + B_2\xi_1\xi_2 + B_3\xi_2^2$$

so that (1.5.7), making use of (1.5.9), becomes

$$(2A_1\xi_1 + A_2\xi_2)(a_{11}\xi_1 + a_{12}\xi_2) + (2A_3\xi_2 + A_2\xi_1)(a_{21}\xi_1 + a_{22}\xi_2)$$
$$= B_1\xi_1^2 + B_2\xi_1\xi_2 + B_3\xi_2^2.$$

From (1.5.8) it follows that

$$\left.\begin{array}{l} 2A_1a_{11} + A_2a_{21} = B_1 \\ 2A_1a_{12} + A_2(a_{11} + a_{22}) + 2A_3a_{21} = B_2 \\ A_2a_{12} + 2A_3a_{22} = B_3. \end{array}\right\} \qquad (1.5.11)$$

Therefore, the determinant must satisfy

$$D \equiv \begin{vmatrix} 2a_{11} & a_{21} & 0 \\ 2a_{12} & a_{11} + a_{12} & 2a_{21} \\ 0 & a_{12} & 2a_{22} \end{vmatrix}$$

$$= 4\{a_{11}[(a_{11}a_{22} - a_{21}a_{12}) + a_{22}^2] - a_{12}a_{21}a_{22}\} \neq 0.$$

Let us make the opposite assumption that $D = 0$. Then we would have to have

$$(a_{11} + a_{22})(a_{11}a_{22} - a_{12}a_{21}) = 0 \qquad (1.5.12)$$

as can easily be calculated. This cannot be the case because of (1.5.10), as the product on the left side of Eq. (1.5.12) would have to become smaller than zero. The equality sign in (1.5.12) is therefore incorrect. The determinant D must be different from zero, (1.5.11) always has a solution, and the form V can indeed be found if U is prescribed.

It may be readily seen that V must be positive definite if U is assumed to be negative definite, because if V were negative, it would agree with the sign of $dV/dt = U$ and (1.5.5) could not have an asymptotic stable stationary solution under these conditions, which disagrees with our assumption. Furthermore, V cannot be positive semidefinite because if dV/dt is always negative and if V became zero only at one point, it would have to become negative, which is not compatible with the assumption that it is positive semidefinite. The only possibility that remains is that V is positive definite, as was postulated.

Let us now use the positive definite form V constructed in this way as a Lyapunov function for the *unabbreviated* Eqs. (1.5.6). For this purpose let us form

$$dV/dt = (\partial V/\partial \xi_i)\dot{\xi}_i.$$

We obtain the following according to (1.5.7):

$$\frac{dV}{dt} = \frac{\partial V}{\partial \xi_i} a_{ik}\xi_k + \frac{\partial V}{\partial \xi_i} \Phi_i = U + \frac{\partial V}{\partial \xi_i} \Phi_i. \qquad (1.5.13)$$

Since the development of $(\partial V/\partial \xi_i)\Phi$ in (1.5.13) starts with terms of the third or higher degree in the ξ_i, and because U is negative definite according to our assumption, dV/dt is negative definite in this case also for $|\xi_i| < h$, regardless of the properties of Φ_i. This means that the assumptions for asymptotic stability of $\xi_i = 0, i = 1, 2, \ldots, n$, are satisfied according to Theorem 2, because V is assumed to be positive definite. Apparently the non-linear terms Φ_i have no influence on the stability. As Theorem 5 states, under these conditions stability can be predicted using only the abbreviated Eqs. (1.5.5), that is, using the equations of the first approximation.

A special case occurs when there are no characteristic exponents with positive real parts (no characteristic multipliers with magnitudes greater than one) for the equations of the first approximation, but when a few of the characteristic exponents have *zero real part* (or if one of the characteristic multipliers has the magnitude one). Lyapunov showed that a so-called *critical case* exists in this case, in which the equations of the first approximation do not indicate whether stability or instability exists. In this case, the analysis must be performed using the unabbreviated Eqs. (1.5.2) and taking the non-linear terms into account.

We will not consider the critical cases any further. They are treated in the book by Malkin mentioned previously. It must be stated, however, that the critical cases are by no means exceptions, but do indeed occur in problems of practical interest.

Exercises 1.5

1. Given $V = \xi_1^2 + \xi_2^2 - 2\xi_1\xi_2 \cos t, \qquad n = 2,$

$$W = \xi_1^2 + 2\xi_1\xi_2 + 2\xi_2^2 + \xi_3^2, \qquad n = 3,$$

$$W = \xi_1^2 + \xi_2^2 + 2\xi_1\xi_2 + \xi_3^3, \qquad n = 3.$$

Determine which of these functions are positive definite and which are only positive semidefinite.

2. Is the function

$$V = (3 + \sin t\xi_1^2)(\xi_1^2 + \xi_2^2), \qquad n = 2,$$

positive definite and does it have an infinitely small upper limit for $|\xi_k|$, $k = 1, 2$, being sufficiently small?

3. Given $\dot{\xi}_1 = \xi_2$, $\dot{\xi}_2 = -3\xi_1 + \xi_2$, and $V = 2\xi_1^2 + \xi_2^2$. Calculate $\dot{V}$ using $\dot{\xi}_i = a_{ik}\xi_k$. Are V and $\dot{V}$ definite?

4. For the following systems of variational equations investigate the stability of the respective trivial solutions using Lyapunov's method:

   (i) $\dot{\xi}_1 = -(\xi_1 - 2\xi_2)(1 - \xi_1^2 - 2\xi_2^2)$,

   $\dot{\xi}_2 = -(\xi_2 + \xi_1)(1 - \xi_1^2 + 2\xi_2^2)$.

   (ii) $A\dot{\xi}_1 = (B - C)\xi_2(k + \xi_3)$,

   $B\dot{\xi}_2 = (C - A)\xi_1(k + \xi_3)$,

   $C\dot{\xi}_3 = (A - B)\xi_1\xi_2$.

   Assume $0 < A < C < B$, and (α) $k + \xi_3 > 0$, (β) $k + \xi_3 = 0$.

   (iii) $\dot{\xi}_1 = a\xi_1^3 - \xi_2$,

   $\dot{\xi}_2 = \xi_1 + a\xi_2^3$.

   Note: a may be either positive or negative.

   (iv) $\dot{\xi}_1 = \xi_2 - 3\xi_3 - \xi_1(\xi_2 - 2\xi_3)^2$,

   $\dot{\xi}_2 = -2\xi_1 + 3\xi_3 - \xi_2(\xi_1 + \xi_3)^2$,

   $\dot{\xi}_3 = 2\xi_1 - \xi_2 - \xi_3$.

   Note: use $V = \alpha\xi_1^2 + \xi_2^2 + \beta\xi_3^2$, $\alpha > 0$, $\beta > 0$, as a Lyapunov function.

5. Investigate the stability of the trivial solution of the following systems in the first approximation:

   (i) $\dot{\xi}_1 = 2\xi_2 + \xi_1^3 - \xi_1^4\xi_2$,

   $\dot{\xi}_2 = -2\xi_1 + \xi_2^3$.

   (ii) $\dot{\xi}_1 = -\xi_1 + 2\xi_2 - \xi_1^3 + \xi_1^2\xi_2$,

   $\dot{\xi}_2 = 2\xi_1 - \xi_2 + \xi_2^3$.

   Should the critical case turn up, decide finally on stability using Lyapunov's method.

6. Investigate the stability of the trivial solution of the system

   $\dot{\xi}_1 = \xi_1 - \xi_2 + \xi_3 + \xi_1 r^2$,

   $\dot{\xi}_2 = \xi_2 + \xi_1 + 2\xi_3 + 2\xi_2 r^2$,

   $\dot{\xi}_3 = \xi_3 - \xi_1 - 2\xi_2 + 3\xi_3 r^2$,

   $r^2 = \xi_1^2 + \xi_2^2 + \xi_3^2$,

   using Lyapunov's direct method. Show that in this case stability can be investigated in the first approximation.

7. Investigate the stability of the trivial solution of the system

$$\dot{\xi}_1 = 2\xi_1 + \xi_2 + 2\xi_1^3 + 2\xi_1\xi_2^2,$$

$$\dot{\xi}_2 = -\xi_1 + 2\xi_2 + \xi_2\xi_1^2 + \xi_2^3,$$

in the first approximation if possible, and using Lyapunov's direct method.

1.6
The Energy Method

1.6.1 Conservative, Scleronomic Systems

For conservative scleronomic systems with finite degrees of freedom it is common to apply the so-called energy method for the assessment of the stability of a particular motion of the system. Very frequently, the particular 'motion' is a stationary (equilibrium) position of the system. Then, the theorem of Dirichlet[1] holds which states that *the equilibrium is stable if the potential energy of the system assumes a minimum in the equilibrium position.*

Another, related theorem[2] asserts that *the equilibrium is stable if the second variation of the potential energy of the system is positive definite at the equilibrium position.*

In the following it will be shown that the two statements are essentially identical, and that they actually result from an application of Lyapunov's direct method when one uses the energy of the system as a Lyapunov function.

First, consider a dynamic, mechanical system with n degrees of freedom. It is described by the system of second-order differential equations.

$$m_i \ddot{x}_i = f_i(x_1, x_2, \ldots, x), \qquad i = 1, 2, \ldots, n, \qquad (1.6.1)$$

where the x_i are the coordinates and the m_i the masses. The f_i are given functions which have the dimensions of forces. Since the system is assumed to be conservative,

$$f_i(x_1, x_2, \ldots, x_n) = -\frac{\partial U}{\partial x_i}, \qquad (1.6.2)$$

where U is the potential energy.

Multiplying both sides in (1.6.1) by x_i and then integrating over the time t, one obtains the equation

$$\int m_i \ddot{x}_i \dot{x}_i \, dt = \int f_i(x_1, x_2, \ldots, x_n) x_i \, dt. \qquad (1.6.3)$$

---

[1] G. Hamel, *Theoretische Mechanick*, p. 268, Springer, Berlin, 1949.
[2] H. L. Langhaar, *Energy Methods in Applied Mechanics*, p. 31, John Wiley and Sons, New York, 1962.

90

By virtue of (1.6.2), this equation changes into

$$\int m_i \ddot{x}_i \dot{x}_i \, dt = - \int \frac{\partial U}{\partial x_i} \frac{dx_i}{dt} \, dt = - \int \frac{\partial U}{\partial x_i} \, dx_i. \tag{1.6.4}$$

Let the integrations be carried out on both sides of (1.6.4). Then, one arrives at

$$\tfrac{1}{2} m_i \dot{x}_i^2 = -U + C, \tag{1.6.5}$$

where C is a constant of integration.

Yet,

$$T = \tfrac{1}{2} m_i \dot{x}_i^2 \tag{1.6.6}$$

is the kinetic energy of the system. Hence, (1.6.5) yields

$$T + U = C. \tag{1.6.7}$$

The quantity on the left-hand side of (1.6.7) is the total energy E of the system. Therefore, it follows from (1.6.7) that

$$E = T + U = C = \text{const.} \tag{1.6.8}$$

Consequently,

$$\dot{E} = 0. \tag{1.6.9}$$

Let E be chosen as the Lyapunov function. Then, if

$$E > 0, \text{ positive, definite,} \tag{1.6.10}$$

is required to hold, the motion of the mechanical system described by (1.6.1) would be stable by virtue of Theorem 1 in Section 1.5.2. Since the kinetic energy T is positive definite according to (1.6.6), the required positive definiteness of the energy E is warranted by the positive definiteness of the potential energy U. Therefore, for conservative, scleronomic systems, it is sufficient for stability that the potential energy U is positive definite.

Let the unperturbed motion x_i^0, whose stability is to be investigated, be an equilibrium position of the system, which is defined by the condition

$$f_i(x_1^0, x_2^0, \ldots, x_n^0) = 0. \tag{1.6.11}$$

Then, by virtue of (1.6.2),

$$\frac{\partial U}{\partial x_i} = 0 \quad \text{for} \quad x_i \equiv x_i^0. \tag{1.6.12}$$

Hence, positive definiteness of U is identical with U having a minimum at the equilibrium position. Therefore, for conservative scleronomic systems, it is sufficient that the potential energy U has a minimum at the equilibrium position in order to make that equilibrium position stable. This statement is equivalent to Dirichlet's stability theorem.

Secondly, consider a more general case, where the equations describing the state of the mechanical system are variational equations. The basic equations

(1.6.1), (1.6.2) are again the starting point. But now set

$$x_i = x_i^0 + \xi_i, \qquad (1.6.13)$$

where x_i^0 are the coordinates of the unperturbed motion and the ξ_i are the variations caused by a perturbation.

Using (1.6.13) in (1.6.1) yields

$$m_i \ddot{x}_i^0 + m_i \ddot{\xi}_i = f_i(x_1^0 + \xi_1, \ldots, x_n^0 + \xi_n). \qquad (1.6.14)$$

Developing the right-hand side of (1.6.14) into a Taylor series yields

$$m_i \ddot{x}_i^0 + m_i \ddot{\xi}_i = f_i(x_1^0, x_2^0, \ldots, x_n^0) + \frac{\partial f_i}{\partial x_k}\bigg|_{x_k = x_k^0} \xi_k. \qquad (1.6.15)$$

Yet, x_i^0 satisfies (1.6.1). Therefore,

$$m_i \ddot{x}_i^0 = f_i(x_1^0, x_2^0, \ldots, x_n^0)$$

and

$$m_i \ddot{\xi}_i = \frac{\partial f_i}{\partial x_k}\bigg|_{x_k = x_k^0} \xi_k \qquad (1.6.16)$$

remains from (1.6.15), which are the variational equations.

Multiplying in (1.6.16) both sides with $\dot{\xi}_i$ and integrating over time t yields

$$\int m_i \ddot{\xi}_i \dot{\xi}_i \, dt = \frac{\partial f_i}{\partial x_k}\bigg|_{x_k = x_k^0} \int \xi_k \dot{\xi}_i \, dt. \qquad (1.6.17)$$

Performing the integration leads to

$$\frac{1}{2} m_i \dot{\xi}_i^2 = \frac{\partial f_i}{\partial x_k}\bigg|_{x_k = x_k^0} \xi_k \xi_i + C \qquad (1.6.18)$$

where C is a constant of integration.

Rearranging in (1.6.18) yields

$$\frac{1}{2} m_i \dot{\xi}_i^2 + \frac{\partial f_i}{\partial x_k}\bigg|_{x_k = x_k^0} \xi_k \xi_i = C. \qquad (1.6.19)$$

Let

$$L = \frac{1}{2} m_i \dot{\xi}_i^2 - \frac{\partial f_i}{\partial x_k}\bigg|_{x_k = x_k^0} \xi_k \xi_i \qquad (1.6.20)$$

and choose this expression as Lyapunov function. Then, by virtue of (1.6.19),

$$\dot{L} = 0. \qquad (1.6.21)$$

For stability of the unperturbed motion x_i^0, one has to require that L be positive definite. The expression

$$T = \tfrac{1}{2} m_i \dot{\xi}_i^2, \qquad (1.6.22)$$

which is a kinetic energy, is obviously positive definite. Therefore, it is sufficient for L to be positive definite, that the expression

$$V = -\frac{\partial f_i}{\partial x_k}\bigg|_{x_k = x_k^0} \xi_k \xi_i \qquad (1.6.23)$$

be positive definite. Observing (1.6.2) in (1.6.23) leads to the requirement

$$V = \frac{\partial^2 U}{\partial x_i \partial x_k}\bigg|_{x_k = x_k^0} \xi_k \xi_i > 0 \qquad (1.6.24)$$

positive definite.

Expression V in (1.6.24) is a quadratic form in the ξ_i. For V to be positive definite, it is necessary and sufficient that the determinant of its coefficients be positive definite. That means,

$$\det\left(\frac{\partial^2 U}{\partial x_i \partial x_k}\right)_{x_k = x_k^0} > 0 \qquad (1.6.24a)$$

positive definite, must hold.

This condition can be brought into another form which one encounters frequently in the literature. Consider the Hamiltonian H. For a conservative, scleronomic system, H is the total energy, i.e.,

$$H = T + U. \qquad (1.6.25)$$

Moreover, assume that one is working with generalized coordinates q_i instead of the x_i. Then, (1.6.24) can be replaced by

$$\det\left(\frac{\partial^2 U}{\partial q_i \partial q_k}\right) > 0.$$

Also,

$$T = T(\dot{q}_i) \qquad (1.6.26)$$

so that

$$\frac{\partial^2 H}{\partial q_i \partial q_k} = \frac{\partial^2 T}{\partial q_i \partial q_k} + \frac{\partial^2 U}{\partial q_i \partial q_k}$$

which holds by virtue of (1.6.25), becomes due to (1.6.26) simply

$$\frac{\partial^2 H}{\partial q_i \partial q_k} \equiv \frac{\partial^2 U}{\partial q_i \partial q_k}$$

Therefore, condition (1.6.24a) can be substituted by

$$\det\left(\frac{\partial^2 H}{\partial q_i \partial q_k}\right) > 0 \qquad (1.6.24b)$$

positive definite. That is the other form of stability condition (1.6.24a), to which reference will be made in the following section.

The preceding results shall now be viewed under the assumption that the unperturbed solution x_i^0 corresponds to an equilibrium position. Then,

$$f_i(x_1^0, x_2^0, \ldots, x_n^0) = 0$$

by definition of an equilibrium position. Hence, by virtue of (1.6.2), also

$$\frac{\partial U}{\partial x_i}\bigg|_{x_i = x_i^0} = 0. \tag{1.6.27}$$

Moreover, it is only a matter of definition to assume that

$$U_0 = U\big|_{x_i = x_i^0} = 0. \tag{1.6.27a}$$

The potential energy is given as

$$U = U\big|_{x_i = x_i^0} + \frac{\partial U}{\partial x_i}\bigg|_{x_i = x_i^0}\xi_i + \frac{1}{2}\frac{\partial^2 U}{\partial x_i\,\partial x_k}\bigg|_{x_i = x_i^0}\xi_i\xi_k + \ldots \tag{1.6.27b}$$

This expression corresponds to

$$U = U_0 + \delta U + \tfrac{1}{2}\delta^2 U + \ldots \tag{1.6.28}$$

By virtue of (1.6.27) and (1.6.27a) it follows from (1.6.27b) and (1.6.28) that

$$U = \frac{1}{2}\frac{\partial^2 U}{\partial x_i\,\partial x_k}\bigg|_{x_i = x_i^0}\xi_i\xi_k = \tfrac{1}{2}\delta^2 U. \tag{1.6.29}$$

Applying the stability condition (1.6.24) to (1.6.29) leads to two conclusions:
For stability of the equilibrium,
  1) the potential energy should be positive definite,
  2) the second variation of the potential energy should be positive definite,
and both conditions, which are identical, should be satisfied at the equilibrium position. With these conclusions, Dirichlet's theorem is again reconfirmed.

1.6.2 Non-autonomous, Non-conservative Systems

Let from now on the application of generalized coordinates q_i instead of natural (Cartesian) coordinates x_i be postulated. Then,

$$U = U(q_i) \tag{1.6.30}$$

shall be the potential energy and

$$T = T(\dot{q}_i^2) \tag{1.6.31}$$

the kinetic energy. Moreover, let the expression

$$H = T + U \tag{1.6.32}$$

be called the Hamiltonian, which in the case of non-autonomous, non-conservative systems is not any longer the total energy E of the mechanical system.

As has been shown in Section 1.6.1, an application of the energy method to the stability analysis of conservative systems is quite straightforward and common. Condition

$$\partial U/\partial q_i = 0 \tag{1.6.33}$$

yields the equilibrium positions, and the criterion

$$\det\left(\frac{\partial^2 U}{\partial q_i \, \partial q_k}\right) = 0 \tag{1.6.34}$$

is applied to detect the instabilities of the system. In the following, it will be shown, how such conditions can be generalized in order to make an application of the energy method applicable to non-autonomous, non-conservative systems. Although the stability investigations for such systems can be performed by other means, it may still be desirable to make also an energy approach available. Such an approach has the advantage of showing that all systems, non-autonomous, non-conservative, conservative ones, can be ruled by one common procedure and demonstrating that conservative systems are just special cases of a more general set of mechanical systems.

The equations of Hamilton for a non-conservative system read

$$\frac{\partial H}{\partial q_i} - N_i = -\dot{p}_i, \qquad \frac{\partial H}{\partial p_i} = \dot{q}_i. \tag{1.6.35}$$

In (1.6.35), H is the Hamiltonian, N_i are the non-conservative force components, q_i the already mentioned generalized coordinates, and p_i the generalized impulses. The latter are defined by

$$p_i = \frac{\partial L}{\partial \dot{q}_i} \tag{1.6.36}$$

where L is the Lagrangian.

Using (1.6.35), the time derivative of the Hamiltonian can be obtained as

$$\dot{H} = \frac{\partial H}{\partial H_i} \, q_i + \frac{\partial H}{\partial p_i} \, \dot{p}_i = N_i \dot{q}_i. \tag{1.6.37}$$

This relationship can be integrated with respect to time to yield

$$H(q_i, p_i, t) = \int_{t_0}^{t} N_i \dot{q}_i \, dt + E_k \tag{1.6.38}$$

where the E_k, $k = 1, 2, 3, \ldots$ are constants of integration. Since the $\dot{q}_i$ can be expressed in terms of the p_i (or vice versa), equation (1.6.38) could serve to draw a phase space diagram in the (p_i, q_i)-space. This might, however, be quite unfeasible insofar as the diagram were explicitly depending on time and would therefore be presentable only in the form of a motion picture. However, let at least the idea of (1.6.38) being a phase space diagram be kept in mind.

Consider a certain, fixed moment of time $t = t_0$. At that instant, multiply the first equation in (1.6.35) by $\dot{q}_i$, the second one by $\dot{p}_i$, and add these two products. The result is, after a simple modification of that sum,

$$\left(\frac{\partial H}{\partial q_i} - N_i\right) + \frac{\partial H}{\partial p_j} \frac{\partial p_j}{\partial q_i} = \Phi(q_i, p_j, P) = 0. \tag{1.6.39}$$

In (1.6.39) it has been indicated that the quantity Φ thus obtained does not only depend on the q_i and p_i but also on certain structural parameters P which affect deeply the very nature of the system. Assigning fixed values to dp_j/dq_i, (1.6.39) could be viewed as the hyper-isocline of the phase space diagram at the moment of time t_0. However, the phase space diagram is actually a 'motion picture'. Therefore, a hyper-surface, being an isocline at $t = t_0$, would not remain an isocline at other instances of time different from $t = t_0$. Yet, it will in spite of this be useful to view (1.6.39) as an information-bearing characteristic of the system, being termed *heterocline* in the following.

It is worthwhile to consider (1.6.39) under the assumption that the system being investigated were a conservative one. Then, $N \equiv 0$. In addition, it is sufficient to use the static method for the stability investigation of a conservative system. In other words, the stability of the equilibrium positions may be analyzed. Therefore, also $p_i \equiv 0$, and $H = p^2/2m + U(q_i)$ changes into U. Taking all these facts into account, relationship (1.6.39) degenerates into $\partial U/\partial q_i = 0$. The conclusion is that condition (1.6.33) is a special case of (1.6.39) for a conservative, static system, as predicted.

A system is stable if the values of its structural parameters lie in a certain appropriate domain. Some of these parameters (or all of them) may be destabilizing, i.e., if its values exceed certain limits the system may lose its stability. In order to evaluate the stability of a system it is, therefore, important to observe the effect which the variation of a destabilizing parameter may have on the stability degree of the system. This can be done by viewing the phase-space diagram. The variation of the structural parameters causes the diagram to deform and the deformation velocity is a measure for the stability behaviour of the system. If this velocity is zero, the system is invariant against a parameter change and consequently stable. It remains stable if the effect of a parameter change is limited and the deformation velocity is of a finite magnitude. The system becomes unstable and the phase space diagram suffers 'snap-through' of critical points if the deformation velocity becomes infinite.

A means of deriving an analytical expression for the deformation velocity consists in taking the total differential of Φ, i.e.

$$d\Phi = \frac{\partial \Phi}{\partial q_i} \, dq_i + \frac{\partial \Phi}{\partial P} \, dP = 0. \qquad (1.6.40)$$

In (1.6.40) it has been assumed for the sake of simplicity that the system depends on a single destabilizing parameter P only. If there were more than one present, the subsequent consideration could easily be generalized accordingly. A more detailed version of (1.6.40) based on (1.6.39) is

$$\left[\left(\frac{\partial^2 H}{\partial q_k \partial q_i} - \frac{\partial N_k}{\partial q_i} + \frac{\partial}{\partial q_i} \frac{\partial H}{\partial p_j} \frac{dp}{dq_k} \right) \right] dq_i + \left[\frac{\partial^2 H}{\partial q_k \partial P} - \frac{\partial N_k}{\partial P} \right] dP = 0. \qquad (1.6.41)$$

In deriving (1.6.41), the fact has been taken into account that neither $\partial H/\partial p_j$ nor dp_j/dq_k depend explicitly on any structural parameter P.

From (1.6.41) the deformation velocity of the phase space diagram is obtained as

$$\frac{dq_i}{dP} = -\frac{\Delta_i}{\Delta} \tag{1.6.42}$$

where

$$\Delta = \det\left[\frac{\partial^2 H}{\partial q_k \partial q_i} - \frac{\partial N_k}{\partial q_i} + \frac{\partial}{\partial q_i}\left(\frac{\partial H}{\partial p_j}\frac{dp_j}{dq_k}\right)\right] \tag{1.6.43}$$

and Δ_i is the determinant which follows from Δ if the ith column of Δ is replaced by the vector

$$b_k = \frac{\partial^2 H}{\partial q_k \partial P} - \frac{\partial N_k}{\partial P}. \tag{1.6.44}$$

In the case of invariance of the system against dP, Eq. (1.6.42) yields

$$\Delta_i = 0 \tag{1.6.45}$$

as a stability criterion. If on the other hand, the system becomes unstable due to a change of the structural parameter P, Eq. (1.6.42) indicates that

$$\Delta = \det\left[\frac{\partial^2 H}{\partial q_k \partial q_i} - \frac{\partial N_k}{\partial q_i} + \frac{\partial}{\partial q_i}\left(\frac{\partial H}{\partial p_j}\frac{dp_j}{dq_k}\right)\right] = 0 \tag{1.6.46}$$

is the instability criterion. This is obvious as in the case of (1.6.45), $dq_i/dP = 0$, and in the case of (1.6.42), $dq_i/dP = \infty$.

It may also be mentioned that relationship (1.6.42) can be integrated to yield the curves

$$q_i = q_i(P). \tag{1.6.47}$$

These are known as deflection-load curves in structural engineering or as resonance curves in the mechanics of vibrations and are a good tool to depict the general stability behaviour of the system. For example, the post-buckling aspect of the system can be studied using (1.6.47).

Let the important criterion (1.6.46) be discussed for the case of a conservative system. As already pointed out, one may then restrict oneself also to the static aspect of the system. Hence, $N_i \equiv 0$, $p_i \equiv 0$, and H can be substituted by $U(q_i)$. Consequently, (1.6.46) is reduced to $\det([\partial^2 U/\partial q_k]\partial q_i) = 0$, which is (1.6.46). In this way it has become obvious that the celebrated stability condition (1.6.34) of the stability theory of static systems is only a special case of the more general theory presented here for any kind of dynamic, non-conservative systems.

Now, it will be shown how the results obtained previously can be applied in a systematic way to the stability evaluation of non-conservative systems.

The *first step* consists in specifying the generalized coordinates q_i, generalized impulses p_i, components N_i of the non-conservative forces, and the structural parameter(s), say P.

The *second step* involves calculating the determinant (1.6.43)

$$\Delta = \det\left[\frac{\partial^2 H}{\partial q_k \, q_i} - \frac{\partial N_k}{\partial q_i} + \frac{\partial}{\partial q_i}\left(\frac{\partial H}{\partial p_j} \frac{\mathrm{d}p_j}{\mathrm{d}q_k}\right)\right],$$

the vector b_i (see (1.6.44)), and the determinant

$$\Delta_i = \begin{vmatrix} a_{11} a_{12} & \cdots & a_{1,i-1} & b_1 & a_{1,i+1} & \ldots & a_{1n} \\ a_{21} a_{22} & \cdots & a_{2,i-1} & b_2 & a_{2,i+1} & \cdots & a_{2n} \\ \vdots & & & & & & \\ a_{n1} a_{n2} & \cdots & a_{n,i-1} & b_n & a_{n,i+1} & \cdots & a_{nn} \end{vmatrix} \qquad (1.6.48)$$

where

$$a_{ki} = \frac{\partial^2 H}{\partial q_k \, \partial q_i} - \frac{\partial N_k}{\partial q_i} + \frac{\partial}{\partial q_i}\left(\frac{\partial H}{\partial p_j} \frac{\mathrm{d}p_j}{\mathrm{d}q_k}\right). \qquad (1.6.49)$$

The *third step* leads to the application of (1.6.43) and (1.6.48). If one is interested in the onset of instability, one will apply (1.6.46), i.e. $\Delta = 0$, and derive from this criterion the conditions for which stability is lost.

If one is concerned with invariance, for example the existence of a limit cycle in the phase space diagram, then one may apply (1.6.45), ie. $\Delta_i = 0$. The result will be an indication of the conditions required for invariance of the system to the change of a certain structural parameter.

If information on the post-buckling behaviour of the system is required, one may integrate the relationship (1.6.42), i.e.

$$\frac{\mathrm{d}q_i}{\mathrm{d}P} = -\frac{\Delta_i}{\Delta}$$

in order to obtain the 'deflection-load' curve.

A final remark should not be forgotten. All the conditions mentioned, (1.6.42), (1.6.45), and (1.6.46), are underdetermined. For an application of these conditions, q_i ought to be known explicitly as a function of time. The same is required of p_i, $\dot{q}_i$, respectively, which can then be specified as functions of q_i. However, knowledge of these quantities would be obtained only by fully integrating the Hamilton equations (1.6.35). Then, the problem would be completely solved, and no specific stability theory needed. Hence, one may conclude that a meaningful application of the here presented stability theory must be based on additional, plausible assumptions on q_i and p_i. Considering the fact that a motion of the system caused by a perturbation, and possibly leading to instability, will essentially have the character of a vibration, appropriate assumptions on q_i and p_i can be made without great difficulties. In this way, the theory developed here will at least be sufficient for yielding approximate information on stability or instability. This will be shown in subsequent examples.

The stability of non-linear, non-conservative systems has been treated in the mechanics of vibrations. Therefore, the energy approach, as developed in this book, will be applied to certain vibrating systems. The results thus obtained can then be compared with those obtained earlier in a different and independent way, and available in the literature. In such a manner, one may gain confidence in the applicability and efficiency of the energy approach proposed here. Following this, the theory may be applied to a structural system subjected to follower forces, for the treatment of which the theory has specifically been established.

The forced and therefore non-conservative vibration of a mass–spring system with a non-linear characteristic is given by

$$m\ddot{q} + f(q) = mB \cos \omega t. \tag{1.6.50}$$

In (1.6.50) m is the mass of the vibrating masspoint, q its deflection from the equilibrium position, f a non-linear function of q, mB the amplitude, ω the frequency of the driving force, t the time, and the dot denotes differentiation with respect to time.

For this system,

$$T = \tfrac{1}{2} m\dot{q}^2 \tag{1.6.51}$$

is the kinetic energy,

$$U = \int f(q) \, dq = \psi(q) \tag{1.6.52}$$

the potential energy, and therefore

$$L = T - U = \tfrac{1}{2} m\dot{q}^2 - \psi(q) \tag{1.6.53}$$

the Lagrangian. Hence,

$$p = \frac{\partial L}{\partial \dot{q}} = m\dot{q} \tag{1.6.54}$$

is the generalized impulse. Thus,

$$H(p, q) = \frac{1}{2} \frac{p^2}{m} + \psi(q) \tag{1.6.55}$$

is the Hamiltonian of the system. The non-conservative force involved is

$$N = mB \cos \omega t, \tag{1.6.56}$$

and the destabilizing parameter is supposed to be the frequency ω of the driving force.

It is reasonable to assume that due to the overwhelming influence of the driving force, after some transition period, the motion of the masspoint will be given by

$$q = A \cos \omega t, \tag{1.6.57}$$

where A is the yet undetermined amplitude. From (1.6.57) follows

$$\dot{q} = -\omega A \sin \omega t = -\sqrt{\omega^2 A^2 - \psi^2 q^2}.$$

Hence,

$$\frac{d\dot{q}}{dq} = \frac{\omega^2 q}{\sqrt{\omega^2 A^2 - \omega^2 q^2}} = -\frac{\omega^2 q}{\dot{q}}. \tag{1.6.59}$$

By virtue of (1.6.54) and (1.6.59)

$$\frac{dp}{dq} = m\frac{d\dot{q}}{dq} = -m\omega^2 \frac{q}{\dot{q}}. \tag{1.6.60}$$

Moreover,

$$\frac{\partial H}{\partial p} = \frac{p}{m} = \dot{q} \tag{1.6.61}$$

as follows from (1.6.54) and (1.6.55). Therefore, (1.6.60) and (1.6.61) yield

$$\frac{\partial H}{\partial p}\frac{dp}{dq} = -m\omega^2 q, \tag{1.6.62}$$

which leads to

$$\frac{\partial}{\partial q}\left(\frac{\partial H}{\partial p}\frac{dp}{dq}\right) = -m\omega^2. \tag{1.6.63}$$

Using (1.6.55) and (1.6.62), the heterocline Φ is obtained according to (1.6.39) as

$$\Phi = \left(\frac{\partial\psi}{\partial q} - mB\cos\omega t\right) - m\omega^2 q = 0. \tag{1.6.64}$$

The determinant Δ consists in this case of a single element which according to (1.6.43), (1.6.55), (1.6.56) and (1.6.63) reads

$$\Delta = \frac{\partial^2\psi}{\partial q^2} - m\omega^2. \tag{1.6.65}$$

The determinant Δ_i consists again of a single element. This follows from (1.6.44), (1.6.55), (1.6.56) and (1.6.48) as

$$\Delta_i \equiv b = -\frac{\partial N}{\partial\omega} = mtB\sin\omega t. \tag{1.6.66}$$

The resonance curve of the vibration is to be derived from (1.6.42), i.e.

$$\frac{dq}{d\omega} = -\frac{\Delta_i}{\Delta}. \tag{1.6.67}$$

Using (1.6.65) and (1.6.66) in (1.6.67) yields

$$\frac{dq}{d\omega} = -\frac{mtB\sin\omega t}{(\partial^2\psi/\partial q^2) - m\omega^2}. \tag{1.6.68}$$

By virtue of (1.6.57), (1.6.68) can be changed into

$$-tA\sin\omega t = -\frac{mtB\sin\omega t}{(\partial^2\psi/\partial q^2) - m\omega^2}. \tag{1.6.69}$$

From (1.6.69) follows

$$-\frac{\partial^2\psi}{\partial q^2} A \sin \omega t + m\omega^2 \sin t = - mB \sin \omega t. \qquad (1.6.70)$$

Using again (1.6.57)), $(\partial^2\psi/\partial q^2)\, A \sin \omega t$ can be rewritten to yield

$$\frac{\partial^2\psi}{\partial q^2} A \sin \omega t = - F_1(A)\sin \omega t - \sum_{k=2}^{n} F_k(A)\sin k\omega t. \qquad (1.6.71)$$

In (1.6.71), $F_1, F_2, \ldots, F_n$ are known functions of the argument A. Substituting (1.6.71) into (1.6.70) yields

$$F_1(A)\sin \omega t + \sum_{k=2}^{n} F_k(A)\sin k\omega t + m\omega^2 A \sin \omega t = - mB \sin \omega t. \qquad (1.6.72)$$

Comparing coefficients of $\sin \omega t$ in (1.6.72), and neglecting higher harmonics, yields finally the resonance curve

$$F_1(A) + m\omega^2 A + mB = 0. \qquad (1.6.73)$$

In an (A, ω)-plane, (1.6.73) corresponds to the curve

$$C_1(A, \omega) + B = 0, \qquad C_1(A, \omega) = \frac{1}{m} F_1(A) + \omega^2 A. \qquad (1.6.74)$$

Instability occurs according to (1.6.46) and (1.6.65) if

$$\frac{\partial^2\psi}{\partial q^2} - m\omega^2 = 0, \qquad q\frac{\partial^2\psi}{\partial q^2} - m\omega^2\, q = 0$$

respectively.

Using (1.6.57), first

$$q\frac{\partial^2\psi}{\partial a^2} = G_1(A)\cos \omega t + \sum_{k=2}^{r} G_k(A)\cos k\omega t \qquad (1.6.76)$$

is obtained, where r is an appropriate number depending on the nature of the terms involved in ψ. Subsequently, using (1.6.57) and (1.6.76) in (1.6.75), one arrives at

$$G_1(A)\cos \omega t + \sum_{k=2}^{r} G_k(A)\cos k\omega t - m\omega^2 A \cos \omega t = 0. \qquad (1.6.77)$$

Comparing coefficients of $\cos \omega t$ in (1.6.77) and neglecting higher harmonics yields

$$G_1(A) - m\omega^2 A = 0. \qquad (1.6.78)$$

Letting $G_1/A = G_1^*(A)$, (1.6.78) can be transformed into

$$G_1^*(A) - m\omega^2 = c^2(A, \omega) = 0, \qquad (1.6.79)$$

which can be interpreted as another curve in the (A, ω)-plane. The intersection of curves (1.6.78) and (1.6.79) determines the point $(A_{crit}, \omega_{crit})$, whose co-ordinates represent the critical values of amplitude A and driving frequency ω at which instability of the vibration occurs.

Next, take Van der Pol's problem as an example for systems exhibiting the phenomenon of relaxation oscillations. This problem will serve to demonstrate the potential of the theory to provide information on the existence of a limit cycle.

The equation of motion of Van der Pol's problem is known to be

$$m\ddot{q} + kq - m\varepsilon(1 - q^2)\dot{q} = 0. \tag{1.6.80}$$

Hence,

$$H = \frac{1}{2}\frac{p^2}{m} + \frac{1}{2}kq^2 \tag{1.6.81}$$

$$N = m\varepsilon(1 - q^2)\dot{q} \tag{1.6.82}$$

$$\frac{\partial H}{\partial p}\frac{dp}{dq} = -m\omega_0^2 q \tag{1.6.83}$$

assuming a solution of the type

$$q = A \sin \omega_0 t, \ \omega_0^2 = k/m. \tag{1.6.84}$$

Taking ε as the parameter P, relationship (1.6.41) assumes for this case the form

$$\left[\frac{\partial^2 H}{\partial q^2} - \frac{\partial N}{\partial q} + \frac{\partial}{\partial q}\left(\frac{\partial H}{\partial p}\frac{dp}{dq}\right)\right]dq + \left[\frac{\partial^2 H}{\partial q\,d\varepsilon} - \frac{\partial N}{\partial \varepsilon}\right]d\varepsilon = 0. \tag{1.6.85}$$

Using (1.6.81), (1.6.82) and (1.6.83) in (1.6.85) yields

$$\left[k - \frac{\partial N}{\partial q} - m\omega^2\right]dq - \frac{\partial N}{\partial \varepsilon}d\varepsilon = 0.$$

But $k - m\omega_0^2 = 0$ by definition. Hence,

$$\frac{\partial N}{\partial q}dq + \frac{\partial N}{\partial \varepsilon}d\varepsilon = 0 \tag{1.6.86}$$

must be required. Multiplying (1.6.86) by dq, integrating, and using the mean value theorem of integration yields

$$\left(\frac{\partial N}{\partial q}\right)_{av}\int (dq)^2 + (d\varepsilon)_{av}\int \frac{\partial N}{\partial \varepsilon}dq = 0. \tag{1.6.87}$$

If the system were insensitive against changes of the parameter $d\varepsilon$, then $dq = 0$ and even more

$$\int (dq)^2 = 0.$$

But then, from (1.6.87) followed

$$\int \frac{\partial N}{\partial \varepsilon}dq = 0 \tag{1.6.88}$$

in the case of invariance.

102

Using (1.6.82) in (1.6.88) yields

$$m \int (1 - q^2)\dot{q}\, dq = m \int (1 - q^2)\dot{q}^2\, dt = 0.$$

Applying (1.6.84) and extending the integration over a period, yields

$$\int_0^{2\pi} [A^2 \omega_0^2 \cos^2 \omega_0 t - A^4 \omega_0^2 \cos^2 \omega_0 t \sin^2 \omega_0 t]\, d\omega_0 t = 0$$

and

$$\pi - A^2 \frac{\pi}{4} = 0.$$

From the preceding equation it follows that $A = 2$, indicating a limit cycle with this amplitude can be expected to exist[1]. Finally consider the hinged rod with a compressive load, as shown in Fig. 34. The energy approach, sucessfully tested in the previous examples, will now be applied to that non-conservative system involving the follower force P.

From the figure it can be deduced that

$$T = \tfrac{1}{2} ml^2 (3\dot{\phi}_1^2 + 2\dot{\phi}_1\dot{\phi}_2 + \dot{\phi}_2^2) \tag{1.6.89}$$

is the kinetic energy and

$$U = \tfrac{1}{2} c (2\phi_1^2 - 2\phi_1\phi_2 + \phi_2^2) \tag{1.6.90}$$

the potential energy. Hence

$$L = \tfrac{1}{2} ml^2 (3\dot{\phi}_1^2 + 2\dot{\phi}_1\dot{\phi}_2 + \dot{\phi}_2^2) - \tfrac{1}{2} c (2\phi_1^2 - 2\phi_1\phi_2 + \phi_2^2) \tag{1.6.91}$$

is the Lagrangian, and

$$p_1 = \frac{\partial L}{\partial \dot{\phi}_1} = ml^2 (3\dot{\phi}_1 + \dot{\phi}_2) \qquad p_2 = \frac{\partial L}{\partial \dot{\phi}_2} = ml^2 (\dot{\phi}_2 + \dot{\phi}_1) \tag{1.6.92}$$

Fig. 34. Hinged rod with tangential, compressive load P

[1]Den Hartog, J. P., *Mechanical Vibrations*, pp. 363–365, McGraw-Hill, New York, 4th ed., 1956.

are the generalised impulses. From (1.6.92) follows

$$\dot{\phi}_1 = \frac{p_1 - p_2}{2ml^2} \qquad \dot{\phi}_2 = \frac{3p_2 - p_1}{2ml^2}, \tag{1.6.94}$$

$$T = \frac{1}{4ml^2}(p_1^2 - 2p_1p_2 + 3p_2^2). \tag{1.6.94}$$

Therefore, the Hamiltonian $H = T + U$ reads

$$H = \frac{1}{4ml^2}(p_1^2 - 2p_1p_2 + 3p_2^2) + \tfrac{1}{2}c(2\phi_1^2 - 2\phi_1\phi_2 + \phi_2^2). \tag{1.6.95}$$

The components of the non-conservative force are

$$N_1 = Pl(\phi_1 - \phi_2), \qquad N_2 = 0. \tag{1.6.96}$$

The motion of the system caused by a perturbation is assumed to be given by

$$\phi_1 = A_1 e^{\omega t}, \qquad \phi_2 = A_2 e^{\omega t}. \tag{1.6.97}$$

Therefore,

$$\frac{d\dot{\phi}_1}{d\phi_1} = \omega, \qquad i = 1, 2, \tag{1.6.98}$$

$$p_1 = ml^2\omega(3\phi_1 + \phi_2), \qquad p_2 = ml^2\omega(\phi_1 + \phi_2),$$

and

$$\left.\begin{array}{ll} \dfrac{dp_1}{d\phi_1} = 3ml^2\omega, & \dfrac{dp_1}{d\phi_2} = ml^2\omega, \\[2mm] \dfrac{dp_2}{d\phi_1} = ml^2\omega, & \dfrac{dp_2}{d\phi_2} = ml^2\omega. \end{array}\right\} \tag{1.6.99}$$

Moreover,

$$\left.\begin{array}{l} \dfrac{\partial H}{\partial p_1} = \dfrac{1}{2ml^2}(p_1 - p_2) = \omega\phi_1, \\[2mm] \dfrac{\partial H}{\partial p_2} = \dfrac{1}{2ml^2}(3p_2 - p_1) = \omega\phi_2. \end{array}\right\} \tag{1.6.100}$$

Therefore,

$$\frac{\partial H}{\partial p_j}\frac{dp_j}{d\phi_1} = \frac{\partial H}{\partial p_1}\frac{dp_1}{d\phi_1} + \frac{\partial H}{\partial p_2}\frac{dp_2}{d\phi_1} = 3ml^2\omega^2\phi_1 + ml^2\omega^2\phi_2,$$

$$\frac{\partial H}{\partial p_j}\frac{dp_j}{d\phi_2} = \frac{\partial H}{\partial p_1}\frac{dp_1}{d\phi_2} + \frac{\partial H}{\partial q_2}\frac{dp_2}{d\phi_2} = ml^2\omega^2\phi_1 + ml^2\omega^2\phi_2, \tag{1.6.101}$$

104

and

$$\frac{\partial}{\partial\phi_1}\left(\frac{\partial H}{\partial p_j}\frac{\mathrm{d}p_j}{\mathrm{d}\phi_1}\right) = 3ml^2\omega^2, \frac{\partial}{\partial\phi_1}\left(\frac{\partial H}{\partial p_j}\frac{\mathrm{d}p_j}{\mathrm{d}\phi_2}\right) = ml^2\omega^2, \left.\begin{array}{c} \\ \\ \end{array}\right\}$$

$$\frac{\partial}{\partial\phi_2}\left(\frac{\partial H}{\partial p_j}\frac{\mathrm{d}p_j}{\mathrm{d}\phi_1}\right) = ml^2\omega^2, \frac{\partial}{\partial\phi_2}\left(\frac{\partial H}{\partial p_j}\frac{\mathrm{d}p_j}{\mathrm{d}\phi_2}\right) = ml^2\omega^2. \qquad (1.6.102)$$

For a stability investigation, condition (1.6.46) has to be checked. But,

$$\Delta = \det(a_{ki}),$$

and the a_{ki} are given by (1.6.49). Using (1.6.95), (1.6.96), and (1.6.102) in (1.6.49) and observing that $q_i = \phi_i$, yields

$$\left.\begin{array}{l} a_{11} = 2c - Pl + 3ml^2\omega^2, \\ a_{12} = -c + Pl + ml^2\omega^2, \\ a_{21} = -c + ml^2\omega^2, \\ a_{22} = c + ml^2\omega_2. \end{array}\right\} \qquad (1.6.103)$$

Hence,

$$\Delta = \left| \begin{array}{cc} 2c - Pl + 3ml^2\omega^2 & -c + Pl + ml^2\omega^2 \\ -c + ml^2\omega^2 & c + ml^2\omega^2 \end{array} \right| \qquad (1.6.104)$$

and from condition (1.6.46), i.e. $\Delta = 0$, follows the characteristic equation

$$\det(\mathbf{A} - \lambda\mathbf{C} = 0, \qquad (1.6.105)$$

where

$$\lambda = -\frac{ml^2\omega^2}{c} \qquad p = \frac{Pl}{c}$$

and $\mathbf{A}$ and $\mathbf{C}$ are the matrices

$$\mathbf{A} = \begin{pmatrix} 2 - p & -1 + p \\ 2 - 1 & 1 \end{pmatrix} \qquad \mathbf{C} = \begin{pmatrix} 3 & 1 \\ 1 & 1 \end{pmatrix}. \qquad (1.6.107)$$

Expanding (1.6.105) yields

$$\lambda^2 - \lambda\left(\frac{7}{2} - p\right) + \frac{1}{2} = 0$$

and

$$\lambda = \frac{7 - 2p}{4} \pm \left[\left(\frac{7 - 2p}{4}\right)^2 - \frac{1}{2}\right]^{1/2}. \qquad (1.6.108)$$

The quantity λ is proportional to ω^2. A flutter frequency exists and the structural system becomes unstable by flutter if λ turns out to be complex. According to (1.6.108) that is the case for

$$\left(\frac{7 - 2p}{4}\right)^1 - \frac{1}{2} < 0.$$

In the limiting case, i.e. at the stability boundary,

$$\left(\frac{7-2p}{4}\right)^2 - \frac{1}{2} = 0. \qquad (1.6.109)$$

Solving (1.6.109) for p yields the flutter load

$$p_{fl} = \frac{7 \pm \sqrt{8}}{2}, \qquad P_{fl} = \frac{7 \pm \sqrt{8}}{2}\frac{c}{l}. \qquad (1.6.110)$$

This value is in complete agreement with the flutter load calculated in Section 2.3.5, if in formula (2.3.83) $\alpha \equiv 1$ is being substituted.

The conclusion is that also for non-conservative systems with follower forces, the proposed energy approach to stability can successfully be applied.

Exercises 1.6

1. Given Duffing's problem with $L = T - U = \frac{1}{2}m\dot{q}^2 - \psi(q)$, $\psi(q) = \frac{1}{2}aq^2 + \frac{1}{4}bq^4$ where a and b are constants. Show that

$$A\omega_0^2 + \frac{3}{4}hA^3 - \omega^2 A = B, \qquad \omega_0^2 = \frac{a}{m}, \qquad h = \frac{b}{m}$$

   is the resonance curve and

$$\omega_0^2 + \frac{9}{4}hA^2 - \omega^2 = 0$$

   the curve corresponding to (1.6.79). Find the critical values A_{crit} and ω_{crit} by bringing these two curves to intersection. Note, m is the mass of the oscillator.

2. The vibration of a system with non-linear damping is governed by the differential equation

$$m\ddot{q} + \frac{1}{3}\dot{q}^3 + kq = p_0 \sin \omega t$$

   where m is the mass, q the deflection, k the spring constant, p_0 the amplitude of the driving force, ω the driving frequency, and t the time. Using the energy approach, find the resonance curve relating the amplitude A of the system with the driving frequency ω.

1.7
Mathematical Approximation Methods

Very often the differential equations of mechanical systems whose stability is being investigated are linear but have varying coefficients. Often they are even non-linear, or they can be partial differential equations. This can also be true for the variational equations. This leads to complicated problems. Since the central problem of stability theory is to estimate the solutions of the variational equations and to determine their behaviour when parameters are changed, it will be necessary to use approximation methods in such difficult cases. These approximations have been developed in applied mathematics, mathematical physics (especially astronomy) or in control theory. We will now consider a few of these approximation methods.

1.7.1 Isocline Method

This method is primarily used to construct *phase curves* in the phase plane in the case of variational equations

$$\dot{\xi}_1 = f_1(\xi_2, \xi_2), \qquad \dot{\xi}_2 = f_2(\xi_1, \xi_2)$$

with non-linear, time-independent functions. The expression

$$d\xi_2/d\xi_1 = f_2(\xi_1, \xi_2)/f_1(\xi_1, \xi_2) = J(\xi_1, \xi_2)$$

is formed and the curves $J(\xi_1, \xi_2) = \text{const}$ are constructed in the ξ_1, ξ_2 plane, the phase plane. These curves are called *isoclines* because they correspond to directional elements having a constant inclination. If a family of isoclines has been constructed in the phase plane and if the corresponding directional elements have been plotted, this leads to an entire field of directional elements with which families of phase curves can be graphically constructed. An example of this procedure, as well as an example of determining stability from the phase curves, has already been given in Section 1.4.1 and Fig. 24.

The book by Kauderer (2.10) contains similar procedures leading to the construction of phase curves from directional elements. These procedures were advanced by Liénard and Schäfer.

1.7.2 Methods of Small Oscillations

In this case we assume that a first-order set of differential equations is given for the mechanical system. For example, it can be the Hamiltonian system

$$\dot{q}_r = \partial H/\partial p_r \qquad\qquad \dot{p}_r = -\partial H/\partial q_r$$
$$H = H(q_1,\ldots,q_f,p_1,\ldots,p_f,t), \qquad r = 1,2,,\ldots,f.$$

Introducing the new variables

$$q_1,\ldots,q_f = y_1,\ldots,y_f \quad \text{and} \quad p_1,\ldots,p_f = y_{f+1},\ldots,y_{2f}.$$

We obtain

$$\dot{y}_i = f_i(y_1,\ldots,y_{2f},t), \qquad\qquad i = 1,2,\ldots,2f. \qquad\qquad (1.7.1)$$

Let us investigate the stability of an 'unperturbed' solution y_i^0. Just as was done in Section 1.2, the solutions y_i adjacent to y_i^0 are assumed to have the form

$$y_i(t) = y_i^0(t) + \xi_i(t).$$

By expanding the f_i into a Taylor series and neglecting terms of higher order, the following linear variational equations are obtained from (1.7.1):

$$\dot{\xi}_i = (\partial f_i/\partial y_k)_{y_r^0}\xi_k, \qquad\qquad i,k = 1,2,\ldots,2f. \qquad\qquad (1.7.2)$$

This method is very often used, but can only be applied if the functions f_i are analytic and can be developed into a Taylor series. In addition, the linearized Eqs. (1.7.2) only determine the stability of the unperturbed solutions y_i^0 if the conditions of Lyapunov's Theorem 5 given in 1.5.3 are satisfied.

1.7.3 Variation of Constants

This is a method derived from the theory of linear differential equations. Let us assume that the system of variational equations is given by

$$\dot{\xi}_i = a_{ik}\xi_k + \Phi_i. \qquad\qquad (1.7.3)$$

It consists of the linear system

$$\dot{\xi}_i = a_{ik}\xi_k \qquad\qquad (1.7.4)$$

and the non-linear perturbation terms of Φ_i. In the simpler case of linear differential equations, the Φ_i would only be functions of the independent variables, and the abbreviated system (1.7.4) becomes inhomogeneous because of them. Additional difficulties arise because, in the case assumed here, the Φ_i are non-linear functions of the independent as well as of the dependent variables.

Let us assume that the fundamental system $\xi_i^\nu(i,\nu = 1,\ldots,n)$ corresponds to (1.7.4). The trial solution $\xi_i = c_\nu\xi_i^\nu$ of (1.7.3) is assumed, and the c_ν are variables instead of constants. We therefore have

$$\dot{\xi}_i = \dot{c}_\nu \xi_i^\nu + c_\nu \dot{\xi}_i^\nu. \tag{1.7.5}$$

If this is substituted in (1.7.3), it follows that

$$\dot{c}_\nu \xi_i^\nu + c_\nu \dot{\xi}_i^\nu = a_{ik} c_\nu \xi_k^\nu + \Phi_i. \tag{1.7.6}$$

Because of (1.7.4) we have

$$\dot{\xi}_i^\nu = a_{ik} \xi_k^\nu, \tag{1.7.7}$$

which results in the following if this result is substituted into (1.7.6):

$$\dot{c}_\nu \xi_i^\nu = \Phi_i(c_1, \ldots, c_n, \xi_1^1, \ldots, \xi_1^n, \xi_2^1, \ldots, \xi_2^n, \ldots, \xi_n^n, t). \tag{1.7.8}$$

A non-linear system of differential equations has thus been found for the c_k which can be solved for the c_k. The final result is

$$\dot{c}_k = \Psi_k(c_1, \ldots, c_n, \xi_1^1, \ldots, \xi_1^n, \xi_2^1, \ldots \xi_2^n, \ldots, \xi_n^n, t), \tag{1.7.9}$$

where Ψ_k is a non-linear function that results from the solution of (1.7.8).

Equation (1.7.9) does not appear to be a simplification, because we departed from the non-linear system (1.7.3) in the ξ_k and obtained a non-linear system (1.7.9) in the c_k, which could be possibly just as complicated.

Let us carry out the following transformation that is different in each case:

$$c_k = H_k(K_1, \ldots, K_n), \qquad k = 1, \ldots, n. \tag{1.7.10}$$

Equation (1.7.9) becomes

$$(\partial H_k/\partial K_i)\dot{K}_i = \Psi_k[H_1(K_1, \ldots, K_n), \ldots, H_n(K_1, \ldots, K_n), \xi_1^1, \ldots, \xi_n^n, t]. \tag{1.7.11}$$

If (1.7.11) is solved for the $\dot{K}_k$, we obtain the following equation in the place of (1.7.9):

$$\dot{K}_k = G_k(K_1, \ldots, K_n, \xi_1^1, \ldots, \xi_n^n, t). \tag{1.7.12}$$

It should be possible to find certain *average values* for the $\dot{K}_k$ over a time interval T

$$[\dot{K}_k]_M = (1/T) \int_0^T G_k \, dt, \tag{1.7.13}$$

where the quantities $K_1, \ldots, K_n$ appearing in G_k can be held constant. In this way an approximation for the function K_k

$$K_k = [\dot{K}_k]_M t + c_k^* \tag{1.7.14}$$

is found. This approximation is justified when the K_k change only slightly in the corresponding time interval T, as has been assumed. The c_k^* represent constants of integration.

If this method is applied to non-linear vibration problems, it is called the *method of slowly changing amplitude* and the time interval T is a period.

Based on (1.7.14), (1.7.10), and the trial solution $\xi_i = c_\nu \xi_i^\nu$, the following expression is found for the desired approximate solution ξ_i:

$$\xi_i = H_\nu([\dot{K}_1]_M t + c_1^*, \ldots, [\dot{K}_n]_M t + c_n^*)\xi_1^\nu.$$

This is the end of the calculation. The trick is to find a suitable transformation (1.7.10), so that the average value (1.7.13) can be formed in such a way that the K_k can be assumed to be constant over the time T.

Example: Let us consider the non-linear differential equation $\ddot{x} + \omega_0^2 x + hx^3 = 0$ with the initial conditions $x(0) = A_0, \dot{x}(0) = 0$. With $\xi_1 = x, \xi_2 = \dot{x}$, it is transformed into

$$\dot{\xi}_1 = \xi_2$$

$$\dot{\xi}_2 = -\omega_0^2 \xi_1 - h\xi_1^3,$$

in which $\Phi_1 = 0$, $\Phi_2 = -h\xi^3$. The linearized system has the fundamental solutions

$$\xi_1^1 = \cos \omega_0 t, \qquad \xi_1^2 = -\sin \omega_0 t, \qquad \xi_2^1 = -\omega_0 \sin \omega_0 t, \qquad \xi_2^2 = -\omega_0 \cos \omega_0 t,$$

so that

$$\left.\begin{array}{l}\xi_1 = c_1 \cos \omega_0 t - c_2 \sin \omega_0 t, \\[1mm] \xi_2 = -\omega_0 c_1 \sin \omega_0 t - \omega_0 c_2 \cos \omega_0 t\end{array}\right\} \qquad (1.7.15)$$

is obtained.

By means of the transformation

$$c_1 = A \cos \Theta, \qquad c_2 = A \sin \Theta, \qquad (1.7.16)$$

the new quantities A and Θ are introduced to replace c_1 and c_2. From (1.7.8) it follows that

$$\dot{c}_1 \cos \omega_0 t - \dot{c}_2 \sin \omega_0 t = 0,$$

$$-\omega_0 \dot{c}_1 \sin \omega_0 t - \omega_0 \dot{c}_2 \cos \omega_0 t = -h(c_1 \xi_1^1 + c_2 \xi_1^2)^3,$$

which, together with (1.7.16), become

$$(\dot{A} \cos \Theta - A \sin \Theta \dot{\Theta}) \cos \omega_0 t - (\dot{A} \sin \Theta + A \cos \Theta \dot{\Theta}) \sin \omega_0 t = 0$$

$$(\dot{A} \cos \Theta - A \sin \Theta \dot{\Theta}) \sin \omega_0 t + (\dot{A} \sin \Theta + A \cos \Theta \dot{\Theta}) \cos \omega_0 t$$
$$= (hA^3/\omega_0)(\cos \Theta \cdot \cos \omega_0 t - \sin \Theta \cdot \sin \omega_0 t)^3.$$

If trigonometric transformations are applied to this expression and the abbreviation $\Phi = \omega_0 t + \Theta$ is introduced, this system can be transformed into

$$\dot{A} \cos \phi - \dot{\Theta} A \sin \phi = 0$$

$$\dot{A} \sin \phi - \dot{\Theta} A \cos \phi = (hA^3/\omega_0) \cos^4 \phi.$$

Solving with respect to $\dot{A}$ and $\dot{\Theta}$, we obtain

$$\dot{A} = (hA^3/\omega_0) \cos^3 \phi \sin \phi, \qquad \dot{\Theta} = (hA^2/\omega_0) \cos^4 \phi,$$

which corresponds to Eqs. (1.7.12). Let us obtain the average values of the slowly varying quantities A and Θ. We find

$$[\dot{A}]_M = \frac{1}{2\pi} \int_0^{2\pi} \frac{hA^3}{\omega_0} \cos^3 \phi \sin \phi \, d\phi = 0,$$

$$[\dot{\Theta}]_M = \frac{1}{2\pi} \int_0^{2\pi} \frac{hA^2}{\omega_0} \cos^4 \phi \, d\phi = \frac{3hA^2}{8\omega_0},$$

so that the following expression is obtained for (1.7.14):

$$A = c_1^* \qquad \Theta = (3hA^2/8\omega_0)t + c_2^*. \tag{1.7.17}$$

If (1.7.17) is substituted in (1.7.16) and the result substituted in (1.7.15), it follows that

$$\xi_1 = c_1^* \cos\left(\frac{3hA^2}{8\omega_2} t + c_2^*\right) \cos \omega_0 t - c_1^* \sin\left(\frac{3hA^2}{8\omega_0} t + c_2^*\right) \sin \omega_0 t,$$

$$\xi_2 = -\omega_0 \left[c_1^* \cos\left(\frac{3hA^2}{8\omega_0} t + c_2^*\right) \sin \omega_0 t + c_1^* \sin\left(\frac{3hA^2}{8\omega_0} t + c_2^*\right) \cos \omega_0 t \right].$$

These equations are transformed into the following expression by means of trigonometric transformations:

$$\xi_1 = c_1^* \cos\left[(\omega_0 + 3hA^2/8\omega_0)t + c_2^*\right]$$

$$\xi_2 = -\omega_0 c_1^* \sin\left[(\omega_0 + 3hA^2/8\omega_0)t + c_2^*\right].$$

The integration constants c_k^* can now be determined using the initial conditions $\xi_1(0) = A_0$, $\xi_2(0) = 0$. This leads to $c_2^* = 0$, $c_1^* = A_0$. The following approximate solution, well known in the theory of vibrations, is thus obtained:

$$\xi_1 \equiv x = A_0 \cos\left[(\omega_0 + 3hA_0^2/8\omega_0)t\right].$$

1.7.4 Perturbation Method

This approximation method was initially resolved in astronomy but has been widely applied in other branches of physics. The fundamental concepts of this method were probably advanced by Poisson.

Let us assume that the motion of a body in the solar system is given by

$$\dot{q}_i = F_{i0} + \mu^k F_{ik} \qquad i = 1, 2, \ldots, n; \qquad k = 1, 2, 3, \ldots, \tag{1.7.18}$$

where the F_{i0} are due to the influence of the sun and therefore represent the most important terms of the system of differential equations. The terms $\mu^k F_{ik}$ represent the much smaller influences of the other planets and can therefore be considered as perturbations of the motion which is primarily caused by the dominant attraction of the sun. Therefore, we are justified in assuming the following trial solution for (1.7.18):

$$q_i = \mu^k q_{ik}, \qquad k = 0, 1, 2, \ldots, \tag{1.7.19}$$

If (1.7.19) is substituted into (1.7.18) and if *terms corresponding to equal powers of μ are set equal,* a calculation method is obtained which will now be described in more detail.

Let us consider a more general physical problem for which the canonical system

$$\dot{p}_i = -\partial H/\partial q_i, \qquad \dot{q}_i = \partial H/\partial p_i$$

holds. Assume that the Hamiltonian function H is given by

$$H = H_0 + \mu^k H_k, \qquad k = 1, 2, \ldots$$

Here H_0 is again the dominant term and the $\mu^k H_k$, $k = 1, 2, \ldots$ represent the perturbations. The obvious trial solution $q_i = \mu^k q_{ik}$, $p_i = \mu^k p_{ik}$ will be used. From this point of view, the assumption that the terms $\mu^k F_{ik}$ or $\mu^k H_k$, respectively, are indeed small perturbations is essential. This is certainly the case if μ is a small parameter. The perturbation method is therefore also often called the *method of the small parameter.* However, it has been extended far beyond its original application, and under certain circumstances it can also be used if μ is no longer a small parameter or if such a parameter does not exist and must be artificially introduced into the calculation. We will show this in examples. The perturbation method was developed especially by Poincaré, who used it mostly to find periodic solutions and to investigate their stability. We will not treat this subject further because it belongs in the field of non-linear vibrations.

The description of the method is as follows.

A system of variational equations

$$\dot{\xi}_1 = a_{ik}\xi_k + \mu\Phi_i(\xi_1, \xi_2, \ldots, \xi_n, t), \qquad i, k = 1, 2, \ldots, n,$$

is given where Φ_i represents the non-linear terms. The trial solution for ξ_i is assumed in the form of a series $\xi_i = \xi_{i0} + \mu^r \xi_{ir}$, $r = 1, 2, \ldots$. This series converges if the parameter μ is sufficiently small. If it is substituted into the system of variational equations, one obtains

$$\dot{\xi}_{i0} + \mu^r \dot{\xi}_{ir} = a_{ik}(\xi_{k0} + \mu^r \xi_{kr}) + \mu\Phi_i(\xi_{s0} + \mu^r \xi_{sr}, t), \qquad s = 1, 2, \ldots, n.$$

By setting terms corresponding to equal powers of μ equal to each other, the following new system is obtained:

$$\dot{\xi}_{i0} = a_{ik}\xi_{k0},$$

$$\dot{\xi}_{ir} = a_{ik}\xi_{kr} + \Psi_i(\xi_{s0}, \ldots, \xi_{sm}, t), \qquad i, s, r = 1, 2, \ldots, n, m < r.$$

The system of the *generating solution* ξ_{i0} is shown on the first line. This solution can easily be found from the system because it is linear. The system of differential equations for the other terms ξ_{ir} of the trial solution is shown on the second line. It is also linear, which means that it is indeed possible to calculate the ξ_{ir}, because Ψ_i only contains known functions $\xi_{s0}, \ldots, \xi_{sm}$ of t that were determined in previous calculations. This is because $m < r$.

Therefore, Ψ_i is a known function of t so that the system of the ξ_{ir} is inhomogeneous but no longer non-linear.

Example: The differential equation

$$\dot{q} = -aq - \mu bq^2$$

was selected for the calculation. a and b are constants, and μ is the small parameter. The initial condition is assumed to be $q = K$ for $t = 0$. By means of the trial solution

$$q(t) = q_0(t) + \mu q_1(t) + \mu^2 q_2(t) + \ldots,$$

one obtains

$$\dot{q}_0 + \mu \dot{q}_1 + \mu^2 \dot{q}_2 + \ldots = -aq_0 - \mu aq_1 - \mu^2 aq_2 - \ldots - \mu bq_0^2 - 2\mu^2 bq_0 q_1 - \ldots$$

The generating solution q_0 satisfies the equation

$$\dot{q}_0 = -aq_0.$$

The other terms of the trial solution satisfy the equations

$$\dot{q}_1 = -bq_0^2 - aq_1, \qquad \dot{q}_2 = -2bq_0 q_1 - aq_2, \ldots.$$

The equation $\dot{q}_0 = -aq_0$ together with the initial condition $q_0 = K$ for $t = 0$ results in

$$q_0 = K\,e^{-at}.$$

The equations $\dot{q}_1 = -bq_0^2 - aq_1 = -bK^2\,e^{-2at} - aq_1$, together with the initial condition $q_1 = 0$ for $t = 0$, result in

$$q_1 = (bK^2/a)(e^{-2at} - e^{-at}),$$

from

$$\dot{q}_2 = -2bq_0 q_1 - aq_2 = -(2b^2 K^3/a)(e^{-3at} - e^{-2at}) - aq_2$$

and together with $q_2 = 0$ for $t = 0$, we obtain

$$q_2 = (b^2 K^3/a^2)(e^{-3at} - 2e^{-2at} + e^{-at}), \ldots,$$

etc.

The terms $q_0, q_1, q_2, \ldots$ obtained step by step finally lead to the following expression using the trial solution:

$$q = K\,e^{-at}[1 + (\mu bK/a)(e^{-at} - 1) + (\mu bK/a)^2(e^{-at} - 1)^2 + \ldots].$$

The *convergence proof* for solutions of this type may now be given. We may use the methods of Poincaré for this purpose. For simplicity, we will present the proof for a system of differential equation given by

$$\dot{q}_1 = F_1(q_1 q_2, \mu, t) = \sum c_{1\alpha\beta\gamma} q_1^\alpha q_2^\beta \mu^\nu$$

$$\dot{q}_2 = F_2(q_1, q_2, \mu, t) = \sum c_{2\alpha\beta\gamma} q_1^\alpha q_2^\beta \mu^\nu,$$

$$(1.7.20)$$

with two independent variables, q_1, q_2. This is a representation due to N. Minorsky. It is assumed that the F_i are analytic and can therefore be represented as power series of q_1, q_2, and μ. The coefficients $c_{i\alpha\beta\gamma}$ are continuous functions of t in the interval $J = [t_0, t_1]$ and it is assumed that the series converge for $|q_i|, |\mu| < r$.

In addition, the series are assumed not to contain constant terms, so that the system (1.7.20) has the particular solution $q_i = 0$ for $\mu = 0$.

In order to obtain the solution, the following trial solution is used:

$$q_i = \mu^k q_{ik}(t), \qquad i = 1, 2, \qquad k = 1, 2, \ldots, \qquad (1.7.21)$$

where $q_{i1}(t_0) = 0$ is assumed to hold. From (1.7.20) and with (1.7.21) it follows that

$$\dot{q}_{11} = c_{1100}q_{11}(t) + c_{1010}q_{21}(t) + c_{1001},$$
$$\dot{q}_{21} = c_{2100}q_{11}(t) + c_{2010}q_{21}(t) + c_{2001}. \qquad (1.7.22)$$

The $q_{11}(t), q_{21}(t)$ can be determined with the initial conditions $q_{11}(t_0) = q_{21}(t_0) = 0$. The following differential equations are found for the other functions of the trial solution.

$$\dot{q}_{1n} = c_{1100}q_{1n}(t) + c_{1010}q_{2n}(t) + \Psi_{1n},$$
$$\dot{q}_{2n} = c_{2100}q_{1n}(t) + c_{2010}q_{2n}(t) + \Psi_{2n}, \qquad (1.7.23)$$

in which the Ψ_{in} are certain polynomials that depend on $c_{i\alpha\beta\gamma}$ and q_{im} if $m < n$. All the q_{in} can be determined step by step from Eqs. (1.7.23). The q_{in} and their derivatives are continuous in the interval J.

In order to show that the series (1.7.21) converge, let us use the auxiliary system

$$\dot{Q}_i = F_i^*(Q_1, Q_2, \mu, t) = \sum c_{i\alpha\beta\gamma}^* Q_1^\alpha Q_2^\beta \mu^\nu \qquad i = 1, 2, \qquad (1.7.24)$$

where the $c_{i\alpha\beta\gamma}^*$ are majorants of the coefficients $c_{i\alpha\beta\gamma}$ in the interval J.

The trial solution

$$Q_i = \mu^k Q_{ik}(t), \qquad Q_{i1}(t_0) = 0 \qquad (1.7.25)$$

is again assumed for (1.7.24). This first leads to the system

$$\dot{Q}_{11} = c_{1100}^* Q_{11} + c_{1010}^* Q_{21} + c_{1001}^*,$$
$$\dot{Q}_{21} = c_{2100}^* Q_{11} + c_{2010}^* Q_{21} + c_{2001}^*. \qquad (1.7.26)$$

The majorants Q_{i1} to q_{i1} and $\dot{Q}_i$ to $\dot{q}_{i1}$ are obtained for the initial conditions $Q_i(t_0) = 0$. The majorants Q_{in} to q_{in} as well as $\dot{Q}_{in}$ to $\dot{q}_{in}$ are obtained in a completely analogous way. It is sufficient to show that the series $Q_i = \mu^k Q_{ik}(t)$ and their derivatives are uniformly convergent in the interval J if the majorant $c_{i\alpha\beta\gamma}^*$ is appropriately chosen and if μ is sufficiently small.

Let M be the upper limit of the F_i in J assuming that the necessary condition

$|q_i|, |\mu| < r$ for the convergence of the F_i is satisfied. The quantity

$$\frac{M(q_1 + q_2 + \mu)}{1 - (q_1 + q_2 + \mu)/r} \tag{1.7.27}$$

is then a majorant for the F_i. The fact that the power series F_i do not have a constant term has been taken into account. The books by Petrovski[1] and Smirnov[2] show why (1.7.27) is a majorant. The system of differential equations

$$\dot{Q}_1 = \dot{Q}_2 = \frac{M(Q_1 + Q_2 + \mu)[1 + (Q_1 + Q_2 + \mu)/r]}{1 - (Q_1 + Q_2 + \mu)/r} \tag{1.7.28}$$

can be used for the majorants. It is only necessary to show that this system has a solution that vanishes for $t = t_0$ and can be developed in powers in μ in the interval J, provided that $|\mu|$ is sufficiently small.

If we set

$$Q_1 + Q_2 + \mu = r\tau$$

(1.7.28) is transformed into the differential equation

$$\dot{\tau} = 2M\tau(1 + \tau)(1 - \tau)$$

where τ is real and the initial conditions are $\tau = \mu/r$ for $t = t_0$. The equation

$$[(1 - \tau)/\tau(1 + \tau)] \, d\tau = 2M \, dt$$

can be integrated and yields the result $c[\tau/(1 + \tau)^2] = e^{2Mt}$. The initial condition determines the constant of integration c as follows:

$$c = [(r + \mu)^2/\mu r] \, e^{2Mt_0}.$$

The final result is

$$\tau/(1 + \tau)^2 = [\mu r/(r + \mu)^2] \, e^{2M(t - t_0)}.$$

It is seen that τ is given by $\tau = \alpha(1 + \tau)^2$, where $\alpha = \mu r e^{2M(t - t_0)}/(r + \mu)^2$ or it is given by

$$\tau^2 + [(2\alpha - 1)/\alpha]\tau + 1 = 0.$$

This means that τ can only be an analytic function of α if $[(2\alpha - 1)/2\alpha]^2 > 1$ or if $\alpha < \frac{1}{4}$, respectively. Owing to the properties of α, the following condition is obtained:

$$\mu r \, e^{2M(t - t_0)}/(r + \mu)^2 < \frac{1}{4}$$

which can be satisfied, provided that $|\mu|$ can be selected appropriately. This guarantees the convergence of the series (1.7.25) and their derivatives. These series really represent solutions, because they satisfy the differential equations

[1]Petrovski, J. G., *Lectures on Partial Differential Equations*, pp. 27–99. Teubner, Leipzig, 1955.
[2]Smirnov, W. J., *Higher Mathematics*, Vol. IV, pp. 323. VEB Deutscher Verlag der Wissenschaften, Berlin, 1961.

and also are convergent. Consequently, the series (1.7.21) are also convergent solutions of (1.7.20), provided that $|\mu|$ is sufficiently small.

Let us now consider a case where the perturbation method is applied and *a parameter is introduced artificially*. Also, the trial solution, which is an expansion with respect to powers of the parameter, converges for arbitrary values of the parameter, not only small ones.

The system of variational equations

$$\dot{\xi}_i = a_{ik}(t)\xi_k; \qquad i, k = 1, 2, \ldots, n, \qquad (1.7.29)$$

is cast into the new form

$$\dot{\xi}_i = \mu a_{ik}(t)\xi_k \qquad (1.7.30)$$

making use of the arbitrary parameter μ. The original form is obtained for $\mu = 1$. The following trial solution for (1.7.30) is selected

$$\xi_i = \xi_{i0} + \mu^r \xi_{ir}, \qquad i = 1, 2, \ldots, n, \qquad r = 1, 2, \ldots. \qquad (1.7.31)$$

Substituting (1.7.31) into (1.7.30), the following result is obtained:

$$\dot{\xi}_{i0} + \mu^r \dot{\xi}_{ir} = \mu a_{ik}(t)[\xi_{k0} + \mu^s \xi_{ks}].$$

From this is follows that

$$\dot{\xi}_{i0} = 0, \qquad \xi_{i0} = \text{const} = c_{i0},$$

$$\dot{\xi}_{ir} = a_{ik}(t)\xi_{k,r-1},$$ (1.7.32)

$$\xi_{ir} = \int_0^t a_{ik}(t)\xi_{k,r-1}\,dt, \qquad i, k = 1, 2, \ldots, \qquad n, r = 1, 2, \ldots.$$

For $t \geqslant 0$ let us assume that $|a_{ik}(t)| \leqslant a$, $i, k = 1, 2, \ldots, n$. Therefore

$$|\dot{\xi}_{i1}| = \left| \int_0^t a_{ik}\xi_{k0}\,dt \right| \leqslant \left| \sum_{k=1}^n c_{k0} \right| \int_0^t a\,dt$$

and

$$|\xi_{i2}| = \left| \int_0^t a_{ik}\xi_{k1}\,dt \right| \leqslant \left| \sum_{k=1}^n c_{k0} \right|_n \int_0^t a\left\{ \int_0^t a\,dt \right\}dt.$$

Partial integration of the result results in

$$\int_0^t a\left\{ \int_0^t a\,dt \right\}dt = \tfrac{1}{2}\left[\int_0^t a\,dt \right]^2.$$

Therefore, we obtain the result

$$|\xi_{i2}| \leqslant (n/2)\left| \sum_{k=1}^n c_{k0} \right|\left[\int_0^t a\,dt \right]^2,$$

and the general result is

$$|\xi_{is}| \leqslant \frac{n^{s-1}}{s!}\left| \sum_{k=1}^n c_{k0} \right|\left[\int_0^t a\,dt \right]^s, \qquad s = 1, 2, 3, \ldots. \qquad (1.7.33)$$

Substitution of (1.7.32) and (1.7.33) into (1.7.31) results in

$$|\xi_i| \leqslant |\xi_{i0}| + |\mu^r \xi_{ir}| \leqslant |c_{i0}| + \frac{\left|\sum\limits_{k=1}^{n} c_{k0}\right|}{n} \sum_{r=1}^{\infty} \frac{n^r}{r!} |\mu^r| \left[\int_0^t a\, dt\right] r$$

or

$$|\xi_i| \leqslant |c_{i0}| + \frac{\left|\sum\limits_{k=1}^{n} c_{k0}\right|}{n} (e^x - 1), \qquad x = n|\mu| \int_0^t a\, dt \qquad (1.7.34)$$

respectively.

It follows from (1.7.34) that the trial solution (1.7.31) is absolutely convergent for any value of the parameter μ, and therefore also for $\mu = 1$. The solution for the original problem (1.7.21) can be obtained from (1.7.31) by making the parameter μ equal to one in the series development (1.7.31) obtained by the perturbation method theory.

The way in which the perturbation method can be used for *stability investigations* will now be shown.

Example: Consider the differential equation

$$\ddot{y} + p(t)y = 0,$$

where $p(t)$ is a *periodic* function. The notation $\xi_1 = y, \xi_2 = \dot{y}$ is introduced which results in the system

$$\dot{\xi}_1 = \xi_2,$$

$$\dot{\xi}_2 = -p\xi_1.$$

It can be written in the form

$$\dot{\xi}_1 = p_{ik}\xi_k, \qquad i, k = 1, 2, \qquad (1.7.35)$$

where $p_{11} = 0, p_{12} = 1, p_{21} = -p, p_{22} = 0$. The case considered previously is obtained from

$$\dot{\xi}_1 = \mu p_{ik}\xi_k. \qquad (1.7.36)$$

The only difference is that the periodic coefficients $p_{ik}(t)$ have taken the place of the general coefficients $a_{ik}(t)$. Equation (1.7.36) becomes (1.7.35) for $\mu = 1$.

Let us consider the results given in Section 1.3.3. It was shown there that the characteristic equation

$$\det(a_{ri} - \rho\delta_{ri}) = \begin{vmatrix} a_{11} - \rho & a_{12} \\ a_{21} & a_{22} - \rho \end{vmatrix} = 0$$

must be used for the investigation of stability. We have the equations $a_{ik} = \xi_i^{(k)}(\omega)$. The quantity ω is the common period of the coefficient p_{ik}. The functions $\xi_i^{(k)}$ represent a fundamental solution system for (1.7.35) or (1.7.36) which satisfy the initial conditions $\xi_i^{(k)}(0) = \delta_{ik}$.

In the present case we have

$$p_{11} + p_{22} = 0.$$

Therefore, the Wronskian determinant satisfies

$$W(t) = W(0) \exp\left(\int_0^\sigma p_{ii} \, dt \right) = W(0) = \text{const.}$$

The equation $W(t) = \det[\xi_i^{(k)}(t)]$ holds, and because

$$W(t) = W(0) = \text{const,}$$

it is found that

$$\begin{aligned}
W(\omega) &= \xi_1^{(1)}(\omega)\xi_2^{(2)}(\omega) - \xi_1^{(2)}(\omega)\xi_2^{(1)}(\omega) \\
&= \xi_1^{(1)}(0)\xi_2^{(2)}(0) - \xi_1^{(2)}(0)\xi_2^{(1)}(0) \\
&= \delta_{11}\delta_{22} - \delta_{12}\delta_{21} = 1.
\end{aligned}$$

The characteristic equation

$$\begin{vmatrix} a_{11} - \rho & a_{12} \\ a_{21} & a_{22} - \rho \end{vmatrix} = \begin{vmatrix} \xi_1^{(1)}(\omega) - \rho & \xi_1^{(2)}(\omega) \\ \xi_2^{(1)}(\omega) & \xi_2^{(2)}(\omega) - \rho \end{vmatrix} = 0$$

results in

$$\rho^2 - [\xi_1^{(1)}(\omega) + \xi_2^{(2)}(\omega)]\rho + [\xi_1^{(1)}(\omega)\xi_2^{(2)}(\omega) - \xi_1^{(2)}(\omega)\xi_2^{(1)}(\omega)] = 0$$

which is also equal to

$$\rho^2 - a_1\rho + 1 = 0, \qquad a_1 = \xi^{(1)}(\omega) + \xi_2^{(2)}(\omega).$$

The roots $\rho_{1,2}$ of this quadratic equation are the characteristic multipliers. The absolute magnitude of these multipliers determines whether stability exists. According to Section 1.3.3, the $\rho_{1,2}$ are related to $a_1/2$ in such a way that the quantity

$$\tfrac{1}{2}[\xi_1^{(1)}(\omega) + \xi_2^{(2)}(\omega)]$$

can also be used to determine stability. In this case, one would have to know $\xi_1^{(1)}(\omega), \xi_2^{(2)}(\omega)$. The latter functions are obtained according to the perturbation calculation method using (1.7.36).

First, the relationship $\xi_2^{(2)} = \dot{\xi}_1^{(2)}$ is used. We can set

$$\tfrac{1}{2}[\xi_1^{(1)}(\omega) + \xi_2^{(2)}(\omega)] = \tfrac{1}{2}[\xi_1^{(1)}(\omega) + \dot{\xi}_1^{(2)}(\omega)].$$

It is permissible to use the two particular solutions $\xi_1^{(1)}$ and $\xi_1^{(2)}$. Since $\xi_1^{(k)}(0) = \delta_{ik}$ and $\dot{\xi}_1^{(1)} = \xi_2^{(1)}$, $\dot{\xi}_1^{(2)} = \xi_2^{(2)}$, the initial conditions are given by

$$\xi_1^{(1)}(0) = 1, \qquad \dot{\xi}_1^{(1)}(0) = 0, \qquad \xi_1^{(2)}(0) = 0, \qquad \dot{\xi}_1^{(2)}(0) = 1.$$

The following trial solutions are assumed for (1.7.36):

$$\left. \begin{aligned}
\xi_1^{(1)} &= \xi_{10}^{(1)} + \mu\xi_{11}^{(1)} + \mu^2\xi_{12}^{(1)} + \ldots, \\
\xi_1^{(2)} &= \xi_{10}^{(2)} + \mu\xi_{11}^{(2)} + \mu^2\xi_{12}^{(2)} + \ldots,
\end{aligned} \right\} \tag{1.7.37}$$

which must satisfy the differential equation $\ddot{\xi}_1^{(r)} = -p\mu\xi_1^{(r)}$, $r = 1, 2$, as well as the initial conditions mentioned above. We require that

$$\xi_{10}^{(1)}(0) = 1, \qquad \dot{\xi}_{10}^{(1)}(0) = 0, \qquad \xi_{10}^{(2)}(0) = 0, \qquad \dot{\xi}_{10}^{(2)}(0) = 1,$$

$$\xi_{1k}^{(1)}(0) = \dot{\xi}_{1k}^{(1)}(0) = \xi_{1k}^{(2)}(0) = \dot{\xi}_{1k}^{(2)}(0) = 0, \qquad k = 1, 2, \ldots.$$

If (1.7.37) is substituted into (1.7.36), the following differential equations are obtained:

$$\ddot{\xi}_{10}^{(r)} = 0,$$

$$\ddot{\xi}_{1k}^{(r)} = -p\xi_{1,k-1}^{(r)}, \qquad r = 1, 2, \qquad k = 1, 2, 3, \ldots.$$

Because of the initial conditions, the following result is obtained:

$$\xi_{10}^{(1)} = 1, \qquad \xi_{1k}^{(1)} = \int_0^t \left\{ \int_0^t -p\xi_{1,k-1}^{(1)} \, dt \right\} dt,$$

$$\xi_{10}^{(2)} = t, \dot{\xi}_{1k}^{(2)} = \int_0^t -p\xi_{1,k-1}^{(2)} \, dt, \qquad k = 1, 2, \ldots. \tag{1.7.38}$$

The terms of the series expansion (1.7.37) can be obtained step by step in this way.

It is already known that the series (1.7.37) converge for any value of μ, and therefore also for $\mu = 1$. The desired fundamental solutions of the original system (1.7.35) can be obtained from (1.7.37) for $\mu = 1$. If $\mu = 1$ and $\xi_{10}^{(1)} = 1, \xi_{10}^{(2)} = t$ and if (1.7.37) is substituted into $\frac{1}{2}[\xi_1^{(1)}(\omega) + \dot{\xi}_1^{(2)}(\omega)]$, the result

$$\tfrac{1}{2}[\xi_1^{(1)}(\omega) + \dot{\xi}_1^{(2)}(\omega)] = 1 + \tfrac{1}{2} \sum_{k=1}^{\infty} [\xi_{1k}^{(1)}(\omega) + \dot{\xi}_{1k}^{(2)}(\omega)] \tag{1.7.39}$$

is obtained. Based on (1.7.38) an estimation of the magnitude of the characteristic multipliers can be made, which is equivalent to a statement about stability.

Let $p(t) \leqslant 0$. It then follows from (1.7.38) that

$$\xi_{11}^{(1)} > 0, \qquad \dot{\xi}_{11}^{(2)} > 0.$$

Also $\xi_{11}^{(2)} > 0$ because according to the second inequality above, $\xi_{11}^{(2)}$ increases and is zero for $t = 0$ according to the initial conditions.

The result $\xi_{1k}^{(1)} > 0, \dot{\xi}_{1k}^{(2)} > 0, \xi_{1k}^{(2)} > 0$ for $k > 1$ is obtained in a similar way. Owing to this fact and because of (1.7.39), the final result

$$\tfrac{1}{2}|\xi_1^{(1)}(\omega) + \xi_2^{(2)}(\omega)| = \tfrac{1}{2}|\xi_1^{(1)}(\omega) + \dot{\xi}_1^{(2)}(\omega)| > 1$$

is obtained. As was shown in Section 1.3.3, this means that the two roots $\rho_{1,2}$ of the characteristic equation become real. Their magnitudes are larger or smaller than one, respectively. It can be concluded from this, instability is present! In the same way it can be shown that

$$\tfrac{1}{2}|\xi_1^{(1)}(\omega) + \xi_2^{(2)}(\omega)| < 1$$

holds for $p(t) \geqslant 0$ and $\omega \int_0^\omega p \, dt \leqslant 4$ so that stability prevails.

The so-called *secular terms* occur often when the perturbation method theory is applied to non-linear vibration problems. These are the terms of the solution series whose amplitude increases indefinitely with time. Important characteristics of the process in question can be obtained by setting up conditions for eliminating the secular terms. For example, this can be done by physical reasoning such as considering the energy balance of the mechanical process represented by the differential equation.

Example: Let us assume that the motion of a point mass suspended from a non-linear spring is governed by the differential equation $\ddot{\chi} + \omega_0^2\chi + \mu\chi^3 = 0$ and that the initial conditions are $\chi = A$, and $\dot{\chi} = 0$ for $t = 0$. If it is assumed that the magnitude of μ is sufficiently small, it is possible to use the trial solution

$$\chi = \chi_0 + \mu\chi_1 + \mu^2\chi_2 + \dots, \qquad \omega^2 = \omega_0^2 + \mu f_1(A) + \mu^2 f_2(A) + \dots,$$

The initial conditions $\chi_0 = A$, $\dot{\chi}_0 = 0$, $\chi_k = \dot{\chi}_k = 0$ for $k \geqslant 1, t = 0$ can be specified. If this trial solution is substituted into the differential equation, the following differential equation for the generating solution is obtained:

$$\ddot{\chi}_0 + \omega^2\chi_0 = 0.$$

It follows that $\chi_0 = A \cos \omega t$ because of the initial conditions.

If all the terms connected with μ are considered in the next step, the following is obtained:

$$\ddot{\chi}_1 + \omega^2\chi_1 = f_1\chi_0 - \chi_0^3 = \left(f_1 A - \frac{3A^3}{4}\right) \cos \omega t - \frac{A^3}{4} \cos 3\omega t,$$

and the general solution is

$$\chi_1 = c_1 \cos \omega t + c_2 \sin \omega t + \frac{\omega t}{2} \left(f_1 A - \frac{3A^3}{4}\right) \sin \omega t + \frac{A^3}{32\omega^2} \cos 3\omega t.$$

The third term is the secular term, which vanishes for

$$f_1 A - (3A^3)/4 = 0.$$

Since $A = 0$ would be trivial, the secular term vanishes if $f_1 = (3A^2)/4$. The initial conditions $\chi_1(0) = \dot{\chi}_1(0) = 0$ determine the integration constants as follows:

$$c_1 = -A^3/32\omega^2, \qquad c_2 = 0,$$

and the final result is

$$\chi_1 = -(A^3/32\omega^2)(\cos \omega t - \cos 3\omega t).$$

In the third step, the terms containing μ^2 result in the equation

$$\ddot{\chi}_2 + \omega^2\chi_2 = f_1\chi_1 + f_2\chi_0 - 3\chi_0^2\chi_1$$

$$= -\frac{3A^2}{4} \frac{A^3}{32\omega^2} (\cos \omega t - \cos 3\omega t)$$

$$+ f_2 A \cos \omega t + 3A^2 \frac{A^3}{32\omega^2} \cos^2 \omega t(\cos \omega t - \cos 3\omega t).$$

The general solution of χ_2 can be calculated from this, which again contains two constants of integration and a secular term. The secular term vanishes if $f_2 = -(3A^4)/128\omega^2$. Taking the boundary conditions into account, the result obtained is

$$\chi_2 = -(A^5/1024\ \omega^4)(\cos \omega t - \cos 5\omega t).$$

If the calculation is terminated at this point, the following approximate solution for the non-linear vibration process described by the original differential equation is obtained:

$$\chi = A \cos \omega t - (\mu A^3/32\omega^2)(\cos \omega t - \cos 3\omega t)$$
$$- (\mu^2 A^5/1024\ \omega^4)(\cos \omega t - \cos 5\omega t).$$

The quantities f_1 and f_2 were determined by making the secular terms vanish. With the trial solution for ω^2, the following expression is found, which specifies the variation of frequency with amplitude:

$$\omega^2 = \omega_0^2 + (\mu 3 A^2/4) - (\mu^2 3 A^4/128\omega^2).$$

Let us consider the perturbation method theory for the case in which the functions Φ_i connected with μ appearing in the expression

$$\dot\xi_i = a_{ik}\xi_k + \mu\Phi_i(\xi_1, \xi_2, \ldots, \xi_n, \mu t)$$

are no longer analytic. The trial solution $\xi_i = \xi_{i0} + \mu^r\xi_{ir}$ generally used in perturbation method theory is no longer valid. The more general trial solutions must be used

$$\xi_i = \sum_{s=0}^{\infty} \xi_{i,s},$$

whose terms can be obtained from the following iteration formula:

$$\dot\xi_{i,s} - a_{ik}\xi_{k,s} = \mu\Phi_i(\xi_{1,s-1}, \ldots, \xi_{n,s-1}, \mu, t_0).$$

Functional analysis must be used in every individual case to determine whether the successive approximation method for ξ_i is convergent or not.

1.7.5 The Methods of Galerkin and Ritz

Often ordinary boundary value problems

$$\left.\begin{array}{l} D[\chi(t)] = F(t), \\ U[\chi(t)]_R = H(t)_R \end{array}\right\} \tag{1.7.40}$$

are encountered in stability studies, where D and U are operators and F and H are functions. The subscript R refers to boundary values. The second line of (1.7.40) contains the boundary conditions.

Let us restrict ourselves to *homogeneous problems* and let us also assume that the operator D has a special form. We shall consider a case of great practical importance, which is the boundary value problem

$$L[\chi(t)] + N[\chi(t)] - c\chi(t) = 0, \left.\begin{array}{c}\\ \end{array}\right\}$$
$$U[\chi(t)]_R = 0 \quad (1.7.41)$$

in which L is a *linear and positive definite* differential operator that is *self-adjoint* with respect to the boundary conditions. N is a non-linear differential operator in the most general case. In particular, N can also be a non-self-adjoint, linear operator or a non-linear function.

Let us consider the following linear auxiliary problem corresponding to (1.7.41)

$$L(\phi(t)) - \lambda\phi(t) = 0, \left.\begin{array}{c}\\ \end{array}\right\}$$
$$U(\phi(t))_R = 0. \quad (1.7.42)$$

It is called an *abbreviated problem*. It has the *same boundary conditions* as the 'unabbreviated problem' (1.7.41), which is very important.

Because of the assumptions made for L, $\lambda = 0$ is not an eigenvalue of the abbreviated problem. This means that a Green function $G(t, \tau)$ with respect to the boundary conditions exists. Therefore $L(\phi) = \lambda\phi$ for ϕ. The expressions

$$\phi = \lambda L^{-1}(\phi), \quad (1.7.43)$$

or

$$\phi = \lambda \int_a^b G(t, \tau)\phi(\tau)\, d\tau \quad (1.7.44)$$

are formed. The inverse operator L^{-1} of L exists, which is closely related to the Green function of L. This fact can be seen from a comparison of (1.7.43) with (1.7.44). In other words: *the integral Eq. (1.7.44) is equivalent to the problem (1.7.42)*.

The abbreviated problem has an infinite number of eigenvalues $\lambda_i, i = 1, 2, 3,$ ..., which are all positive and real. The corresponding eigenfunctions are a *closed, orthogonal, and normalized* function system. The Green function is *symmetrical and positive definite* and can be expanded into the series

$$G(t, \tau) = \sum_{i=1}^{\infty} \frac{\phi_i(t)\phi_i(\tau)}{\lambda_i} \quad (1.7.45)$$

which is absolutely and uniformly *convergent* according to the theorem of Mercer. The eigenfunctions $\xi_i(t)$ are orthogonal and normalized and therefore satisfy the conditions

$$(\phi_i, \phi_k) = \delta_{ik},$$
$$\{\phi_i, L(\phi_k)\} = [\phi_i, \phi_k] = \lambda_k\delta_{ik} \quad (1.7.46)$$

The following notation has been used.

The expression

$$(\phi_i, \phi_k) = \int_a^b \phi_i \phi_k \, \mathrm{d}t$$

is the 'scalar product' of the functions ϕ_i, ϕ_k. The expression

$$[\phi_i, \phi_k] = \{\phi_i, L(\phi_k)\} = \int_a^b \phi_i L(\phi_k) \, \mathrm{d}t$$

is the 'generalized scalar product' of the same functions. If the operator L is self-adjoint, it means that the relationships

$$\{\phi_i, L(\phi_k)\} = \{\phi_k, L(\phi_i)\}$$

or
$$[\phi_i, \phi_k] = [\phi_k, \phi_i] \tag{1.7.47}$$

are satisfied.

Since the abbreviated and the unabbreviated problem satisfy the same boundary conditions, the solutions $\chi(t)$ of the unabbreviated problem (1.7.41) satisfy the boundary conditions of the abbreviated problem (1.7.42). Since the eigenfunctions $\phi_i(t)$ of (1.7.42) are a closed system of functions in the space of functions, which are as often differentiable as operator L requires it and which satisfy the boundary conditions $U(\phi(t)) = 0$, the expansion theorem can be applied. The series expansion

$$\chi(t) = \sum_{k=1}^{\infty} \rho_k \phi_k(t), \tag{1.7.48}$$

which is absolutely and uniformly convergent, can be formed.

The *Galerkin method* consists of substituting the series (1.7.48) into the differential equation instead of $\chi(t)$. All terms are multiplied by $\phi_i(t)$ and integration is performed over the interval prescribed by the boundary values a, b, of t. In this way the following algebraic and infinite system of equations is found:

$$\sum_{k=1}^{\infty} \rho_k \int_a^b \phi_i L(\phi_k) \, \mathrm{d}t - c \sum_{k=1}^{\infty} \rho_k \int_a^b \phi_i \phi_k \, \mathrm{d}t$$

$$= - \int_a^b \phi_i N \left(\sum_k \rho_k \phi_k \right) \mathrm{d}t, \qquad i = 1, 2, 3, \ldots, \tag{1.7.49''}$$

which determines the expansion coefficients ρ_k. It can also be written in the form

$$\sum_k \rho_k [\phi_i, \phi_k] - c \sum_k \rho_k (\phi_i, \phi_k) = -(\phi_i, N), \qquad i = 1, 2, 3, \ldots. \tag{1.7.49'}$$

Equation (1.7.46) is then applied. Several simplifications occur and the result is

$$\rho_i = - \frac{(\phi_i, N)}{\lambda_i - c} = - \frac{1}{\lambda_i - c} \int_a^b \phi_i N \left(\sum_{k=1}^{\infty} \rho_k \phi_k \right) \mathrm{d}t, \qquad i = 1, 2, 3, \ldots. \tag{1.7.49}$$

Since it is not possible to solve immediately the infinite system of equations (1.7.49), the Galerkin method stipulates that the infinite series $\Sigma_k \rho_k \phi_k$ on the right-hand side of (1.7.49) is truncated at the term having $k = n$ and finite order. One also makes the system of equations finite: The subscript i runs from one to n. This leads to the *Galerkin equations*

$$\rho_i^* = -\frac{1}{\lambda_i - c} \int_a^b \phi_i N \left(\sum_{k=1}^n \rho_k^* \phi_k \right) dt, \qquad i = 1, 2, \ldots, n, \quad (1.7.50)$$

from which the approximation values ρ_i^* can be calculated. According to (1.7.48), the approximation

$$\chi(t) = \sum_{k=1}^n \rho_k^* \phi_k(t) \qquad (1.7.51)$$

is formed from them. This is an approximation for the desired solution $\chi(t)$ of (1.7.41), and depending on the specific problem under consideration, it may be a solution only in the 'weak sense'.

By letting n increase at each step of the calculation, a sequence of successive approximation is carried out. The convergence of the sequence must be investigated in each individual case.

Let us consider the simplest case where the operator N is not self-adjoint but is *linear*, so that

$$N \equiv L^*,$$

$$\int_a^b \phi_i N \left(\sum_k \rho_k \phi_k \right) dt \equiv \sum_k \rho_k \int_a^b \phi_i L^*(\phi_k) \, dt = \sum \rho_k [\phi_i, L^*(\phi_k)].$$

Equation (1.7.49) then becomes

$$\rho_i(\lambda_i - c) + \sum_k \rho_k [\phi_i, L^*(\phi_k)] = 0, \qquad i = 1, 2, 3, \ldots,$$

or

$$\sum_{k=1}^\infty a_{ik} \rho_k = 0,$$

$$\qquad (1.7.52)$$

$$a_{ik} = (\lambda_i - c)\, \delta_{ik} + [\phi_i, L^*(\phi_k)], \qquad i = 1, 2, 3, \ldots.$$

Equations (1.7.52) again represent an infinite algebraic system of equations which is now *linear* for determining the ρ_k. The ρ_k can be obtained by successive approximation according to the Galerkin method. The theory of infinite determinants of von Koch can be used to determine whether the successive approximation converges toward the unique solution of (1.7.52) and whether the Galerkin method is applicable[1].

---

[1]Leipholz, H., *Stability of Elastic Systems*, Sijthoff et Nordhoff, Alphen aan den Rijn, The Netherlands, pp. 171–177, and 177–184, 1980.

Let us now consider the more general case where N is a *non-linear* operator or is a non-linear function.

If $L(\phi) = \lambda\phi$ possesses the eigenfunctions ϕ_i and the eigenvalues λ_i, $L(\phi) - c\phi = \varkappa\phi$ will have the same eigenfunctions ϕ_i and the eigenvalues $\varkappa_i = \lambda_i - c$. If $L(\phi)$ with the given boundary conditions has the inverse

$$L^{-1}(u) = \int_a^b G(t, \tau)u \, d\tau$$

with the Green function

$$G(t, \tau) = \sum_k \frac{\phi_k(t)\phi_k(\tau)}{\lambda_k}$$

then $L(\phi) - c\phi = L^*(\phi)$ has the inverse

$$L^{*-1}(u) = \int_a^b G^*(t, \tau)u \, d\tau \qquad (1.7.53)$$

for the same boundary conditions and the Green function is

$$G^*(t, \tau) = \sum_k \frac{\phi_k(t)\phi_k(\tau)}{\lambda_k - c}.$$

We have assumed that $c \neq \lambda_k$ holds so that G^* indeed exists. The case $c = \lambda_k$ must be examined separately which will not be done here.

It follows from $L(\chi) + N(\chi) - c\chi = 0$ that

$$L^*(\chi(t)) = -N[\chi(t)],$$

$$\chi(t) = -L^{*-1}[N(\chi)]$$

$$\chi(t) = -\int_a^b G^*(t, \tau)N[\chi(\tau)] \, d\tau. \qquad (1.7.54)$$

Let us again consider (1.7.49) and let us rewrite this equation with another integration variable. We then obtain

$$\rho_i = -\int_a^b \frac{\phi_i(\tau)}{\lambda_i - c} N\left(\sum_k \rho_k\phi_k(\tau)\right) d\tau, \qquad i = 1, 2, 3, \ldots.$$

Each of these ρ_i is now multiplied by $\phi_i(t)$, and the result is summed over i.
This result in

$$\sum_i \rho_i\phi_i(t) = -\int_a^b \sum_i \frac{\phi_i(t)\phi_i(\tau)}{\lambda_i - c} N\left(\sum_k \rho_k\phi_k(\tau)\right) d\tau. \qquad (1.7.55)$$

When this result is compared with (1.7.48) and (1.7.53), it is found that (1.7.55) is the non-linear integral equation (1.7.54). It follows that the successive approximation method according to the Galerkin equations is completely equivalent to the successive approximation method with the integral equation (1.7.54) if

$$\chi^{(n)}(t) = -\int_a^b G^{*(n)}(t, \tau)N\left(\sum_{k=1}^n \rho_k\phi_k(\tau)\right) d\tau$$

$$G^{*(n)}(t, \tau) = \sum_{k=1}^n \frac{\phi_k(t)\phi_k(\tau)}{\lambda_k - c}.$$

(1.7.56)

A solution of the system of Eqs. (1.7.49) can be found and the solution of (1.7.41) can be written in the form (1.7.48) if it can be shown that there is a solution for the integral equation (1.7.54). If this is the case, the Galerkin method can be used and then consists of solving the reduced system of Eqs. (1.7.50), provided that it may also be shown that the approximation (1.7.56) is close to the solution of (1.7.54). Therefore, everything has been reduced to the investigation of non-linear integral equations. The works by Leipholz[1] and Krasnoselski[2] contain a detailed treatment of the subject, especially a discussion of the Galerkin method convergence. These works show that the Galerkin method does indeed converge for special non-linear operators or functions N. A very special case exists if c is equal to zero and the operator N is a *function of the independent variable t*. We then have the problem

$$L[\chi(t)] = f(t),$$
$$U[\chi(t)]_R = 0$$

(1.7.57)

and the Galerkin method can be transformed into the Ritz method. From this it can be seen *that the Ritz method is a special case of the much more general Galerkin method.*

It can be shown that obtaining a solution of (1.7.57) is the same as finding a function $\chi(t)$ which makes the functional

$$\Phi(\chi) = \{L(\chi), \chi\} - 2(\chi, f)$$

(1.7.58)

a minimum. If the term $\{L(\chi_0), \chi_0\}$, where χ_0 is the unique solution of (1.7.57), is added and subtracted to the right-hand side of (1.7.58), the following is obtained, taking (1.7.47) and the relationship $L(\chi_0) = f$ into account:

or

$$\Phi(\chi) = \{L(\chi - \chi_0), \chi - \chi_0\} - \{L(\chi_0), \chi_0\}$$
$$\Phi(\chi) = [(\chi - \chi_0), (\chi - \chi_0)] - [\chi_0, \chi_0].$$

(1.7.59)

The relationship (1.7.59) shows that $\Phi(\chi)$ is a minimum when the first term is zero, which is the case for $\chi = \chi_0$. The solution of (1.7.57) makes the functions (1.7.58) a minimum, as we have postulated.

We may also show that the function χ_0, which makes (1.7.58) a minimum is also a solution of (1.7.57). Let α be an arbitrary number, and ϕ an arbitrary

[1]Leipholz, H., *Acta Mech.* 1, 339–353 (1965).
[2]Krasnoselski, M. A., *Topological Methods in the Theory of Non-Linear Integral Equations*, Oxford University Press.

function. Since χ_0 makes $\Phi(\chi)$ a minimum based on our assumptions, we have

$$\Phi(\chi_0 + \alpha\phi) - \Phi(\chi_0) \geqslant 0.$$

If we take (1.7.58) and (1.7.47) into account, the quadratic form in α may be rewritten as follows:

$$2\alpha\{L(\chi_0) - f, \phi\} + \alpha^2\{L(\phi)\phi\} \geqslant 0.$$

It is not negative. Its discriminant must therefore satisfy

$$\{L(\chi_0) - f, \phi\}^2 \leqslant 0,$$

and this may only be true for $L(\chi_0) = f$, i.e., χ_0 is a solution of (1.7.57), which was postulated previously.

Instead of solving (1.7.57), the extremal problem $\Phi(\chi) = \min$ may be solved. It is assumed that L is positive definite [Eq.(1.7.57) has only *one* solution] and self-adjoint [because (1.7.47) is employed].

The *Ritz method* consists of substituting the series trial solution (1.7.48) into $\Phi(\chi)$, resulting in

$$\Phi(\chi) = \left\{ \sum_k \rho_k L(\phi_k), \sum_i \rho_i\phi_i \right\} - 2\left(\sum_k \rho_k\phi_k, f \right).$$

If (1.7.47) is employed, this relationship may be rewritten as follows:

$$\Phi(\chi) = \rho_i^2\{\phi_i, L(\phi_i)\} + 2 \sum_{k \neq i} \{\phi_i, L(\phi_k)\}\rho_i\rho_k - 2\rho_i(\phi_i, f) - 2 \sum_{k \neq i} \rho_k(\phi_k, f)$$

in which one arbitrary index i has been emphasized. The desired solution may be obtained by finding the minimum of $\Phi(\chi)$. The equation

$$\partial\Phi/\partial\rho_i = 0, \qquad i = 1, 2, 3, \ldots,$$

is thus formed. This leads to

$$\partial\Phi/\partial\rho_i = 2 \sum_k \rho_k\{\phi_i, L(\phi_k)\} - 2(\phi_i, f) = 0,$$

so that the infinite system of equations

$$\sum_{k=1}^{\infty} \rho_k[\phi_i, \phi_k] = (\phi_i, f), \qquad i = 1, 2, 3, \ldots,$$

is obtained for determining the coefficients ρ_k. In formal terms, it is completely analogous to the system (1.7.49') if we set $c = 0$ and if the function $-f$ is substituted for the operator N. This exact assumption was employed for the transition from (1.7.41) to (1.7.57). This shows that the Ritz and Galerkin methods are closely related. The Ritz method is a special case of the Galerkin method, which is only valid for certain special assumptions.

If the Ritz method is continued, a finite system of equations is used instead of an infinite system. An iteration method using

$$\sum_{k=1}^{n} \rho_k[\phi_i, \phi_k] = (\phi_i, f), \qquad i = 1, 2, 3, \ldots, n$$

is employed. This determines the series coefficients ρ_k of the solution (1.7.48) to any desired degree of accuracy.

The principles of mechanics may be used to derive the methods of Galerkin and Ritz. Let us first consider the most general principle, which is the principle of virtual displacements, and let us assume a deformable solid body. In a discussion of the motion or equilibrium of a body, the theory of stability assumes that the body undergoes a perturbation. In mathematical terms, this perturbation may be treated as a virtual displacement δw. The *principle of virtual displacements* results in an equation specifying the time behaviour of the displacement vector w. If the magnitude of w is bounded for all times, the stability of the corresponding equilibrium position prevails.

The principle of virtual displacements results in the relationship

$$- \int_v \mu \ddot{w} \, \delta w \, dV - \delta(\Pi_i + \Pi_a) + \int_0 k_a \, \delta w \, dO + \sum_i K_{ai} \, \delta w_i = 0.$$

The following notation is used: μ is the mass density, w is the displacement vector, Π_i is the potential of internal monogenic forces, Π_a potential of external polygenic forces, V the volume, O the elastic system surface, k_a the polygenic external surface forces, K_{ai} polygenic external single forces, δw_i the displacement vectors of the points at which the latter are applied, and the double overdots refer to double differentiation with respect to time.

Monogenic forces may be derived from a scalar function. Conservative forces represent one example of these forces, since they may be derived from scalar potential functions, and their virtual work is the variation of the corresponding potential. Polygenic forces, on the other hand, cannot be derived from a scalar function, and their virtual work is not the variation of a scalar function. Friction forces represent an example of polygenic forces, as well as the so-called 'follower' forces, which change their direction with the deformation of the body along which they act. The concepts of monogenic and polygenic forces are explained in detail in the study of Lanczos (6).

The potential energy is given by $\Pi_i + \Pi_a = A$, so that the principle of virtual displacements may also be written in the following form:

$$\delta A = - \int_v \ddot{\mu} w \, \delta w \, dV + \int_0 k_a \, \delta w \, dO + \sum_{i=1}^{n} K_{ai} \, \delta w_i. \qquad (1.7.60)$$

If we note that $T = \int_v \frac{1}{2} \cdot \mu w^2 \, dV$ is the kinetic energy and that the Lagrangian L and the Hamiltonian H satisfy the following relationships

$$L = T - A, \qquad H = T + A,$$

the following new equations may be obtained in place of (1.7.60). If

$$\delta T = \int_v \mu \dot{w} \, \delta \dot{w} \, dV$$

is subtracted from both sides of (1.7.60) and if multiplication by -1 is performed, it follows that

$$\delta L = \int_v \mu(\ddot{w}\,\delta w + \dot{w}\,\delta\dot{w})\,\mathrm{d}V - \int_0 k_a\,\delta w\,\mathrm{d}O - \sum_{i=1}^n K_{ai}\,\delta w_i,$$

or

$$\delta L = \frac{\mathrm{d}}{\mathrm{d}t}\int_v \mu\dot{w}\,\delta w\,\mathrm{d}V - \int_0 k_a\,\delta w\,\mathrm{d}O - \sum_{i=1}^n K_{ai}\,\delta w_i. \qquad (1.7.61)$$

On the other hand, if δT is added on both sides of (1.7.60), the result

$$\delta H = \int_v \mu(\dot{w}\,\delta\dot{w} - \ddot{w}\,\delta w)\,\mathrm{d}V + \int_0 k_a\,\delta w\,\mathrm{d}O + \sum_{i=1}^n K_{ai}\,\delta w_i. \qquad (1.7.62)$$

Let us now consider two special cases. It is first assumed that the *stability of an equilibrium position* is being investigated ($\ddot{w} = 0$), and that no polygenic forces are present ($k_a = 0$, $K_{ai} = 0$). Equation (1.7.60) then becomes $\delta A = 0$, A = extremum, and the *principle of minimum potential energy* is obtained. Let us now consider the *stability of elastic motion* where there are no polygenic forces (this is the case when only weight forces or 'undirectional' forces are applied). Equation (1.7.61) then becomes

$$\delta L = (\mathrm{d}/\mathrm{d}t)\int_v \mu w\,\delta\dot{w}\,\mathrm{d}V.$$

An additional calculation results in

$$\delta \int_{t_1}^{t_2} (T - A)\,\mathrm{d}t = 0$$

which is the *Hamilton principle*

$$\int_{t_1}^{t_2} (T - A)\,\mathrm{d}t = \text{extremum.}$$

Both of these special cases result in *variational problems* which can be solved by a *direct method* — that is, the Ritz method. From the standpoint of mechanics, it may be concluded that the Ritz method can be applied to stability investigations of mechanical processes if there are *no polygenic* forces present. These polygenic forces make it impossible to formulate the classical variational problems. Therefore, the Ritz method has a limited range of applicability. We shall very shortly establish the fact that the Galerkin method can be obtained from (1.7.62). It has thus been determined that the Galerkin method is more general and comprehensive, because it can also be applied when polygenic forces are present. This means that the problem is no longer an ordinary variational problem. Let us start with (1.7.62) which may be written in the following form:

$$\delta H = \delta A + \delta T = Z.$$

The quantity Z is an abbreviation for the terms on the right side of (1.7.62).

In the most general case, we have

$$\delta A = -\delta T + Z. \tag{1.7.63}$$

On the other hand, the special case of equilibrium position stability for only monogenic forces corresponds to

$$\delta A = 0. \tag{1.7.64}$$

Let us assume the trial solution $w = \Sigma_k \rho_k w_k$ of the Galerkin or Ritz method. It will be assumed from this point on that forces and displacements are directed along the same line, so that a scalar calculation may be performed. It follows from (1.7.64) that

$$(\partial A / \partial \rho_k) \, \delta \rho_k = 0.$$

Owing to the arbitrary nature of the variations $\delta \rho_k$, the system of Ritz equations

$$\partial A / \partial \rho_k = 0, \qquad k = 1, 2, 3, \ldots,$$

is obtained. If the same trial solution is used, it follows from (1.7.63) that

$$\frac{\partial A}{\partial \rho_i} \delta \rho_i = -\frac{\partial T}{\partial \rho_i} \delta \rho_i + \int_v \sum_k \sum_j L_j^{(1)}(w) L_j^{(2)}(w_k) \, \delta \rho_k \, \mathrm{d}V$$

$$+ \sum_k \sum_j R_j^{(1)}(w) R_i^{(2)}(w_k) \, \delta \rho_k,$$

where $L_j^{(i)}, i = 1, 2$ are linear differential operators, and $R_j^{(i)}, i = 1, 2$, are linear boundary value expressions. Owing to the fact that the variations are arbitrary, the following system of equations is obtained:

$$\frac{\partial A}{\partial \rho_i} = -\frac{\partial T}{\partial \rho_i} + \int_v \sum_k \rho_k L_j^{(1)}(w_k) L_j^{(2)}(w_i) \, \mathrm{d}V$$

$$+ \sum_k \rho_k R_j^{(1)}(w_k) R_j^{(2)}(w_i), \qquad i = 1, 2, 3, \ldots,$$

which corresponds to the *Galerkin equations* if the coordinate functions w_k of the trial solution are appropriately selected.

It is apparent that the second system of equations is more general than the Ritz system of equations, where all terms on the right-hand side are zero. It may again be concluded that the Ritz method can only be applied when certain special assumptions have been made, which make it possible to set terms on the right-hand side equal to zero. On the other hand, in general it is necessary to use the Galerkin method.

The Galerkin method consists of substituting H in (1.7.62) and carrying out the variation, resulting in

$$\int_v D(w) \, \delta w \, \mathrm{d}V + R(w, \delta w) = 0. \tag{1.7.65}$$

D is a differential operator, and R is a boundary value expression. If the trial solution $w = \Sigma_k \rho_k w_k$ is substituted in (1.6.65), the following is obtained:

$$\sum_j \int_v D\left[\sum_k \rho_k w_k\right] w_j \, \delta\rho_j \, dV + R\left[\sum_r \rho_r w_r, \delta\rho_j\right] = 0. \qquad (1.7.66)$$

The *restricted Galerkin method* consists of selecting the coordinate functions w_k of the trial solution in such a way that *all boundary conditions* are satisfied. Thus

$$R\left[\sum_r \rho_r w_r, \delta\rho_j\right] = 0,$$

and all the boundary terms disappear. Only the following expression remains from (1.7.66):

$$\sum_j \int_v D\left[\sum_k \rho_k w_k\right] w_j \, \delta\rho_j \, dV = 0.$$

Because of the fact that the variations $\delta\rho_j$ are arbitrary, the following system of equations

$$\int_v D\left[\sum_k \rho_k w_k\right] w_j \, dV = 0, \qquad j = 1, 2, 3, \ldots, \qquad (1.7.67)$$

is obtained for determining the series coefficient ρ_k. If

$$D(w) = L(w) + N(w) - cw$$

holds, the above expression becomes the system of Eqs. (1.7.49″) if it is assumed that the integration is performed over a one-dimensional region.

The *general Galerkin method* may also be employed if the dynamic boundary conditions are complicated, and if it is difficult to find coordinate functions w_k which satisfy the dynamic boundary conditions. In this case the functions w_k are used which satisfy as many, but *not all*, dynamic boundary conditions. The calculation must then start with Eq. (1.7.66), and all boundary terms are included which do not vanish, because the corresponding w_k is used. An example of this method will be presented shortly. This example will demonstrate the fact that the methods of Galerkin and Ritz are extremely important in stability investigations, especially in the mechanics of deformable, rigid bodies.

Example: Let us consider the case of a homogeneous column with constant cross section, clamped at the bottom. It is loaded by a 'subtangential' compressive force (Fig. 35). This is a 'follower' force which is always inclined to the vertical by the angle $\alpha\phi_l$ as the column deforms. The quantity α is a parameter which satisfies the condition $0 < \alpha \leqslant 1$, and ϕ_l is the angle between the vertical and the tangent of the column axis at its tip. Let us investigate the column *stability of the trivial equilibrium position* (straight column axis). *Transverse vibrations* are superimposed on the equilibrium position, and whether or not the deflections of the column axis remain small is determined.

Fig. 35. Buckling column with sub-tangential compressive force

The load P_k is the buckling load or critical load at which the deflections increase indefinitely. If there are several such loads, P_k, we shall restrict ourselves to the smallest of these loads because in practice it is usually sufficient to determine this smallest critical load.

According to the principle of d'Alembert, the inertia forces applied to the column elements are taken into account. The sectional method is applied, and the following relationship is obtained for the equilibrium of the internal and external moments.

$$EJy''(\chi, t) = P_v[y(l, t) - y(\chi, t)] - P_H(l - \chi) - \int_\chi^l \mu\ddot{y}(\xi, t)(\xi - \chi)\,d\xi. \tag{1.7.68}$$

EJ is the constant bending stiffness, μ the constant linear mass density of the column, and $P_v = P \cos \alpha\phi_l$, $P_H = P \sin \alpha\phi_l$. It can be assumed that $\alpha\phi_l$ is a small angle, so that $\cos \alpha\phi_l \approx 1$ and $\sin \alpha\phi_l \approx \alpha\phi_l$ holds. The calculation may be therefore continued with $P_v \approx P$, $P_H \approx P\alpha\phi_l$. In addition, the prime indicates the derivative with respect to χ, and the overdot the derivative with respect to time t.

The following partial differential equation is obtained from (1.7.68) by twofold differentiation with respect to χ:

$$EJ\frac{\partial^4 y(\chi, t)}{\partial \chi^4} + P\frac{\partial^2 y(\chi, t)}{\partial \chi^2} + \mu\frac{\partial^2 y(\chi, t)}{\partial t^2} = 0.$$

The trial solution $y(\chi, t) = e^{\omega t} f(\chi)$ results in the following ordinary differential equation:

$$EJf^{IV} + Pf'' + \mu\omega^2 f = 0. \tag{1.7.69}$$

132

In addition, the boundary conditions

$$f(0) = f'(0) = f''(l) = 0, \qquad f'''(l) = (\alpha - 1)(P/EJx)f'(l) \qquad (1.7.70)$$

must be taken into account. Furthermore, $\phi_l = f'(l)$ was employed. The boundary value problem given by (1.7.69) and (1.7.70) will now be solved by the Galerkin method. The following approximate trial solution for $f(x)$ is used:

$$f_1(\chi) = \rho_1 f_1(\chi) + \rho_2 f_2(\chi)$$

$$f_1(\chi) = 5 - \cos 2\pi(\chi/l) - 4 \cos \pi(\chi/l), \qquad (1.7.71)$$

$$f_2(\chi) = 28 - \cos \tfrac{3}{2} \pi(\chi/l) - 27 \cos (\pi/2)(\chi/l).$$

The functions f_1 and f_2 do *not* satisfy all the boundary conditions. The fourth condition of (1.7.70) is violated. False results would be obtained if

$$D(f) = EJf^{IV} + Pf'' + \mu\omega^2 f$$

were used in the calculation and if the Galerkin equations by (1.7.67) were used according to

$$\int_0^l D[\rho_1 f_1(\chi) + \rho_2 f_2(\chi)] f_i(\chi) \, d\chi = 0, \qquad i = 1, 2, \ldots . \qquad (1.7.72)$$

This would be permissible if f_1 and f_2 satisfied *all* boundary conditions. Because this is not the case, the principle of virtual displacements must be used in order to find so-called extended equations of Galerkin. The non-zero boundary values of the selected trial functions f_1 and f_2 must be included in the calculation in order to obtain these extended equations.

Because $k_a \equiv 0$,

$$\sum_{i-l}^{n} K_{ai} \, \delta w_i \equiv K_{a1} \, \delta w_1 \equiv - P\alpha\phi_l \, \delta w(t, l), \qquad w(t, \chi) \equiv y(t, \chi),$$

$$\phi_l \equiv y'(t, l),$$

Eq. (1.7.62) yields the following equation:

$$\delta H = \int_0^l \mu(\dot{y} \, \delta\dot{y} - \ddot{y} \, \delta y) \, d\chi - P\alpha y'(t, l) \, \delta y(t, l). \qquad (1.7.73)$$

The relationship

$$H = \int_0^l \left[\frac{\mu}{2} \dot{y}^2 + \frac{EJ}{2} y''^2 - \frac{P}{2} y'^2 \right] d\chi = \int_0^l F \, d\chi$$

holds for an elastic column subjected to the given load. If the variation of H is carried out, the result is

$$\delta H = \int_0^l = \left[\frac{\partial F}{\partial y} - \frac{\partial}{\partial \chi} \left(\frac{\partial F}{\partial y'} \right) + \frac{\partial^2}{\partial \chi^2} \left(\frac{\partial F}{\partial y''} \right) \right] \delta y \, d\chi$$

$$+ \left[\frac{\partial F}{\partial y'} \delta y + \frac{\partial F}{\partial y''} \delta y' - \frac{\partial}{\partial x} \frac{\partial F}{\partial y''} \delta y \right]_0^l + \int_0^l \frac{\partial F}{\partial \dot{y}} \delta\dot{y} \, d\chi.$$

If this result and $F = (\mu \dot{y}^2 + EJy''^2 - Py'^2)/2$ are substituted into (1.7.73), the following result is obtained:

$$\int_0^l [(Py'' + EJy^{IV})\, \delta y + \mu \dot{y}\, \delta \dot{y}]\, d\chi + [-Py'\, \delta y + EJy''\, \delta y' - EJy'''\, \delta y]_0^l$$

$$= \int_0^l \mu(\dot{y}\, \delta \dot{y} - \ddot{y}\, dy)\, d\chi - P\alpha y'(t,l)\, \delta y(t,l). \qquad (1.7.74)$$

The trial solution $y(\chi, t) = e^{\omega t} f(\chi)$ is used and (1.7.74) becomes

$$\int_0^l (Pf'' + EJf^{IV} + \mu\omega^2 f)\, \delta f\, d\chi + [-Pf'\, \delta f + EJf''\, \delta f - EJf'''\, \delta f]_0^l$$

$$+ P\alpha f'(l)\, \delta f(l) = 0. \qquad (1.7.75)$$

In the Galerkin method, if functions $f_k, k = 1, 2$, are used, which satisfy all the conditions (1.7.70), then all terms in (1.7.75) outside of the integral sign would vanish and it would be possible to use (1.7.72). Since one of the selected functions satisfies only the three first conditions (1.7.70), the following expression remains from (1.7.75):

$$\int_0^l (Pf'' + EJf^{IV} + \mu\omega^2 f)\, \delta f\, d\chi + (\alpha - 1)Pf'(l) - EJf'''(l)\, \delta f(l) = 0.$$

If the approximation $f^*(\chi) = \rho_1 f_1(\chi) + \rho_2 f_2(\chi)$ is substituted in this relationship for $f(\chi)$, the result is

$$\int_0^l \{[\rho_1 D(f_1) + \rho_2 D(f_2)]\, (f_1\, \delta\rho_1 + f_2\, \delta\rho_2)\}\, d\chi + (\alpha - 1)P[\rho_1 f_1'(l)$$

$$+ \rho_2 f_2'(l)]\, [f_1(l)\, \delta\rho_1 + f_2(l)\, \delta\rho_2] - EJ[\rho_1 f_1'''(l)$$

$$+ \rho_2 f_2'''(l)]\, [f_1(l)\, \delta\rho_1 + f_2(l)\, \delta\rho_2] = 0.$$

By multiplying and collecting all terms containing $\delta\rho_1$ and $\delta\rho_2$, the following equations are obtained because the $\delta\rho_1, \delta\rho_2$ are arbitrary:

$$\rho_1 \left[\int_0^l D(f_1) f_1\, d\chi + (\alpha - 1)Pf_1'(l)f_1(l) - EJf_1'''(l)f_1(l) \right]$$

$$+ \rho_2 \left[\int_0^l D(f_2) f_1\, d\chi + (\alpha - 1)Pf_2'(l)f_1(l) - EJf_2'''(l)f_1(l) \right] = 0,$$

$$\rho_1 \left[\int_0^l D(f_1) f_2\, d\chi + (\alpha - 1)Pf_1'(l)f_2(l) - EJf_1'''(l)f_2(l) \right]$$

$$+ \rho_2 \left[\int_0^l D(f_2)\, f_2\, d\chi + (\alpha - 1)Pf_2'(l)f_2(l) - EJf_2'''(l)f_2(l) \right] = 0.$$

These are linear, homogeneous, algebraic equations that determine the quantities ρ_1 and ρ_2. If the calculation is continued with the functions f_1, f_2

according to (1.7.71), the following result is obtained:

$$\rho_1\left(16\,\frac{\pi^4 EJ}{l^3} - 10P\,\frac{\pi^2}{l} + 33.5\,\mu\omega^2 l\right)$$

$$+ \rho_2\left(10.5\,\frac{\pi^4 EJ}{l^3} - 21P\,\frac{\pi^2}{l} + 96\alpha P\,\frac{\pi}{l} + 78.6\,\mu\omega^2 l\right) = 0,$$

$$\rho_1\left(10.5\,\frac{\pi^4 EJ}{l^3} - 21P\,\frac{\pi^2}{l} + 78.6\,\mu\omega^2 l\right)$$

$$+ \rho_2\left(25.3\,\frac{\pi^4 EJ}{l^3} - 92.3P\,\frac{\pi^2}{l} + 336\,\alpha P\,\frac{\pi}{l} + 198.3\,\mu\omega^2 l\right) = 0.$$

The determinant of this system of equations must be zero if this homogeneous system is to have unique solutions ρ_1, ρ_2. Introducing the abbreviations

$$a_1 = 33.5l, \qquad a_2 = 198.3l, \qquad b = 78.6l$$

$$c_1 = 16\,\frac{\pi^4 EJ}{l^3} - 10P\,\frac{\pi^2}{l}, \qquad c_2 = 25.3\,\frac{\pi^4 EJ}{l^3} - 92.3P\,\frac{\pi^2}{l} + 336\alpha P\,\frac{\pi}{l},$$

$$d_1 = 10.5\,\frac{\pi^4 EJ}{l^3} - 21P\,\frac{\pi^2}{l} + 96\,\frac{\pi}{l}\,\alpha P, \qquad d_2 = 10.5\,\frac{\pi^4 EJ}{l^3} - 21P\,\frac{\pi^2}{l},$$

this condition is formulated as follows:

$$\det\left[\begin{pmatrix} c_1 & d_1 \\ d_2 & c_2 \end{pmatrix} + \omega^2\begin{pmatrix} \mu a_1 & \mu b \\ \mu b & \mu a_2 \end{pmatrix}\right] = 0. \tag{1.7.76}$$

Equation (1.7.76) is quadratic with respect to ω^2. Setting $\omega^2 = -\lambda$, we obtain

$$\lambda^2 - \lambda\left[\frac{a_1 c_2 + a_2 c_1}{\mu(a_1 a_2 - b^2)} - \frac{b(d_1 + d_2)}{\mu(a_1 a_2 - l^2)}\right] + \frac{c_1 c_2 - d_1 d_2}{\mu^2(a_1 a_2 - l^2)} = 0. \tag{1.7.77}$$

The equilibrium position $y \equiv 0$ of the column is stable if the axial deflections $y(t, \chi)$ of the column carrying out transverse oscillations remain bounded at all times. Since $y(t, \chi) = e^{\omega t} f(\chi)$ holds, this is only the case for imaginary $\omega = \pm(-\lambda)^{1/2}$. For stability, λ must be real and satisfy $\lambda > 0$. The stability is reached when either λ becomes zero or complex.

Equation (1.7.77) describes the characteristics of λ. If

$$c_1 c_2 - d_1 d_2 = 0, \tag{1.7.78}$$

λ can become zero and therefore (1.7.78) is a possible stability boundary. If the discriminant of (1.7.77) is positive, the roots of this equation are real. If it is negative, the roots are conjugate complex. If the discriminant is set equal to zero,

$$\text{Discr} = [a_1 c_2 + a_2 c_1 - b(d_1 + d_2)]^2 - 4(c_1 c_2 - d_1 d_2)(a_1 a_2 - b^2) = 0, \tag{1.7.79}$$

which results in another stability boundary.

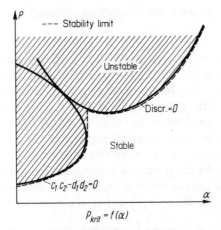

Fig. 36. Stability map

Both conditions (1.7.78) and (1.7.79) can be plotted in the P, α plane in the form of curves. This results in a stability diagram (Fig. 36). The lower limit of the instability region gives the buckling load P as a function of the parameter α.

Very often mechanical processes, the stability of which is being investigated, are described by boundary-value problems that contain partial differential equations or system of partial differential equations. Even though the extension of stability theory from ordinary to partial differential equations would be very desirable, much work has still to be done to overcome the mathematical difficulties involved. The study of Hahn (1.2) contains a brief discussion of such questions and contains comments on the work of the Russian investigator V. J. Zubov.

In general it will again be necessary to use approximation methods which essentially are based on the Galerkin method. This emphasizes its importance.

1.7.6 The Harmonic Balance Method

The method of harmonic balance, first developed by Krylov and Bogoljubov, is an often used approximation method for the treatment of non-linear problems of vibration and control theory. it is described in various textbooks (2.5, 2.12, 2.13) so that we can restrict ourselves to an introduction to the subject.

The harmonic balance method is often called the *method of harmonic linearization*. This harmonic linearization differs from ordinary linearization where the non-linear function is replaced by the first term of a Taylor series (geometrically speaking, a curve is replaced by its tangent at a certain location) in that a linear approximation expression $c\chi(t)$ is determined for a non-linear function $f[\chi(t)]$. However, its proportionality factor is no longer a true constant.

136

Example: Let us consider the differential equation

$$\ddot{\chi} + \omega_0^2 \chi = f(\chi) = \chi^3.$$

For its solution, the basic assumption that it is approximately harmonic is made, i.e., $\chi \approx b_1^* \sin \omega t$. Then $f(\chi)$ is a periodic function which satisfies $f(\chi) \approx B_1 \sin \omega t$. The relationship $\sin \omega t \approx \chi/b_1^*$ follows from the approximate trial solution for χ. Therefore, $f(\chi) \approx (B_1/b_1^*)\chi$ holds corresponding to harmonic linearization. Apparently the factor $c = B_1/b_1^*$ has become a function of the amplitude b_1^*. The great practical importance of the harmonic balance method comes from the fact that by this method of linearization the true character of the non-linear problems is more readily apparent.

Let us continue the calculation in order to clarify the features of the method. The Fourier coefficients B_1 of $f(\chi)$ satisfy the relationship

$$B_1 = (2/T) \int_0^T f[\chi(\xi)] \sin \omega\xi \, d\xi,$$

where T is the period of the process. The result

$$cb_1^* = B_1 = (2/T) \int_0^T f[\chi(\xi)] \sin \omega\xi \, d\xi$$

is obtained, which will be used immediately. If $\chi = b_1^* \sin \omega t$ is now substituted into the linearized differential equation $\ddot{\chi} + \omega_0^2 \chi = c\chi$, the result is

$$b_1^*(\omega_0^2 - \omega^2) = cb_1^*.$$

The first relationship is used to obtain

$$b_1^*(\omega_0^2 - \omega^2) = (2/T) \int_0^T f[\chi(\xi)] \sin \omega\xi \, d\xi.$$

Because $T = 2\pi/\omega$, $f[\chi(\xi)] = \chi^3(\xi) = b_1^{*3} \sin^3 \omega\xi$,

$$b_1^*(\omega_0^2 - \omega^2) = (\omega/\pi) \int_0^{2\pi/\omega} b_1^{*3} \sin^4 \omega\xi \, d\xi$$

is obtained. After integration and transformation, the following equation is obtained:

$$\omega_0^2 - \omega^2 = \tfrac{3}{4} b_1^{*2},$$

which is well known in the theory of vibrations and gives the dependence of amplitude b_1^* on the frequency ω. The approximate solution of the problem has thus been found.

Up to the present we have discussed the simple vibration problem

$$\ddot{\chi} + \omega_0^2 \chi = f(\chi). \tag{1.7.80}$$

ω_0 was a given constant and $f(\chi)$ is a non-linear restoring function that satisfies

$$\int_0^T f[\chi(\xi)] \, d\xi = 0, \tag{1.7.81}$$

thus greatly simplifying the problem. In addition, let us assume that (1.7.80) has a periodic solution that is twice differentiable and has a period T. The boundary conditions corresponding to (1.7.80) are therefore

$$\chi(0) = \chi(T), \qquad \dot{\chi}(0) = \dot{\chi}(T). \tag{1.7.82}$$

The harmonic balance method can be extended to more general problems, for example, the problem

$$D[\chi(t)] = f(\chi),$$

where D is a linear differential operator of higher order with constant coefficients. It can also be applied to

$$\ddot{\chi} + \omega_0^2 \chi = f(\chi, \dot{\chi}).$$

Let us now extend the method in another direction. The following *linear substitution problem* corresponding to (1.7.80) is introduced:

$$\ddot{\chi} + \omega_0^2 \chi = c_1 \chi + c_2 \dot{\chi}. \tag{1.7.83}$$

The general trial solution is

$$\chi = a_1^* \cos \omega t + b_2^* \sin \omega t. \tag{1.7.84}$$

If the latter is substituted into the right-hand side of (1.7.80), $f(\chi)$ becomes a periodic function that can be expanded in Fourier series. Only the first term of the series is retained. Owing to the assumption (1.7.81) there is not a constant term in this series. The following result is thus obtained:

$$f(\chi) = f(a_1^* \cos \omega t + b_1^* \sin \omega t) \approx A_1 \cos \omega t + B_1 \sin \omega t. \tag{1.7.85}$$

This expression should coincide with the right-hand side of the substitution problem so

$$A_1 \cos \omega t + B_1 \sin \omega t = c_1 \chi + c_2 \dot{\chi} \tag{1.7.86}$$

must hold. If the trial solution (1.7.84) is again substituted on the right-hand side of (1.7.86), the following system of equations is obtained:

$$\begin{aligned} A_1 &= c_1 a_1^* + c_2 b_1^* \omega, \\ B_1 &= c_1 b_1^* - c_2 a_1^* \omega. \end{aligned} \tag{1.7.87}$$

The Fourier coefficients A_1 and B_1 satisfy the relationships

$$A_1 = (2/T) \int_0^T f(\chi) \cos \omega \xi \, d\xi, \qquad B_1 = (2/T) \int_0^T f(\chi) \sin \omega \xi \, d\xi,$$

and (1.7.87) becomes

$$(2/T) \int_0^T f(\chi) \cos \omega \xi \, d\xi = c_1 a_1^* + c_2 b_1^* \omega,$$

$$(2/T) \int_0^T f(\chi) \sin \omega \xi \, d\xi = c_1 b_1^* - c_2 a_1^* \omega. \tag{1.7.88}$$

138

Substitution of the trial solution (1.7.84) into the linear substitution problem (1.7.83) results in

$$a_1^*(\omega_0^2 - \omega^2) \cos \omega t + b_1^*(\omega_0^2 - \omega^2) \sin \omega t$$
$$= c_1 a_1^* \cos \omega t + c_1 b_2^* \sin \omega t - c_2 a_1^* \omega \sin \omega t + c_2 b_1^* \omega \cos \omega t.$$

This is only true if the equations

$$a_1^*(\omega_0^2 - \omega^2) = c_1 a_1^* + c_2 b_1^* \omega,$$
$$b_1^*(\omega_0^2 - \omega^2) = c_1 b_1^* - c_2 a_1^* \omega \qquad (1.7.89)$$

are satisfied. Comparison of (1.7.88) and (1.7.89) shows that

$$a_1^*(\omega_0^2 - \omega^2) = (2/T) \int_0^T f(\chi) \cos \omega \xi \, d\xi,$$
$$\qquad (1.7.90)$$
$$b_1^*(\omega_0^2 - \omega^2) = (2/T) \int_0^T f(\chi) \sin \omega \xi \, d\xi$$

must hold. The following fundamental equations of the harmonic balance method are found using the new variables $a_1 = (T/2)^{1/2} a_1^*$:

$$a_1 = -\frac{1}{\omega^2 - \omega_0^2} \int_0^T f[\chi(\xi)] (2/T)^{1/2} \cos \omega \xi \, d\xi,$$
$$\qquad (1.7.91)$$
$$b_1 = -\frac{1}{\omega^2 - \omega_0^2} \int_0^T f[\chi(\xi)] (2/T)^{1/2} \sin \omega \xi \, d\xi.$$

The most important application of the method is the determination of periodic solutions of mechanical processes and the investigation of their stability.

Example[1]: Let us assume that a mechanical process is described by the nonlinear differential equation

$$\ddot{\chi} - (\alpha - \beta)|\chi|)|\dot{\chi}|\dot{\chi} + \omega_0^2(\chi + \gamma \chi^3) = 0.$$

Let us investigate whether a periodic solution exists and whether it is stable.

The following trial solution $\chi = a_1^* \cos \omega t$ is used for this purpose. For purposes of simplicity, $b_1^* = 0$ has been assumed (in contrast to the first example where $a_1^* \equiv 0, b_1^* \neq 0$), so that the following relationships are obtained for the coefficients c_1 and c_2 of the linear substitution problem (1.6.83):

$$c_1 = (2/a_1^* T) \int_0^T f[\chi, \dot{\chi}] \cos \omega \xi \, d\xi,$$

$$c_2 = -(2/a_1^* \omega T) \int_0^T f[\chi, \dot{\chi}] \sin \omega \xi \, d\xi.$$

---

[1]This example is from Magnus, K., *Vibrations*, p. 125. Teubner, Stuttgart, 1961.

The following substitution must be made:

$$\chi \approx a_1^* \cos \omega\xi, \qquad \dot\chi \approx -\omega a_1^* \sin \omega\xi, \qquad T = 2\pi/\omega,$$

as well as

$$f[\chi, \dot\chi] = (\alpha - \beta|\chi|)|\dot\chi|\dot\chi - \omega_0^2 \gamma \chi^3.$$

The result after integration is

$$c_1 = -\tfrac{3}{4}\omega_0^2 \gamma a_1^{*2}, \qquad c_2 = \frac{8}{3}\frac{\alpha\omega}{\pi}a_1^* - \frac{\beta\omega}{\pi}a_1^{*2},$$

and therefore (1.7.83) becomes

$$\ddot\chi + \omega_0^2 \chi = -\frac{3}{4}\omega_0^2 \gamma a_1^{*2}\chi + a_1^* \frac{\omega}{\pi}\left(\frac{8}{3}\alpha - \beta a_1^*\right)\dot\chi.$$

The linear substitution problem is given by the vibration equation

$$\ddot\chi + a_1^* \frac{\omega}{\pi}\left(\beta a_1^* - \frac{8}{3}\alpha\right)\dot\chi + \omega_0^2\left(1 + \frac{3}{4}\gamma a_1^{*2}\right)\chi = 0.$$

When the damping coefficient $d = a_1^*(\omega/\pi)(\beta a_1^* - (8/3)\alpha)$ vanishes, the periodic solution is an undamped oscillation. This is the case for $a_1^* = a_{1\text{st}}^* \equiv (8/3)(\alpha/\beta)$. Simultaneously, the stationary amplitude $a_{1\text{st}}^*$ of this undamped oscillation has been found. Its frequency can be obtained from the coefficients of the third of the linear vibration equation

$$\omega^2 = \omega_0^2(1 + 3/4\gamma a_{1\text{st}}^{*2}) \text{ or } \omega = \omega_0(1 + 3/4\gamma a_{1\text{st}}^{*2})^{\frac{1}{2}}.$$

If the amplitude of the approximate solution satisfies

$$a_1^* > 8\alpha/3\beta = a_{1\text{st}}^*,$$

the damping coefficient d becomes positive and the oscillation is damped. Conversely if $a_1^* < a_{1\text{st}}^*$, d becomes negative and we have an induced oscillation case. This is because stability in the large and instability in the small prevail, and because the periodic solution with the amplitude $a_{1\text{st}}^*$ is described by a stable *limit cycle* in the phase diagram.

The harmonic balance method can also be used to determine the existence and stability of periodic solutions of systems of differential equations such as

$$\dot\xi_i = F_i(\xi_1, \ldots, \xi_n), \qquad i = 1, 2, \ldots, n, \qquad (1.7.92)$$

which correspond to non-linear variational problems.

Let us replace (1.7.92) by the linear substitution system

$$\dot\xi_i = c_{1ik}\xi_k + c_{2ik}\dot\xi_k. \qquad (1.7.93)$$

The method is used to determine the coefficients. If, in particular,

$F_i(\xi_1, \ldots, \xi_n) = \Sigma_k F_{ik}(\xi_k)$ holds, the following result is obtained:

$$c_{1ik} = (1/\pi a_k^*) \int_0^{2\pi} F_{ik}(a_k^* \sin \phi) \sin \phi \, d\phi,$$

$$c_{2ik} = (1/\pi \omega a_k^*) \int_0^{2\pi} F_{ik}(a_k^* \sin \phi) \cos \phi \, d\phi. \tag{1.7.94}$$

The c_{jik} are related to 'description functions,' which are used in the theory of non-linear control systems.

The trial solution $\xi_k = a_k^* e^{\lambda t}$ and the linear substitution system (1.7.93) results in the following homogeneous algebraic system of equations:

$$a_k^*(c_{1ik} + c_{2ik}\lambda - \delta_{ik}\lambda) = 0, \qquad i = 1, 2, \ldots, n.$$

It only has non-trivial solutions for the a_k^* if its determinant is equal to zero. The requirement

$$\det(c_{1ik} + c_{2ik}\lambda - \delta_{ik}\lambda) = 0$$

must be satisfied, which leads to the characteristic equation

$$a_k\lambda^k = 0, \qquad k = 1, 2, \ldots, n,$$

of the problem.

It is therefore possible to apply the stability investigation method of Section 1.3 to the characteristic equation. This leads to results that make it possible to estimate the stability of the problem at hand.

If, for example, the Hurwitz criterion is applied to the coefficients a_k of the characteristic equation given above, the so-called *A-curve method* of Magnus is obtained. More about this method can be found on pp. 171, 172.

Exercises 1.7

1. Given the differential equation $ma^2 \dfrac{d^2q}{dt^2} + b_1 \dfrac{dq}{dt} + mga \sin q = 0$,

   which describes the motion of a damped simple pendulum.
   a) Show that by means of proper transformations, this differential equation can be brought into the form

   $$\frac{dy}{dq} = \frac{F(q)}{y} - \lambda,$$

   where $F(q) = -\sin q$, $\lambda = b_1(ma\sqrt{ag})^{-1}$, $y = \dot{q}(a/g)^{1/2}$
   b) For $\lambda = \frac{1}{4}$, plot the phase curve diagram around the origin of the y, q-plane using the isocline method.
2. Prove that the method of variation of constants yields for differential equation $\ddot{q} + \omega_0^2(1 - m \cos \omega_1 t)q = 0$ the approximate solution

   $$q = A\left(1 + \frac{m}{4} \cos \omega_1 t\right) \cos\left(\omega_0 t - \frac{m\omega_0}{2\omega_1} \sin \omega_1 + \Theta_0\right)$$

   provided $m \ll 1$ and $\omega_1 \ll \omega_0$.

3. Given the differential equation $\ddot{y} + g(\chi)y = 0$, $g(\chi)$ periodic with the period ω, $g(\chi) \geqslant 0$, $g(\chi) \neq 0$. Using the perturbation method, find the condition under which the solutions of this differential equation are stable for $y_1(0) = 1$, $\dot{y}_1(0) = 0$, $y_2(0) = 0$, $\dot{y}_2(0) = 1$, where the $y_i, i = 1, 2$, are the fundamental solutions.

4. The boundary-eigenvalue problem of a membrane in an airstream, compressed by longitudinal forces q and supported along the edges on which q is applied, is given by

$$\mu\ddot{w} + \alpha w^N + qw'' + M^* w' = 0, \ w = w(\chi, t),$$

$$w(0, t) = w(l, t) = w''(0, t) = w''(l, t) = 0,$$

$$\mu, \alpha, q, M^* = \text{constant}.$$

Using Galerkin's method, determine the stability domain in the q, M^*-plane.

5. The boundary-eigenvalue problem of an elastically supported rod with varying flexural rigidity and subject to a compressive longitudinal force is given by

$$[(3 - \chi^2)y'']'' + 60y = -\lambda y'',$$
$$y(-1) = y''(-1) = y(+1) = y''(+1) = 0.$$

Using Ritz' method, find the first critical value of λ.

6. Using the method of harmonic balance, show that

$$\chi = [\tfrac{4}{3}(\omega_0^2 - \omega^2)]^{\frac{1}{2}} \sin \omega t$$

is an approximate solution to the differential equation

$$\ddot{\chi} + \omega_0^2\chi - \chi^3 = 0.$$

7. Using the method of harmonic balance, show that the oscillator, whose motion is given by the differential equation $\ddot{\chi} - (\alpha - \beta\chi^2)\dot{\chi} + \chi = 0$, is able to perform a stable, periodic oscillation with the amplitude $A = 2\sqrt{(\alpha/\beta)}$.

PART 2

Application of Stability Theory
to Selected Problems of Mechanics

2.1
Celestial Mechanics Problems

2.1.1 Stability of Central Force Motion

The equations of motion of central force motion of a point mass are given by

$$m(\ddot{r} - r\dot{\phi}^2) = f(r),$$

$$r\ddot{\phi} + 2\dot{r}\dot{\phi} = (1/r)(\mathrm{d}/\mathrm{d}t)(r^2\dot{\phi}) = 0. \tag{2.2.1}$$

If the attracting force directed to the center satisfies Newton's laws, we have

$$f(r) = (mM\gamma/r^2) = -(mK/r^2). \tag{2.1.2}$$

r and γ are the polar coordinates of motion, m and M are the masses of the two bodies, and γ and K are certain constants.

The Lagrangian of central force motion is given by

$$L = (m/2)(\dot{r}^2 + r^2\dot{\phi}^2) - \int_r^\infty f(r)\,\mathrm{d}r. \tag{2.1.3}$$

If the terms and formulas of Hamiltonian mechanics are not known to the reader, he is referred to the works of Frank and v. Mises (7). Because $\mathrm{d}L/\mathrm{d}\phi = 0$, ϕ is a 'cyclic' coordinate, and one of the Lagrangian equations,

$$\frac{\mathrm{d}L}{\mathrm{d}\phi} - \frac{\mathrm{d}}{\mathrm{d}t}\frac{\partial L}{\partial\dot{\phi}} = 0,$$

results in $\mathrm{d}/\mathrm{d}t(\partial L/\partial\dot{\phi}) = 0$, or finally $\mathrm{d}L/\partial\dot{\phi} = \text{const}$. The generalized momenta are given by

$$p_r = \partial L/\partial\dot{r} = m\dot{r}, \; p_\phi = \partial\dot{\phi}L/\partial = mr^2\dot{\phi}.$$

Owing to the relationship $\partial L/\partial\dot{\phi} = \text{const}$, we have

$$r^2\dot{\phi} = h = \text{const}. \tag{2.1.4}$$

This result can also easily be derived from the second equation of motion (2.1.1).

In central force motion the radius vector r represents the 'non-cyclic' coordinate and the angle ϕ is the 'cyclic' coordinate. Let us investigate the *stability of quasistatic motions.* Quasistatic motions, which are similar to equilibrium, are motions in which the non-cyclic coordinate r remains constant and the cyclic coordinate ϕ increases linearly with time. These motions are circular motions for which $r = r_0 = \text{const}$ holds. Therefore, according to (2.1.4), $\dot{\phi} = h/r_0^2 = \text{const}$, and $\phi = ht/r_0^2 + C$ holds. Thus, r and ϕ satisfy indeed the conditions mentioned above.

Let us now carry out a perturbation of the quasistatic circular motion in such a way that the non-cyclic coordinate r varies while the momentum p_ϕ of the cyclic coordinate ϕ remains constant. This is similar to the perturbation of equilibrium positions. Consequently, let us investigate the *'conditional'* stability of circular motions. We have

$$r = r_0 + \xi, \quad p_\phi = mr^2\dot{\phi} = mh = \text{const},$$

that is, $h = \text{const}$.

Let us first apply the *method of small oscillations*. Substitution of the quantities $r = r_0 + \xi$, $r_0 = \text{const}$, $h = r_0^2\dot{\phi} = \text{const}$ into the first equation of motion (2.1.1) yields the following result,

$$\ddot{\xi} - h^2/(r_0 + \xi)^3 - (1/m)f(r_0 + \xi) = 0, \tag{2.1.5}$$

where ξ is the variation of r. If the terms of this equation are expanded into a series, the following result is obtained:

$$\ddot{\xi} - \frac{h^2}{r_0^3}\left(1 - \frac{3\xi}{r_0} + \frac{6\xi^2}{r_0^2} \cdots \right) - \frac{1}{m}\left[f(r_0) + \left(\frac{\mathrm{d}f(r)}{\mathrm{d}r}\right)_{r=r_0}\xi \right.$$
$$\left. + \frac{1}{2}\left(\frac{\mathrm{d}^2 f(r)}{\mathrm{d}r^2}\right)_{r=r_0}\xi^2 \cdots \right] = 0. \tag{2.1.6}$$

The first equation of (2.1.1) yields the result

$$-r_0\dot{\phi}^2 - (1/m)f(r_0) = 0$$

for $r = r_0$. Since the relationship $r_0\dot{\phi}^2 = h^2/r_0^3$ is obtained from $r_0^2\dot{\phi} = h$, this result is equal to

$$-h^2/r_0^3 - (1/m)f(r_0) = 0. \tag{2.1.7}$$

After substitution of (2.1.7) into (2.1.6) and neglecting the higher-order terms, we obtain

$$\ddot{\xi} + \left[\frac{3h^2}{r_0^4} + \frac{1}{m}\left(\frac{\mathrm{d}f(r)}{\mathrm{d}r}\right)_{r=r_0}\right]\xi = 0. \tag{2.1.8}$$

Therefore, a *linear* variational equation has been obtained which can be rewritten as follows because of (2.1.7):

$$\ddot{\xi} - \left[\frac{3f(r_0)}{mr_0} + \frac{1}{m}\left(\frac{\mathrm{d}f(r)}{\mathrm{d}r}\right)_{r=r_0}\right]\xi = 0 \tag{2.1.9}$$

or with the abbreviation

$$\alpha = \frac{3f(r_0)}{mr_0} + \frac{1}{m}\left(\frac{\mathrm{d}f(r)}{\mathrm{d}r}\right)_{r=r_0} \tag{2.1.10}$$

it can be rewritten as follows:

$$\ddot{\xi} - \alpha\xi = 0.$$

With $\xi = y_1$, $\dot{\xi} = \xi_2$, Eq. (2.1.11) finally results in a system of differential equations

or

$$\left.\begin{aligned}\dot{\xi}_1 &= \xi_2, & \dot{\xi}_2 &= \alpha\xi_1 \\ \dot{\xi}_1 &= a_{ik}\xi_k, & i,k &= 1,2.\end{aligned}\right\} \tag{2.1.12}$$

The system coefficients are $a_{11} = 0$, $a_{12} = 1$, $a_{21} = \alpha$, $a_{22} = 0$ and the characteristic equation follows from

$$\det(a_{ik} - \lambda\,\delta_{ik}) = \begin{vmatrix} -\lambda & 1 \\ \alpha & -\lambda \end{vmatrix} = \lambda^2 - \alpha = 0.$$

The *characteristic exponents* $\lambda_{1,2}$ satisfy

$$\lambda_{1,2} = \pm\alpha^{1/2}.$$

Theorem 2 of Section 1.3.1 states that instability occurs if $\alpha > 0$ because one of the characteristic exponents has a positive real part. In this case, according to Theorem 5 of Section 1.5.3, instability can be predicted with certainty from the linear variational equations. If it is assumed that

$$f(r) = -mK/r_n, \tag{2.1.13}$$

the instability condition is

$$\alpha = \frac{3f(r_0)}{mr_0} + \frac{1}{m}\left(\frac{\mathrm{d}f(r)}{\mathrm{d}r}\right)_{r=r_0} = -\frac{3K}{r_0^{n+1}} + \frac{nK}{r_0^{n+1}} > 0. \tag{2.1.14}$$

Equation (2.1.14) shows that central force motions with a central force of the (2.1.13) are unstable for $n < 3$.

If the linear variational equations were used to determine whether stability is possible for $\alpha \leqslant 0$, a vanishing double root would be found for $\alpha = 0$. Two imaginary roots with zero real part would be found for $\alpha < 0$. Therefore, $\alpha \leqslant 0$ corresponds to the *critical case*, and the question of stability can no longer be decided with certainty from the linear variational equations. It is then necessary to use the non-linear equation (2.1.5). The substitution $\xi = \xi_1$, $\dot{\xi} = \xi_2$, and Eq. (2.1.5) result in the system

$$\dot{\xi}_1 = \xi_2, \quad \dot{\xi}_2 = h^2/(r_0 + \xi_1)^3 + (1/m)f(r_0 + \xi_1). \tag{2.1.15}$$

The *Lyapunov function* is assumed to have the following form:

$$V(\xi_1, \xi_2) = \frac{1}{2}\xi_2^2 + \frac{1}{2}\frac{h^2}{(r_0 + \xi_1)^2} - \frac{1}{m}F(r_0 + \xi_1) - \frac{1}{2}\frac{h^2}{r_0^2} + \frac{1}{m}F(r_0), \tag{2.1.16}$$

where $F(x) = \int f(x)\,dx$. Apparently $V(0,0) = 0$, and due to the second equation in (2.1.15), also $\dot{V} \equiv 0$. The condition for which the function V is definite in the vicinity of the origin must be investigated in order to apply the Lyapunov stability Theorem 1 of Section 1.5.3. $V(\xi_1, 0)$ is expanded with respect to ξ_1 around the origin, which results in

$$V(\xi_1, 0) = V(0, 0) + \left(\frac{\partial V}{\partial \xi_1}(\xi_1, 0)\right)_{\xi_1 = 0} \xi_1 + \frac{1}{2}\left(\frac{\partial^2 V}{\partial \xi_1^2}(\xi_1, 0)\right)_{\xi_1 = 0} \xi_1^2$$

$$+ \frac{1}{3!}\left(\frac{\partial^3 V}{\partial \xi_1^3}(\xi_1, 0)\right)_{\xi_1 = 0} \xi_1^2 + \ldots.$$

$V(\xi_1, \xi_2)$ is positive definite in the vicinity of the origin if $V(\xi_1, 0)$ is. Because $V(0,0) = 0$ the sign of $V(\xi_1, \xi_2)$ depends only on the first- and second-order ξ_1 terms sufficiently close to the origin. Because of (2.1.7))we have

$$\left(\frac{\partial V}{\partial \xi_1}(\xi_1, 0)\right)_{\xi_1 = 0} = -\frac{h^2}{r_0^3} - \frac{1}{m}f(r_0) = 0,$$

so that the first-order ξ_1 term is zero. Therefore, the sign of $V(\xi_1, \xi_2)$ is determined only by the properties of the term $\frac{1}{2}\{\partial^2 V(\xi_1, 0)/\partial \xi_1^2\}_{\xi_1 = 0}\xi_1^2$. In order for $V(\xi_1, \xi_2)$ to be positive definite, the coefficient $\{\partial^2 V(\xi_1, 0)/\partial \xi_1^2\}_{\xi_1 = 0}$ must always be positive.

Therefore

$$\left(\frac{\partial^2 V}{\partial \xi_1^2}(\xi_1, 0)\right)_{\xi_1 = 0} = \frac{3h^2}{r_0^4} - \frac{1}{m}\left(\frac{df(x)}{dx}\right)_{\xi_1 = 0} > 0$$

must be required. For $f(x) = -mK/x^n$ this means that

$$3h^2/r_0^4 - nK/r_0^{n+1} > 0$$

must hold.

Because of (2.1.7), we have $h^2/r_0^3 = -(1/m)f(r_0) = K/r_0^n$, and the last inequality becomes

$$3K/r_0^{n+1} - nK/r_0^{n+1} > 0.$$

Therefore, it can be concluded that $n < 3$ must hold for the stability of a circular motion considered as a quasistatic central force motion. Therefore, the circular motion is stable for the Newtonian law $n = 2$ with respect to sufficiently small 'conditional' perturbations (perturbations of the radius without perturbations of p_ϕ).

It will now be shown that point masses moving along circles can only have *orbital stability*.

The equation $r = r_0 = \text{const}$ holds. $\dot{\phi} = h/r_0^2 = \text{const}$ then follows from (2.1.4) and therefore $v = r_0\dot{\phi} = h/r_0 = \text{const}$. If the initial conditions of the unperturbed motion are r_0, v_0 at the time $t = 0$, $r \equiv r_0$, $v \equiv v_0$ holds at all times.

Let us consider the motion of a point mass along a circular trajectory with the initial conditions $r_0^* = r_0 + \varepsilon_{r0}$, $v_0 = v_0^* + \varepsilon_{v0}$ as an example of perturbed motion. Since $r^* = \text{const}$, $v^* = \text{const}$ also holds for this motion, $r^* = r + \varepsilon_r = r_0^*$, $v^* = v + \varepsilon_v = v_0^*$ holds always so that $r^* - r = r_0^* - r_0 = \varepsilon_r = \varepsilon_{r0}$, $v^* - v = v_0^* - v_0 = \varepsilon_v = \varepsilon_{v0}$. If the initial perturbations ε_{r0}, ε_{v0} have been selected small enough, the perturbations ε_r, ε_v will also remain as small. In particular, the difference ε_r between the radii of the circles is always small enough so that the circles are as close as desired. This implies orbital stability.

On the other hand, there is no stability in the sense of Lyapunov. This is an example that shows that the concept of stability is a function of the selected coordinates. In order to show this, let us consider the problem in Cartesian coordinates.

The unperturbed motion satisfies

$$x = r_0 \cos(v_0 t/r_0), \qquad y = r_0 \sin(v_0 t/r_0),$$

and the perturbed motion satisfies

$$x^* = r_0^* \cos(v_0^* t/r_0^*), \qquad y^* = r_0^* \sin(v_0^* t/r_0^*).$$

The equation

$$x^* - x = r_0^* \cos(v_0^* t/r_0^*) - r_0 \cos(v_0 t/r_0)$$

holds, and let us assume that

$$v_0^*/r_0^* - v_0/r_0 = \varepsilon.$$

Thus

$$x^* - x = r_0^* \cos(v_0 t/r_0 + \varepsilon t) - r_0 \cos(v_0 t/r_0)$$

is obtained.

Let us consider the motion at a certain time $t = \pi/\varepsilon$. At this instant $x^*(\pi/\varepsilon) - x(\pi/\varepsilon) = -(r_0^* + r_0)\cos(v_0\pi/r_0\varepsilon)$ holds and also

$$y^*(\pi/\varepsilon) - y(\pi/\varepsilon) = -(r_0^* + r_0)\sin(v_0\pi/r_0\varepsilon).$$

The Euclidean distance of the image points $P[x^*(t), y^*(t)]$, $Q[x(t), y(t)]$, which is given by

$$d[P(t), Q(t)] = \{[x^*(t) - x(t)]^2 + [(y^*(t) - y(t)]^2\}^{1/2},$$

has the value $d[P(\pi/\varepsilon), Q(\pi/\varepsilon)] = r_0^* + r_0$ at the time $t = \pi/\varepsilon$. This is a finite, large, and constant value which cannot be made as small as desired regardless of how small the initial perturbations $x_0^* - x_0$, $y_0^* - y_0$ have been selected. Therefore, it is not possible to make the distance $d(P, Q)$ arbitrarily small at *all* times by making the initial perturbations sufficiently small. The Lyapunov stability condition is not satisfied and there is no Lyapunov stability.

2.1.2 Stability Investigations of the Three-body Problem

Let us consider the restricted three-body problem for which the mass of one body (for example, the satellite) is made so *small* that it can be neglected with

150

respect to the mass of the other two bodies (for example, earth and moon). In addition, let us consider a *plane problem* for which the small body always moves in the plane common to all three bodies. We shall justify these assumptions in the following.

The acceleration b satisfies the following equation in the inertial system, according to Newton's law

$$b = -k\left(\frac{m_E r_{ES}}{|r_{ES}|^3} \frac{m_M r_{MS}}{|r_{MS}|^3}\right). \tag{2.1.17}$$

m_E and m_M are the masses of the earth and the moon, respectively. The definitions of r_{ES} and r_{MS} follow from Fig. 37. It is useful to use the synodic reference system (x, y, z) to describe the three-body problem from this point on. It rotates around the z axis (Fig. 37) with the angular rate $\omega = 2.66 \times 10^{-6}\,\text{s}^{-1}$.

As a result of the laws of relative motion, the following equation is obtained in the synodic system:

$$b = \partial v/\partial t + 2\omega \times v + \omega \times (\omega \times r) = \partial v/\partial t + 2\omega \times v - \omega^2 r_H. \tag{2.1.18}$$

The following units are selected: the length unit $|a_M|$, the time unit $T/2\pi$ (where T is a sidereal month), and the mass unit $m_E + m_M$. All other notation can be obtained from Fig. 37.

In these units we have $\omega = 1$, $k = 1$ so that the following equation for the motion of the satellite is obtained from (2.1.17) and (2.1.18):

$$\partial v/\partial t + 2e_z \times v = \text{grad } U. \tag{2.1.19}$$

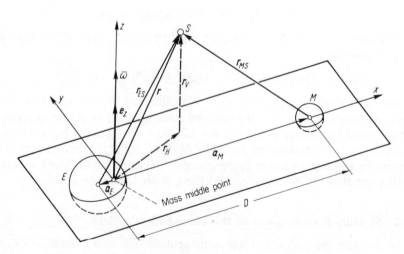

Fig. 37. Three-body problem: E, earth; M, moon; S, satellite

Here e_z is the unit vector in the z direction, and we have

$$U = \frac{1}{2} r_H^2 + \frac{1-\mu}{|r_{ES}|} + \frac{\mu}{|r_{MS}|}, \qquad (2.1.20)$$

where $\mu \approx m_M/m_E = 1/82$.

Let us first show that the motion of the satellite remains in the earth–moon plane (x, y) if the initial values r_0, v_0 are vectors in this plane. In this way we show that the restriction to *plane motion* is a reasonable assumption for the three-body problem.

Scalar multiplication of (2.1.19) with v results in

$$v(\partial v/\partial t - \operatorname{grad} U) = 0. \qquad (2.1.21)$$

This means that the vector $c = \partial v/\partial t - \operatorname{grad} U = 2v \times e_z$ is perpendicular to the vector v and lies in the x, y plane. If the initial values, r_0, v_0 of a motion are selected so that they are vectors in the x, y plane, $\operatorname{grad} U$ is also a vector in the x, y plane at this time, as well as $\partial v/\partial t = c + \operatorname{grad} U$. Therefore, v and r remain in the x, y plane for all times. The motion is plane as predicted.

Equation (2.1.21) can be written in the form

$$v(\partial v/\partial t) = v \operatorname{grad} U = dU/dt.$$

Integration of this expression yields the Jacobi integral

$$v^2 = 2U - J, \qquad (2.1.22)$$

where J is the Jacobi constant.

In the following discussion of the plane three-body problem, the equipotential curves $2U = C_i$ corresponding to different values of the constants C_i, $i = 1, 2, 3, \ldots$, are plotted. These curves are called *Hill limit curves* of zero relative velocity or Darwin curves. Figure 38 shows these curves, which correspond to limits of motion. This can be seen as follows: Let us assume that the satellite begins its motion from the position r_a with the velocity v_a (Fig. 39). From Eq. (2.1.22) it follows that the Jacobi constant of this motion is $J = C_a - v_a^2$ because the satellite began its motion along the equipotential curve $2U = C_a$. When the satellite reaches the curve $2U = C_i$, the square of the magnitude of the velocity is

$$v_i^2 = C_i - J = v_a^2 + C_i - C_a.$$

The equipotential curve $2U = J$ represents a limit to the motion because $v_0^2 = J - J = 0$ holds along it. In general, however, the satellite will not remain in relative equilibrium at the limit curve but instead will move away from it again. For $v_0 = 0$ (which holds along the limit curve), it follows form (2.1.19) that $\partial v_0/\partial t = \operatorname{grad} U$. If $\operatorname{grad} U = 0$ does not also hold by coincidence, $\partial v_0/\partial t \neq 0$ must hold along the limiting curve and therefore $b \neq 0$ must hold. Consequently there is a force acting on the satellite so that this one is repulsed from the limit curves.

Fig. 38. Darwin curves

Libration points:	L_1	L_2	L_3	L_4	L_5
Distance from $E = r_i$:	0.85	1.17	0.99	1.0	1.0
Distance from $M = \rho_i$:	0.15	0.17	1.99	1.0	1.0

The condition grad $U = 0$ leads to singular points of the surface $2U = $ const. The level curves of the surface are the Hill limit curves $2U = C_i$, $i = 1, 2, 3, \ldots$, in the x, y plane. The singular points in the x, y plane correspond to the *libration points* L_k, $k = 1, 2, 3, 4, 5$. The coordinates of these points are the Lagrange particular integrals of the three-body problem. Let us consider the case where the satellite remains in relative equilibrium at one of the libration points where $v = 0$, $\partial v / \partial t = 0$. In addition, the coordinates of the libration point satisfy the condition grad $U = 0$. All the terms in (2.1.19) are simultaneously zero and the system of differential equations is satisfied.

Fig. 39. Hill limit curves

Let us now investigate the stability of the points L_1 to L_5 in the three-body problem for the earth, moon, and satellite.

For this three-body problem we have

$$2U = x^2 + y^2 + \frac{2(1-\mu)}{|r_{ES}|} + \frac{2\mu}{|r_{MS}|} = C, \qquad \begin{aligned} |r_{ES}| &= [(x-x_E)^2 + y^2]^{1/2} \\ |r_{MS}| &= [(x-x_M)^2 + y_2]^{1/2} \end{aligned}$$

and grad $U = 0$ is given by

$$x - (1-\mu)\frac{x-x_E}{|r_{ES}|^3} - \mu\frac{x-x_M}{|r_{MS}|^3} = 0, \qquad y - (1-\mu)\frac{y}{|r_{ES}|^3} - \mu\frac{y}{|r_{MS}|^3} = 0.$$

$$(2.1.23)$$

The libration points L_1 to L_3 lie along the x axis where $y = 0$. The second of Eqs. (2.1.23) is identically satisfied. The x coordinates of the points of the L_i, $i = 1, 2, 3$, can therefore be calculated from the first of Eqs. (2.1.23) with $y = 0$, which results in

$$x - (1-\mu)(x-x_E)^{-2} - \mu(x-x_M)^{-2} = 0. \qquad (2.1.24)$$

Since the origin of the x, y axis system is at the center of gravity of the earth and moon, $x_E = -\mu$, $x_M = 1 - \mu$. If the distance between M and L_2 is called ρ, then $x = 1 - \mu + \rho$. If this is substituted into (2.1.24), the following equation is obtained for determining ρ:

$$\rho^5 + (3-\mu)\rho^4 + (3-2\mu)\rho^3 - \mu\rho^2 - 2\mu\rho - \mu = 0.$$

For $\mu = 1/82$, the one real positive solution is $\rho = 0.17$. The locations of L_1, L_3 along the x axis can also be calculated in this way. The results are given in the table of Fig. 38.

Figure 40 shows the location of the points L_4, L_5. It may be seen, for example, that L_4 has coordinates

$$x = \tfrac{1}{2} - \mu, \quad y = \tfrac{1}{2}\sqrt{3}. \qquad (2.1.25)$$

Fig. 40. Position of the libration point L_4

154

If the libration point locations are known, the Gaussian curvature

$$K = (U_{xx}U_{yy} - U_{xy}^2)/(1 + U_x^2 + U_y^2)^2 \qquad (2.1.26)$$

of the surface U (and therefore also of $2U$) at these points can be found. From

$$U = \frac{1}{2}(x^2 + y^2) + \frac{1 - \mu}{|r_{ES}|} + \frac{\mu}{|r_{MS}|},$$

$$|r_{ES}| = [(x - x_E)^2 + y^2]^{\frac{1}{2}}, \quad |r_{MS}| = [(x - x_M)^2 + y^2]^{\frac{1}{2}},$$

it follows that

$$U_{xx} = \left[1 - (1 - \mu)\frac{1}{|r_{ES}|^3} - \mu\frac{1}{|r_{MS}|^3}\right]$$

$$+ 3\left[(1 - \mu)\frac{(x - x_E)^2}{|r_{ES}|^5} + \mu\frac{(x - x_M)^2}{|r_{MS}|^5}\right],$$

$$U_{yy} = \left[1 - (1 - \mu)\frac{1}{|r_{ES}|^3} - \mu\frac{1}{|r_{MS}|^3}\right] \qquad (2.1.27)$$

$$+ 3y^2\left[(1 - \mu)\frac{1}{|r_{ES}|^5} + \mu\frac{1}{|r_{MS}|^5}\right],$$

$$U_{xy} = 3y\left[(1 - \mu)\frac{x - x_E}{|r_{ES}|^5} + \mu\frac{x - x_M}{|r_{MS}|^5}\right].$$

$y = 0$ holds for L_1 to L_3. The following relationships are obtained for these points from (2.1.27):

$$U_{xy} = 0, \quad U_{yy} = \left[1 - (1 - \mu)\frac{1}{|x - x_E|^3} - \mu\frac{1}{|x - x_M|^3}\right], \qquad U_{xx} = 3 - 2U_{yy}.$$

The sign of K is determined by the sign of

$$U_{xx}U_{yy} = 3U_{yy} - 2U^2yy$$

because $U_{xy} = 0$. For example, for L_1 where $x - x_E = 0.85$, $x - x_M = -0.15$ and because $\mu = 1/82$, we have

$$U_{yy} = 1 - \frac{0.988}{0.85^3} - \frac{0.012}{0.15^3} = -4.15$$

Since U_{yy} is negative, our question is already answered, because K must also be negative. Therefore, L_1 is a *hyperbolic* point of U. The same method shows that L_2, L_3 are also hyperbolic points of U.

Equation (2.1.25) is satisfied for L_4. Furthermore, the coordinates of this point satisfy $|r_{ES}| = |r_{MS}| = 1$, $x - x_E = -(x - x_M) = \frac{1}{2}$ (see Fig. 40) and

$$U_{xx} = \frac{3}{4}, \quad U_{yy} = \frac{9}{4}, \quad U_{xy} = \frac{3}{2}\sqrt{3}\left(\frac{1}{2} - \mu\right). \qquad (2.1.28)$$

We then have

$$U_{xx}U_{yy} - U_{xy}^2 = \frac{3}{4}\frac{9}{4} - \frac{3 \cdot 9}{4}\left(\frac{1}{2} - \mu\right)^2 = \frac{27}{16}\left[1 - 4\left(\frac{1}{2} - \mu\right)^2\right]. \quad (2.1.29)$$

Since $\mu = 1/82$ and $4(\frac{1}{2} - \mu)^2 < 1$, we have $U_{xx}U_{yy} - U_{xy}^2 > 0$ and therefore also $K > 0$ so that L_4 is an *elliptic* point of U. The same is true for point L_5 that is a mirror image of L_4 with respect to the x axis.

Let us now show the manner in which the curvature of U at the singular points (libration points) leads to a first approximation of the stability of these points. It will be shown that the points L_1 to L_3 are unstable according to the definition of Lyapunov, because they are hyperbolic points of U with $K < 0$. It will be seen that the points L_4, L_5 are elliptic points of U with $K > 0$ which are stable in the first approximation.

The following calculation is made to prove this postulate. The scalar form of Eq. (2.1.19) is

$$\ddot{x} - 2\dot{y} = U_x,$$
$$\ddot{y} + 2\dot{x} = U_y. \quad (2.1.30)$$

The libration points L_i, $i = 1, 2, 3, 4, 5$, which have the coordinates $x_i = a_i$, $y_i = b_i$, satisfy grad $U = 0$ and therefore

$$U_x(a_i, b_i) = U_y(a_i, b_i) = 0 \qquad \text{for all } i. \quad (2.1.31)$$

According to (2.1.30), (2.1.31) they also satisfy

$$\left.\begin{array}{l} \ddot{a}_i - 2\dot{b}_i = 0, \\ \ddot{b}_i + 2\dot{a}_i = 0 \end{array}\right\} \qquad \text{for all } i. \quad (2.2.32)$$

The variations ξ, η are introduced in order to investigate the stability of the L_i. We have

$$x = a_i + \xi,$$
$$y = b_i + \eta. \quad (2.1.33)$$

The variational equations for the stability investigation are obtained from (2.1.30) by substitution of (2.1.33). This results in

$$\left.\begin{array}{l} \ddot{a}_i - 2\dot{b}_i + \ddot{\xi} - 2\dot{\eta} = U_x(a_i + \xi, b_i + \eta), \\ \ddot{b}_i + 2\dot{a}_i + \ddot{\eta} + 2\dot{\xi} = U_y(a_i + \xi, b_i + \eta). \end{array}\right\} \quad (2.1.34)$$

Because of (2.1.32) and (2.1.31), and making use of the Taylor expansions of U_x and U_y, we obtain

$$\left.\begin{array}{l} \ddot{\xi} - 2\dot{\eta} = U_{xx}(a_i, b_i)\xi + U_{xy}(a_i, b_i)\eta + X(\xi, \eta), \\ \ddot{\eta} + 2\dot{\xi} = U_{xy}(a_i, b_i)\xi + U_{yy}(a_i, b_i)\eta + Y(\xi, \eta). \end{array}\right\} \quad (2.1.35)$$

This is a non-linear system of equations, the non-linear terms of which are given by X and Y.

Linearized equations corresponding to (2.1.35) given by

$$\ddot{\xi} - 2\dot{\eta} = U_{xx}^{(i)}\xi + U_{xy}^{(i)}\eta,$$

$$\ddot{\eta} + 2\dot{\xi} = U_{xy}^{(i)}\xi + U_{yy}^{(i)}\eta, \qquad (2.1.36)$$

are used in the stability investigation, where

$$U_{xx}^{(i)} = U_{xx}(a_i, b_i), \text{ etc.}$$

The trial solutions

$$\xi = a \, e^{\lambda t}, \ \eta = b \, e^{\lambda t} \qquad (2.1.37)$$

substituted into (2.1.36) result in the system

$$a\lambda^2 - 2b\lambda = aU_{xx}^{(i)} + bU_{xy}^{(i)},$$

$$b\lambda^2 + 2a\lambda = aU_{xy}^{(i)} + bU_{yy}^{(i)} \qquad (2.1.38)$$

In order that this system of equations may have non-trivial solutions for a and b, we must set

$$\begin{vmatrix} \lambda^2 - U_{xx}^{(i)} & -2\lambda - U_{xy}^{(i)} \\ 2\lambda - U_{xy}^{(i)} & \lambda^2 - U_{yy}^{(i)} \end{vmatrix} = 0.$$

This leads to the characteristic equation

$$\lambda^4 + (4 - U_{xx}^{(i)} - U_{yy}^{(i)})\lambda^2 + [U_{xx}^{(i)}U_{yy}^{(i)} - (U_{xy}^{(i)})^2] = 0. \qquad (2.1.39)$$

Because of (2.1.26) and (2.1.31), the Gaussian curvature K of the surface U takes on the following values at the libration points:

$$K^{(i)} = U_{xx}^{(i)}U_{yy}^{(i)} - (U_{xy}^i)^2, \qquad i = 1, 2, 3, 4, 5. \qquad (2.1.40)$$

Therefore, Eq. (2.1.39) can be transformed into

$$\lambda^4 + (4 - U_{xx}^{(i)} - U_{yy}^{(i)})\lambda^2 + K^{(i)} = 0. \qquad (2.1.41)$$

Let us introduce the further abbreviation

$$4 - U_{xx}^{(i)} - U_{yy}^{(i)} = \beta^{(i)} \qquad (2.1.42)$$

The characteristic equation finally becomes

$$\lambda^4 + \beta^{(i)}\lambda^2 + K^{(i)} = 0. \qquad (2.1.43)$$

The roots of this equation are given by

$$\lambda_1 = \left[-\frac{\beta^{(i)}}{2} + \left(\frac{\beta^{(i)2}}{4} - K^{(i)}\right)^{1/2} \right]^{1/2}, \qquad \lambda_2 = -\lambda_1,$$

$$\lambda_3 = \left[-\frac{\beta^{(i)}}{2} - \left(\frac{\beta^{(i)2}}{4} - K^{(i)}\right)^{1/2} \right]^{1/2}, \qquad \lambda_4 = -\lambda_3. \qquad (2.1.44)$$

At the libration points for L_1, L_2, L_3 — that is, for $i = 1, 2, 3, K^{(i)} < 0$ holds. According to (2.1.44) $\lambda_1 > 0$ and is real, so that instability follows from

(2.1.37). According to the theorems derived previously, in this case the result of the stability calculation using the linear equations (2.1.36) is also valid for the non-linear case (2.1.35). This means that for the condition $K^{(i)} < 0$ the libration points L_1, L_2, L_3 are unstable not only in terms of the definition of small oscillations but also in terms of the Lyapunov definition. The negative Gaussian curvature K of the surface U, i.e., the hyperbolic character of U of the points L_1, L_2, L_3 derived from it, has indeed revealed the instability of these points.

For the libration points L_4, L_5, i.e., for $i = 4, 5$, we find $\beta^{(i)} = 1$ from (2.1.42). This can easily be calculated for L_4, where $U_{xx} = 3/4$, $U_{yy} = 9/4$ according to (2.1.28). This also holds for L_5. The characteristic equation (2.1.43) can therefore be simplified as follows:

$$\lambda^4 + \lambda^2 + K^{(i)} = 0, \qquad i = 4, 5. \qquad (2.1.45)$$

Instead of (2.1.44), the following simplification can be made:

$$\left.\begin{array}{ll} \lambda_1 = \{\tfrac{1}{2} [-1 + (1 - 4K^{(i)})^{\frac{1}{2}}]\}^{\frac{1}{2}}, & \lambda_2 = -\lambda_1 \\ \lambda_3 = \{\tfrac{1}{2} [-1 - (1 - 4K^{(i)})^{\frac{1}{2}}]\}^{\frac{1}{2}}, & \lambda_4 = -\lambda_3 \end{array}\right\} \text{ for } i = 4, 5. \quad (2.1.46)$$

From (2.1.46) it may be seen that all λ_k, $k = 1, 2, 3, 4$, are purely imaginary. Because of (2.1.37), small oscillation stability in the first approximation prevails if

$$\tfrac{1}{4} > K^{(i)} > 0, \qquad i = 4, 5, \qquad (2.1.47)$$

is satisfied.

Since we are considering the earth–moon–satellite problem, we have $\mu = 1/82$ so that the following result is obtained from (2.1.40), (2.1.28) for $i = 4$, i.e., from (2.1.29)

$$K^{(4)} = \frac{27}{16} \left[1 - 4 \left(\frac{1}{2} - \frac{1}{82} \right)^2 \right] \approx 0.08.$$

Condition (2.1.47) is satisfied at the elliptical point L_4 because of the positive and sufficiently small Gaussian curvature $K^{(4)}$ of the surface U at this point. Therefore, stability in the first approximation does indeed prevail at this point. This also holds for L_5.

Since the λ_k, $k = 1, 2, 3, 4$, are purely imaginary for (2.1.47), the critical case according to the definition of Lyapunov prevails. This means that the stability of L_4, L_5 obtained from the linearized equations (2.1.36) does not necessarily apply to the general case based on the non-linear equations (2.1.35). In order to obtain results in this case, the stability would have to be investigated again, and the non-linear terms would have to be taken into account. A first step in this direction was made by Leontovic.[1]

---

[1]Leontovic, A. M., *Dokl. Akad. Nauk SSSR* **143**, 525–528 1962.

158

Exercises 2.1

1) Show that Eqs. (2.1.1) are equivalent to the simultaneous system

$$\ddot{x} = f\frac{x}{r}, \quad \ddot{y} = f\frac{y}{r},$$

where x and y are cartesian coordinates with the origin in the center of the motion. Derive the first integral

$$x\dot{y} - y\dot{x} = h,$$

where $h = 2\dot{A}$, and A is the area swept over by the radius vector. Prove this fact.

2) Using Eqs. (2.1.1) show that (i) $\ddot{r} = h^2 r^{-3} + f$ and, using the transformation $u = r^{-1}$ that (ii) $f = -h^2 u^2 (u + d^2 u/d\phi^2)$.

   Let a particle describe a circle with radius a passing through the origin of the axis system. Show by means of the previously derived formula for f that the central force is of the form $f = -k^2 r^5$, where k is an appropriately chosen constant.

   Discuss the stability of this motion assuming that r is perturbed and that the quantity $k = r^2 \dot{\phi}$ is kept constant.

   Hint: the polar equation of the circle is

$$r = 2a \cos \theta = u^{-1}.$$

3) Consider Eqs. (2.1.1)) to describe the motion of a satellite moving around the earth with $f = -Kr^{-2}$. Then, according to an equation derived in problem 2,

$$u + \frac{d^2 u}{d\phi^2} = \frac{K}{h^2}.$$

   Solve this differential equation for u and evaluate the constant of integration from the energy equation.

   Show that the final result is

$$u = \frac{K}{h^2} (1 + e \cos \phi),$$

where

$$e = (1 + 2Eh^2/K^2)^{1/2}$$

and E is the total energy of the motion.

   Describe the various orbits the satellite describes

$$e > 1, \qquad e = 1, \qquad 0 < e < 1, \qquad e = 0.$$

2.2
Problems of the Mechanics of Rigid Bodies and of Systems

2.2.1 Stability of Vibrations

Vibration problems and the associated stability problems have been treated extensively in the literature. For example, we should mention the books by Klotter (2.4), Minorsky (2.1), Kauderer (2.10), and Magnus (2.14). The following limited discussion is given in order to show the way in which the approximate methods derived in Section 1.6 can be applied. Let us restrict ourselves to *non-linear vibrations of one-mass systems*. This makes it possible to deal with the most important questions of modern stability theory without any great difficulty.

Let us assume a vibration process given by the non-linear differential equation (Fig. 41)

$$\ddot{x} + \omega_0^2 x + \mu \Phi(x, \dot{x}) = G(t). \tag{2.2.1}$$

ω_0 is a given constant, μ a small parameter, Φ a non-linear function, and G is a periodic function.

The first possibility is to discuss problem (2.2.1) in the *phase plane*. The phase curves can be obtained with the isocline method, for example. The method can only be recommended when $G \equiv 0$, that is, when the mechanical process is *autonomous*. If, in addition, Φ is only a function of x, the phase curve can be determined exactly by integrating (2.2.1)

$$\dot{x} = \sqrt{2} [E - U(x)]^{1/2} \tag{2.2.2}$$

where

$$U(x) = \tfrac{1}{2} \omega_0^2 x^2 + \mu \int \Phi(x) \, dx.$$

The phase curves are levels of the energy surface $E = \tfrac{1}{2} \dot{x}^2 + U(x)$ in the $\dot{x}$, x plane. The course of these curves makes it possible to determine the stability of stationary points (equilibrium positions) or limit cycles (periodic solutions). If the phase curve equation depends on the parameter, the stability investigation consists of determining the change in the properties of the stationary

159

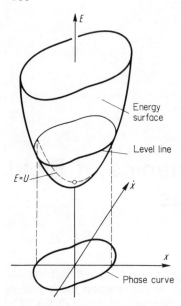

Fig. 41. Connection between energy surface and phase curves

points or limit cycles as the parameter is varied. The 'critical' parameter values are of special interest, because the transition from stable to unstable occurs there.

Example: Let us consider $\ddot{x} + \omega_0^2 x + \mu x^3 = 0$. The phase curves are determined from $\dot{x} = (1/\sqrt{2})(4E - 2\omega_0^2 x^2 - \mu x^4)^{1/2}$ (Fig. 42). The phase diagram shows that the stability of the stationary solution $\dot{x} = x = 0$, as well as that of the periodic solutions, do not change if the parameters ω_0 or μ are changed. The periodic solutions have orbital stability with respect to perturbations of the initial conditions.

The approximation methods of Section 1.6 can also be used to investigate the stability of the solutions of the problem (2.2.1). Let us now consider this method.

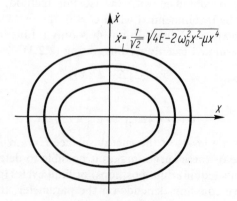

Fig. 42. Phase diagram

Let us reconsider the *method of small oscillations*. It is assumed that (2.2.1) has a periodic solution, which is essentially given by $x_s = A \cos \omega t$. In order to investigate its stability, $x = x_s + \xi$ is substituted into (2.2.1), and the result is

$$\ddot{\xi} + \omega_0^2 \xi + \mu \Phi(x_s + \xi, \dot{x}_s + \dot{\xi}) = G(t) - \ddot{x}_s - \omega_0^2 x_s^2.$$

Linearization of this equation results in

$$\ddot{\xi} + \omega_0^2 \xi + \mu \left[\frac{\partial \Phi(x, \dot{x})}{\partial x} \right]_{\substack{x = x_s \\ x = x_s}} \xi + \mu \left[\frac{\partial \Phi(x, \dot{x})}{\partial \dot{x}} \right]_{\substack{x = x_s \\ x = x_s}} \dot{\xi}$$

$$= G(t) - \ddot{x}_s - \omega_0^2 x_s^2 - \mu \Phi(x_s, \dot{x}_s).$$

The right-hand side is zero according to Eq. (2.2.1), and the following variational equation remains:

$$\ddot{\xi} + \omega_0^2 \xi + \mu \left\{ \left[\frac{\partial \Phi}{\partial x} \right]_{\substack{x = x_s \\ x = x_s}} \xi + \left[\frac{\partial \Phi}{\partial \dot{x}} \right]_{\substack{x = x_s \\ x = x_s}} \dot{\xi} \right\} = 0. \qquad (2.2.3)$$

This is a linear differential equation with variable, *periodic* coefficients. The remainder of the stability investigation follows Section 1.3.3.

Example: Let us again consider $\ddot{x} + \omega_0^2 x + \mu x^3 = 0$. The variational equation is $\ddot{\xi} + \omega_0^2 \xi + 3\mu A^2 (\cos^2 \omega t) \xi = 0$. It can be changed into the following standard form of the Mathieu differential equation by the substitutions of

$$\lambda = (1/\omega^2)(\omega_0^2 + 3\mu A^2/2), \qquad h = (-1/\omega^2)(3\mu A^2/4), \qquad \tau = \omega t,$$

$$d^2\xi/d\tau^2 + (\lambda - 2h \cos 2\tau)\xi = 0.$$

In Section 1.6.4 the approximate solution

$$x = A \cos \omega t \qquad \text{of } \ddot{x} + \omega_0^2 x + \mu x^3 = 0$$

was found. The relationship $\omega_0^2 = \omega^2 + \frac{3}{4}\mu A^2$ was required to hold as a first approximation. Therefore, $\lambda = 1 - h$ would show that this approximation lies along a limiting curve of the Strutt diagram (Fig. 16). It thus follows that x_s was a solution, as can be determined from the phase diagram (Fig. 42), and that it must have orbital stability. It can easily be shown that it is not stable according to the definition of Lyapunov. This is a first indication that the concept of *orbital stability* is very important in vibration problems, because often the Lyapunov stability definition is too restrictive in practical applications.

It is better to use the *harmonic linearization* in such a case, because the linearization described above is often not permissible and may lead to the critical case where the real parts of the characteristic exponents vanish, which means that it certainly is questionable. Harmonic linearization is carried out by first obtaining the linear substitution equations using the harmonic balance as described in Section 1.7.6. The coefficients of the substitution equation are then used to determine stability. This was done for the non-linear differential equation

$$\ddot{x} - (\alpha - \beta \, | \, x \, |) \, | \, \dot{x} \, | \, \dot{x} + \omega_0^2 (x + \gamma x^3) = 0.$$

It was immediately possible to conclude the stability of the periodic solution from the behaviour of the damping coefficients of the substitution equation.

Example: Let us again consider the differential equation $\ddot{x} + \omega_0^2 x + \mu x^3 = 0$, which will be transformed into the substitution equation $\ddot{x} + \omega_0^2 x = cx$. From Section 1.7.6 it follows that

$$c = - \frac{\mu \omega}{A \pi} \int_0^{2\pi/\omega} A^3 \sin^4 \omega \xi \, \mathrm{d}\xi = - \frac{3\mu A^2}{4},$$

so that $\ddot{x} + (\omega_0^2 + 3\mu A^2/4) \; x = 0$ is obtained. This substitution equation corresponds to an oscillation with amplitude A and the frequency $\omega = [\omega_0^2 + 3\mu A^2/4]^{\frac{1}{2}}$ which depends on the amplitude. Apparently the phase curves always have orbital stability because the stability limit $\omega = 0$ can never be reached. On the other hand, oscillations corresponding to adjacent values of A have different frequencies.

This means that image points moving along adjacent phase curves having orbital stability can be displaced by a finite amount at certain times, even though the phase curves may be as close to each other as desired. Therefore, Lyapunov stability does not exist. We have again shown the properties of this differential equation in another way.

It is possible that the harmonic balance method will be suitable for *special stability criteria*, which can be different in each case and are adjusted to the conditions at hand. We will continue this method by means of two examples.

Example 1: Let us first consider a *forced vibration* with the differential equation

$$\ddot{x} + \omega_0^2 x + \mu x^3 = Z \cos \omega t.$$

It can be written in the following form with the notation $L = \mathrm{d}^2/\mathrm{d}t^2 + \omega_0^2$ and $f(x, t) = Z \cos \omega t - \mu x^3$: $L(x) = f(x, t)$. The solution is $x = L^{-1}[f(x, t)]$. The inverse of the operator L is given by

$$L^{-1}(\phi) = - \frac{2}{T} \int_0^t \sum_k \frac{\cos k\omega (t - \xi)}{k^2 \omega^2 - \omega_0^2} \phi(\xi) \, \mathrm{d}\xi$$

and the non-linear integral equation

$$x(t) = - \frac{2}{T} \int_0^t \sum_{k=1}^\infty \frac{\cos k\omega (t - \xi)}{k^2 \omega^2 - \omega_0^2} [Z \cos \omega \xi - \mu x^3 (\xi)] \, \mathrm{d}\xi$$

is obtained where $T = 2\pi/\omega$. According to the harmonic balance method, only the first term in the sum under the integral sign is retained and therefore $k = 1$. The approximation $x = A \cos \omega t$ is assumed for x. This results in

$$A \cos \omega t = - \frac{\omega}{\pi(\omega^2 - \omega_0^2)} \left[\int_0^{2\pi/\omega} Z \cos^2 \omega \xi \, \mathrm{d}\xi - \mu \int_0^{2\pi/\omega} A^3 \cos^4 \omega \xi \, \mathrm{d}\xi \right] \cos \omega t.$$

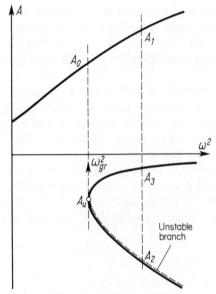

Fig. 43. Resonance curve

The integration is carried out and the coefficients of cos ωt are set equal, which results in the relationship

$$Z = (3\mu/4)A^3 - A(\omega^2 - \omega_0^2). \tag{2.2.4}$$

For constant Z this relationship represents the resonance curve that gives the dependence of the amplitude A on the square of the exciting frequency ω (Fig. 43). As the figure shows, it is possible to have a jump in the amplitude from A_u to A_0 for $\omega = \omega_{gr}$, which indicates that the stability boundary has been reached. Three amplitudes are possible for $\omega > \omega_{gr}$. A decision must be made as to which of the amplitudes correspond to stable or unstable solutions. This is done in the following way.

The resonance curve is a level line of the surface $Z = Z(A, \omega^2)$ generated by (2.2.4). Since $Z = $ const holds for the resonance curve, the following equation must also be satisfied:

$$dZ = (\partial Z/\partial A)\, dA + (\partial Z/\partial \omega^2)\, d\omega^2 = 0.$$

This equation, together with the relationship $\partial Z/\partial \omega^2 = -A$ according to (2.2.4), results in the following equation:

$$dA/d\omega^2 = A(\partial Z/\partial A)^{-1} \tag{2.2.5}$$

Obviously $\partial Z/\partial A > 0$ must hold for stable solutions, which means that the magnitude of the forcing function must increase when A is increased. Conversely, $\partial Z/\partial A < 0$ for unstable solutions. If this is taken into account in (2.2.5), it follows that the stability boundary is given by $\partial Z/\partial A = 0$, that is, by $dA/d\omega^2 = \infty$. The unstable branch of the resonance curve must start at the point A_u where the resonance curve has a vertical tangent. The stable branches follow from (2.2.5) and $\partial Z/\partial A > 0$. In Fig. 43 they are the branches with positive slope. The unstable branch is the one with negative slope, because $\partial Z/\partial A < 0$.

Example 2: Let us consider the differential equation $\ddot{x} + \omega_0 x = av^2(\phi - \dot{x}/v)$. It corresponds to the *bending oscillations of a wing*. The flight velocity is v and the wing simultaneously twists by an amount ϕ. The critical flight velocity v_k is to be determined at which flutter oscillations, and therefore instability phenomena, of the wing begin.[1]

This example is used to show that stability can be determined using the approximate trial solution $x = A\cos \omega_0 t$ of the harmonic balance method and by a consideration of the *energy balance*.

The amount of the damping energy dissipated over a period is given by

$$E_D = \int_0^T av\dot{x}^2 \, dt.$$

The energy supplied by the excitation over a period is given by

$$E_Z = -\int_0^T av^2\phi\dot{x} \, dt.$$

The amplitude of the torsional oscillation ϕ is different from that of the bending oscillation x and there is a phase displacement between them. Therefore, $\phi = bA\cos(\omega_0 t - \psi)$, and the following expressions are calculated:

$$E_D = \int_0^{2\pi} av\omega_0^2 A^2 \sin^2 \omega_0 t \, d(\omega_0 t) = av\omega_0^2 A^2 \pi,$$

$$E_Z = \int_0^{2\pi} av^2 b\omega_0 A^2 \sin \psi \sin^2 \omega_0 t \, d(\omega_0 t) = av^2 b\omega_0 A^2 \sin \psi \pi.$$

The oscillation can only build up to flutter of the wing for $E_Z > E_D$, that is, for $v > \omega_0/b \sin \psi$. The stability boundary is given by $E_Z = E_D$, which means that the critical flight velocity is $v = \omega_0/b \sin \psi$.

The question of *orbital stability* can also be decided by the *variation of constants* method described in Section 1.7.3. Let us consider the phase plane $\dot{A}, A$ and the phase curves in this plane (Fig. 44). The time parameter t increases along the phase curve towards the right in the upper half-plane. In the lower half-plane, it increases towards the left. The points $A = 0$ are *equilibrium points* for the amplitude that correspond to *periodic solutions*. Points such as A_4, A_0, and A_2 are *stable*, and the corresponding phase curves of the periodic solutions in the $\dot{x}$, x plane have orbital stability if the time in the $\dot{A}, A$ plane increases in the direction toward these points. On the other hand, the points A_3, A_1 are *unstable* because time increases in the direction away from them.

---

[1]This example was given by Magnus (2.14, p. 126).

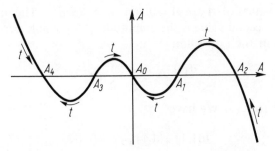

Fig. 44. Phase curve of the vibration amplitude

It can be seen that $d\dot{A}/dA < 0$ holds in the region of stable points and $d\dot{A}/dA > 0$ holds in the region of unstable points. If $[\dot{A}]_M = 0$ is identically satisfied, which is the case for the example $\ddot{x} + \omega_0^2 x + hx^3 = 0$, then the mechanical system has orbital stability of the neutral type.

In the variation of constants method, the average value $[\dot{A}]_M$ is substituted for $\dot{A}$ and the stability is determined from the expression $d[\dot{A}]_M/dA$. This was already carried out for the example mentioned above.

Example: Let us consider the *Van der Pol equation*

$$\ddot{x} - (\alpha - \beta x^2)\dot{x} + x = 0$$

to show the exact calculation. Let us proceed according to Section 1.7.3. For $x = \xi_1$, $\dot{x} = \xi_2$, the differential equation is transformed into the system

$$\dot{\xi}_1 = \xi_2 \qquad \dot{\xi}_2 = (\alpha - \beta\xi_1^2)\xi_2 - \xi_1.$$

The trial solution $\xi_1 = A\cos(t + \theta) = A\cos\psi$, $\xi_2 = -A\sin\psi$ is substituted into the system of differential equations, where A and θ are allowed to be variables. The following is then obtained:

$$\dot{\xi}_1 = \dot{A}\cos\psi - A\sin\psi(1 + \dot{\theta}) = \xi_2 = -A\sin\psi,$$

$$\dot{\xi}_2 = -\dot{A}\sin\psi - A\cos\psi(1 + \dot{\theta}) = (\alpha - \beta\xi_1^2)\xi_2 - \xi_1,$$

$$= (\alpha - \beta A^2\cos^2\psi)(-A\sin\psi) - A\cos\psi,$$

from which it follows that

$$\dot{A}\cos\psi - A\dot{\theta}\sin\psi = 0,$$

$$-\dot{A}\sin\psi - A\dot{\theta}\cos\psi = -A(\alpha - \beta A^2\cos^2\psi)\sin\psi.$$

These are the determining equations for $\dot{A}$, $\dot{\theta}$, from which the following expression is calculated:

$$\dot{A} = A(\alpha - \beta A^2\cos^2\psi)\sin^2\psi.$$

The final result is

$$[\dot{A}]_M = \frac{1}{2\pi}\int_0^{2\pi} A(\alpha - \beta A^2\cos^2\psi)\sin^2\psi \, d\psi = \frac{\alpha A}{2}\left(1 - \frac{\beta}{\alpha}\frac{A^2}{4}\right).$$

The stationary values of A are $A = 0$ and $A_{st} = 2(\alpha/\beta)^{1/2}$. The first value is the trivial one. The second value is the amplitude of a possible periodic solution. The properties of the expression

$$\frac{d[\dot{A}]_M}{dA} = \frac{\alpha}{2} - \frac{3}{8}\beta A^2$$

are determined for A_{st}. We have

$$\{d[\dot{A}]_M/dA\}_{A=A_{st}} = -\alpha,$$

which we shall assume to be negative for positive α. It follows that the corresponding periodic solution has *orbital stability*.

Let us now consider the application of the *perturbation method*. The calculation of *parameter-dependent stability regions* will be shown by means of an example.

Example: Let us consider the differential equation

$$\ddot{x} + \lambda^2[1 + \mu f(t)]x = 0,$$

where λ^2 and μ are parameters and the function $f(t)$ is *periodic* with period ω. It is assumed that this function can be expanded into the series $f_i(t)\mu^i$, $i = 0, 1, 2, 3, \ldots$, according to powers of μ.

The substitutions $\xi_1 = x$, or $\xi_2 = \dot{x}$ transform the differential equation into the system $\dot{\xi}_1 = \xi_2$, $\dot{\xi}_2 = -\lambda^2(1 + \mu f)\xi_1$ or $\xi_i = p_{ik}\xi_k (i, k = 1, 2)$, respectively, where $p_{11} = 0$, $p_{12} = 1$, $p_{21} = -\lambda^2(1 + \mu f)$, $p_{22} = 0$. This is the same situation as in the example in Section 1.7.4 and the results reported there can be used here. The magnitude of $\frac{1}{2}[\xi_1^{(1)}(\omega) + \dot{\xi}_1^{(2)}(\omega)]$ determines stability, where $\xi_1^{(1)}$, $\xi_1^{(2)}$ are fundamental solutions of the system that satisfy the initial conditions:

$$\xi_1^{(1)}(0) = 1, \qquad \dot{\xi}_1^{(1)}(0) = 0, \qquad \xi_1^{(2)}(0) = 0, \qquad \dot{\xi}_1^{(2)}(0) = 1.$$

Since $\xi_1 = x$, we can also use the old notation and can use the magnitude of $K = \frac{1}{2}[x_1(\omega) + \dot{x}_2(\omega)]$ to determine stability, where x_1 and x_2 are two solutions of $\ddot{x} + \lambda^2[1 + \mu f(t)]x = 0$ which satisfy the corresponding initial conditions. There is stability for $|K| < 1$ and instability for $|K| > 1$. The regions of stability and instability are separated by the condition $|K| = 1$. Let us assume that the convergent series expansion

$$x_i = x_{ik}\mu^k, \qquad i = 1, 2, \qquad k = 0, 1, 2, 3, \ldots,$$

is possible for $|\mu| < \varepsilon$. Substituting these series into the differential equation, taking $f = f_k\mu^k$, $k = 0, 1, 2, 3, \ldots$, into account and setting the coefficients of equal powers of μ equal to zero, we obtain the following system of differential equations:

$$\ddot{x}_{i0} = -\lambda^2 x_{i0}, \qquad \ddot{x}_{ik} = -\lambda^2\left(x_{ik} + \sum_{s=0}^{k-1} f_{k-s-1}x_{is}\right), \qquad i = 1, 2, k = 1, 2, 3, \ldots,$$

Fig. 45. Stability map: $\lambda_k(\mu) = [n + a_k(\mu)]\,\pi/\omega$

with the initial conditions

$$x_{10}(0) = 1, \qquad\qquad \dot{x}_{10}(0) = 0,$$

$$x_{20}(0) = 0, \qquad\qquad \dot{x}_{20}(0) = 1, \qquad i = 1, 2,\ k = 1, 2, 3, \ldots.$$

$$x_{ik}(0) = \dot{x}_{ik}(0) = 0,$$

In particular, $x_{10} = \cos \lambda t$, $x_{20} = (1/\lambda)\sin \lambda t$ so that

$$K = \cos \lambda\omega + \tfrac{1}{2}[x_{ik}(\omega) + \dot{x}_{2k}(\omega)]\mu^k, \qquad k = 1, 2, 3, \ldots. \qquad (2.2.6)$$

For $K = (-1)^n$, $\lambda = (n + \alpha)\pi/\omega$, Eq. (2.2.6) results in a sequence of α values that prescribe the stability region boundaries in the λ, μ plane. This sequence is given by the roots $\alpha_k(\mu)$ of the equation

$$(-1)^n = (-1)^n\left(1 - \frac{\alpha^2\pi^2}{2!} + \frac{\alpha^4\pi^4}{4!} - \ldots\right)$$

$$+ \tfrac{1}{2}[x_{11}(\omega) + \dot{x}_{21}(\omega)]_{\lambda=(n+\alpha)\pi/\omega}\cdot\mu + \ldots.$$

The boundaries between the stability and instability regions can also be interpreted as level lines of the surface $K(\lambda, \mu)$ in the λ, μ plane (Fig. 45), which helps to clarify the calculation.

2.2.2 Stability of Control Systems

Stability investigations are important in control theory. In the design and optimization of loops one of the important questions is to design the control system in such a way that it will be stable for all cases, at least in the large. The reader is referred to the books on control theory by Lefschetz (1.4), Letov

(1.10), Solodovnikov (2.5), Popov (2.12), and Oppelt (2.11). In this book it will only be possible to give a short discussion of the topic and to show a few applications of stability theory to control theory. First, a linear problem is discussed using the methods of Section 1.3.3. A non-linear problem will be discussed, which will be linearized using the harmonic balance method, and the method outlined in Section 1.3 will be applied. Finally, a non-linear problem will be investigated using the Lyapunov stability theorems.

Example 1: The *linear control loop* shown in the block diagram of Fig. 46 is given. The equation of motion for the dynamic behaviour of the control loop is

$$T_{s2}\ddot{x} + \dot{x} = s_1 \dot{y}_s + y_s \tag{2.2.7}$$

and the equation of motion for the controller is

$$T_{r2}\ddot{y}_R + T_{r1}\dot{y}_R + y_R = r_0(\omega - x). \tag{2.2.8}$$

In addition, $y_R = -y_s$, so that the following differential equation of the control loop is obtained from (2.2.7) and (2.2.8):

$$T_{s2}T_{r2}x^{IV} + (T_{r1}T_{s2} + T_{r2})\dddot{x} + (T_{r1} + T_{s2})\ddot{x} + (1 + r_0 s_1)\dot{x} + r_0 x$$
$$= r_0 w + r_0 s_1 \dot{w}. \tag{2.2.9}$$

It is assumed that $w = w_0 = $ const holds for the command variable. The stationary solution $x_0 = w_0$ is obtained from (2.2.9) and the stability of this solution will now be discussed. For this purpose, the substitution $x = x_0 + \xi$ is made into (2.2.9). The following variational equation is obtained:

$$T_{s2}T_{r2}\xi^{IV} + (T_{r1}T_{s2} + T_{r2})\dddot{\xi} + (T_{r1} + T_{s2})\ddot{\xi} + (1 + r_0 s_1)\dot{\xi} + r_0 \xi = 0. \tag{2.2.10}$$

It is assumed that the magnitudes of the system parameters T_{s2}, T_{r1}, T_{r2} are given by

$$T_{s2} = 0.10, \qquad T_{r1} = 0.15, \qquad T_{r2} = 0.01$$

in the appropriate dimensions. The parameters s_1 and r_0 must be selected in

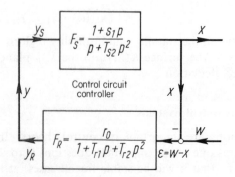

Fig. 46. Block diagram of a linear control loop

Fig. 47. Stability map in the parameter plane

such a way that the stationary solution x_0 is stable. Without first deriving a system of first-order differential equations, we find the following characteristic equation with the trial solution $x = e^{\lambda t}$ and the numerical values T_{s2}, T_{r1}, T_{r2} given above:

$$\lambda^4 + 25\lambda^3 + 250\lambda^2 + a_3\lambda + a_4 = 0, \qquad (2.2.11)$$

where $a_3 = (1 + r_0 s_1) \times 10^3$ and $a_4 = r_0 \times 10^3$. The *Hurwitz matrix* is

$$\mathbf{H} = \begin{bmatrix} 25 & 1 & 0 & 0 \\ a_3 & 250 & 25 & 1 \\ 0 & a_4 & a_3 & 250 \\ 0 & 0 & 0 & a_4 \end{bmatrix}$$

In order that stability may prevail, the principal minors of this matrix must all be positive. This leads to the conditions

$$a_3 > 0, \qquad a_4 > 0, \qquad 6250 a_3 - a_3^2 - 625 a_4 > 0.$$

The stability diagram is shown in Fig. 47. The missing system values

$$r_0 = a_4 \times 10^{-3} = 5,$$
$$s_1 = (a_3 \times 10^{-3} - 1)(1/r_0) = 2/5 = 0.4$$

are obtained by selecting a point from the stability region which is not cross-hatched (for example, P with $a_3 = 3 \times 10^3$, $a_4 = 5 \times 10^3$). If these values are used, the control loop is stable with respect to perturbations of the stationary solution.

Example 2: The *stability of periodic solutions* is often of interest. Let us consider the control loop represented by the block diagram shown in Fig. 48. The following equations are found from the block diagram:

$$x_1 = z - y, \qquad x_2 = F_L x_1, \quad y = F_{NL} x_2.$$

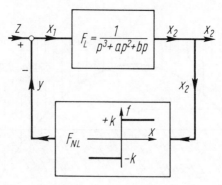

Fig. 48. Block diagram of a non-linear control loop

The frequency response of the control loop is calculated as follows:

$$F = F_L/(1 + F_L F_{NL}),$$

where $x_2 = Fz$. In order to investigate the stability of oscillations that occur in the non-linear system without any external influences, it is sufficient to consider the relationship

$$(1 + F_L F_{NL})x_2 = 0.$$

For

$$F_L = 1/(p^3 + ap^2 + bp)$$

and

$$F_{NL}x_2 \equiv f(x_2)$$

in

$$(p^3 + ap^2 + bp)x_2 + f(x_2) = 0,$$

it is transformed into the differential equation

$$\dddot{x}_2 + a\ddot{x}_2 + b\dot{x}_2 + f(x_2) = 0 \qquad (2.2.12)$$

because $p^n = \mathrm{d}^n/\mathrm{d}t^n$. In this example, $f(x)$ is given by the *non-linear* function $f(x) = k \operatorname{sgn} x$.

The *harmonic linearization* method is applied to (2.2.12) and $f(x_2) = cx_2$ with

$$c = (1/\pi A)\int_0^{2\pi} f(A \sin \phi)\sin \phi \, \mathrm{d}\phi = 4k/\pi A$$

is obtained. The harmonic linearized substitution equation

$$\dddot{x}_2 + a\ddot{x}_2 + b\dot{x}_2 + cx_2 = 0 \qquad (2.2.13)$$

can be used for the rest of the investigation. The book by Popov and Paltov (2.13) contains a very detailed treatment of this equation.

A periodic solution with amplitude A and frequency ω exists if $\lambda = \mathrm{i}\omega$

satisfies the characteristic equation (2.2.13). Therefore

$$(i\omega)^3 + a(i\omega)^2 + bi\omega + c = 0,$$

which leads to

$$X(A, \omega) + iY(A, \omega) = 0. \qquad (2.2.14)$$

The functions

$$X(A, \omega) = -a\omega^2 + c, \qquad Y(A, \omega) = -\omega^3 + b\omega$$

depend on A as well as ω, because at least some of the coefficients in (2.2.13) are functions of A. In our example, this holds true for the coefficient c. Equation (2.2.14) decomposes into the two determining equations

$$X(A, \omega) = 0, \qquad Y(A, \omega) = 0,$$

which must result in positive and real values for A and ω if the system is capable of any oscillations at all.

We obtain

$$-a\omega^2 + (4k/\pi A) = 0, \qquad -\omega^3 + b\omega = 0$$

with the solutions

$$\omega^2 = b, \ A = 4k/\pi ab. \qquad (2.2.15)$$

It is possible for oscillations to exist for $a > 0$, $b > 0$, provided they are stable. The stability is determined as follows. If the $(n-1)$th Hurwitz determinant H_{n-1} satisfies the condition

$$H_{n-1} = 0,$$

the system is on the oscillation boundary. Let us suppose that the amplitude of the oscillation is changed by ΔA. If this oscillation is stable, then a damped process must result for $\Delta A > 0$ and a process with negative damping must result for $\Delta A < 0$. Therefore, the following is a necessary condition for stability of the periodic solution with the amplitude A_0

$$(\partial H_{n-1}/\partial A)_{A=A_0} > 0.$$

Another necessary condition is that all Hurwitz determinants are positive with the exception of H_{n-1}.

In our case this amounts to the condition that the coefficients of (2.2.13) are positive (which is implied from the above condition $a > 0$, $b > 0$, because c is greater than zero in all cases) and that

$$\frac{\partial H_2}{\partial A} = \frac{\partial}{\partial A}(ab - c) = \frac{4k}{\pi A^2} > 0,$$

which is also satisfied in all cases.

It is therefore possible to conclude that for $a > 0$, $b > 0$ it is possible to have stable oscillations with the parameters (2.2.15). The *A-curve method* (Section

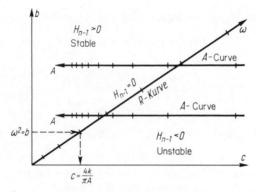

Fig. 49. The A-curve method

1.7.6) gives the same result. The R curve is given by $H_{n-1} \equiv H_2 = 0$ in the parameter plane b, c (Fig. 49). It follows from

$$H_2 = ab - c = 0.$$

that the R curve is the line

$$b = (1/a)c$$

passing through the origin. Along the oscillation boundary the square of the frequency satisfies

$$\omega^2 = a_n(H_{n-3}/H_{n-2}).$$

In our case $a_n = a_3 = c$, $H_{n-2} = H_1 = a$, $H_{n-3} = H_0 = 1$, so that

$$\omega^2 = c/a$$

is obtained. Because $H_{n-1} = H_2 = ab - c = 0$, we have $c/a = b$ so that the well-known relationship $\omega^2 = b$ is obtained in this way. It is therefore possible to plot the ω scale along the R curve shown in Fig. 49.

Since the parameter b is independent of A, the A curves are the horizontal lines, and the A scales are given by $A = 4k/\pi c$. Since the A curves intersect the R curve first in the unstable region and then the stable region as A is increased, $\partial R/\partial A \equiv H_2/\partial A > 0$ holds. The oscillations corresponding to the intersection points of the A curves with the R curves are all stable. These are the same stability conditions found earlier by calculation.

The amplitude $A = 4k/\pi c$ can be read from Fig. 49 for any $\omega^2 = b$. Because $c = ab$, $A = 4k/\pi ab$. There is a stable oscillation with a certain amplitude A for any arbitrary frequency ω. Thus, all the results of the previous calculation have been illustrated graphically.

Example 3: Finally let us consider the application of the *Lyapunov theory* to a *non-linear* problem of control theory (Fig. 50). Let us assume that the dif-

Fig. 50. Course of the non-linear function of a control loop

ferential equations of the control loop are given by

$$\ddot{x} + a\dot{x} + by = c,$$
$$\dot{y} = F(\sigma),$$
$$\sigma = ex + g\dot{x} + h\ddot{x} - ky. \tag{2.2.16}$$

The coefficients are constants $(a, b, c > 0)$ and $F(\sigma) > 0$ for $\sigma \neq 0$.

By means of the transformation $q_1 = x$, $q_2 = \dot{x}$, $q_3 = y$, (2.2.16) is transformed into the system

$$\dot{q}_1 = q_2,$$
$$\dot{q}_2 = -aq_2 - bq_3 + c,$$
$$\dot{q}_3 = F[eq_1 + (g - ah)q_2 - (k + bh)q_3 + ch]. \tag{2.2.17}$$

The following stationary solution is obtained for $\dot{q}_i = 0$, the stability of which will now be investigated:

$$q_1^0 = kc/be, \qquad q_2^0 = 0, \qquad q_3^0 = c/b. \tag{2.2.18}$$

The trial solution $\xi_i = q_i - q_i^0$, $\dot{\xi}_i = \dot{q}_i$ leads to the following variational equations from (2.2.17):

$$\dot{\xi}_1 = \xi_2,$$
$$\dot{\xi}_2 = -(a\xi_2 + b\xi_3),$$
$$\dot{\xi}_3 = F[e\xi_1 + (g - ah)\xi_2 - (k + bh)\xi_3]. \tag{2.2.19}$$

The following *positive definite* function V is selected as the *Lyapunov function:*

$$V = A\left(\frac{a}{b}\xi_2 + \dot{\xi}_3\right)^2 + B\xi_3^2 + C\int_0^{\sigma^*} F(\sigma^*)\,d\sigma^*,$$
$$\sigma^* = e\xi_1 + (g - ah)\xi_2 - (k + bh)\xi_3, \tag{2.2.20}$$

with the constants A, B, C which are assumed to be greater than zero. Let us now form

$$\frac{dV}{dt} = \frac{\partial V}{\partial \xi_1}\dot{\xi}_1 + \frac{\partial V}{\partial \xi_2}\dot{\xi}_2 + \frac{\partial V}{\partial \xi_3}\dot{\xi}_3,$$

which becomes, making use of (2.2.19) and (2.2.20),

$$\frac{dV}{dt} = -\frac{2aA}{b^2}(a\xi_2 + b\xi_3)^2 - C(k + bh)F^2(\sigma^*)$$

$$+ \left[\frac{2aA}{b} + eC - aC(g - ah)\right]F(\sigma^*)\xi_2$$

$$+ [2A + 2B - bC(g - ah)]F(\sigma^*)\xi_3.$$

If we set

$$(2aA/b) + C(e - ag + a^2h) = 0,$$

$$2(A + B) = bC(a - ah),$$

$$k + bh > 0, \qquad\qquad (2.2.21)$$

the remaining terms are

$$dV/dt = -(2aA/b_2)(a\xi_2 + b\xi_3)^2 - C(k + bh)F^2(\sigma^*),$$

which is a *negative semi-definite* function. The stationary solution (2.2.18) is therefore stable according to Theorem 1 of Section 1.5.3 for the assumptions (2.2.21).

If the first two conditions of (2.2.21) are solved for A and B which are assumed to be positive, it follows that

$$A = \frac{bC}{2a}(ag - e - a^2h) > 0, \qquad B = \frac{bC(g - ah) - 2A}{2} = \frac{beC}{2a} > 0.$$

These conditions together with the third condition of (2.2.21) and the fact that $a, b > 0$, leads to the condition

$$ag > e + a^2h,$$

$$e > 0,$$

$$k + bh > 0.$$

The solution (2.2.18) is stable if the parameters of the control loop satisfy these conditions.

2.2.3 Stability of the Gyroscope

Numerous important gyroscope problems of various types occur in technology. The stability of the gyroscope itself or of the gyroscope as a stabilizing element are problems which are often encountered.

Let us describe the gyroscopic motion in the rotating body-fixed reference system x, y, z along the principal axes of inertia. The following vector equation of motion is obtained from the angular momentum theorem:

$$\partial D/\partial t + \omega \times D = M \qquad (2.2.22)$$

$D = (D_x, D_y, D_z)$ is the angular momentum vector, $\omega = (\omega_x, \omega_y, \omega_z)$ the rotation vector, and $M = (M_x, M_y, M_z)$ the torque vector. The angular momentum components satisfy

$$D_x = A\omega_x, \qquad D_y = B\omega_y, \qquad D_z = C\omega_z,$$

where A, B, C are the principal moments of inertia. Therefore, the components of (2.2.22) are

$$A\dot{\omega}_x + (C - B)\omega_y\omega_z = M_x,$$

$$B\dot{\omega}_y + (A - C)\omega_z\omega_x = M_y,$$

$$C\dot{\omega}_z + (B - A)\omega_x\omega_y = M_z. \qquad (2.2.23)$$

Example 1: Let us first consider a special case of gyroscope motion without forces, so that $M \equiv 0$. Therefore, Eq. (2.2.23) results in

$$A\dot{\omega}_x + (C - B)\omega_y\omega_z = 0,$$

$$B\dot{\omega}_y + (A - C)\omega_z\omega_x = 0,$$

$$C\dot{\omega}_z + (B - A)\omega_x\omega_y = 0. \qquad (2.2.24)$$

One *stationary motion* is given by $\omega_{x0} = \text{const}$, $\omega_{y0} = \omega_{x0} = 0$ because these values satisfy the system (2.2.24). Let us investigate the stability of the motion by setting $\omega_x = \omega_{x0} + \xi_1$, $\omega_y = \xi_2$, $\omega_z = \xi_3$. The following non-linear variational equations are obtained from (2.2.24):

$$A\dot{\xi}_1 + (C - B)\xi_2\xi_3 = 0,$$

$$B\dot{\xi}_2 + (A - C)(\omega_{x0} + \xi_1)\xi_3 = 0,$$

$$C\dot{\xi}_3 + (B - A)(\omega_{x0} + \xi_1)\xi_2 = 0. \qquad (2.2.25)$$

If the investigation is carried out by the *small oscillation method*, it is necessary to linearize the problem by setting products of the variations equal to zero. Equation (2.2.25) is therefore simplified as follows:

$$\dot{\xi}_1 = 0,$$

$$\dot{\xi}_2 = [(C - A)/B]\,\omega_{x0}\xi_3,$$

$$\dot{\xi}_3 = [(A - B)/C]\,\omega_{x0}\xi_2. \qquad (2.2.26)$$

The linearized system (2.2.26) has the characteristic equation

$$\begin{vmatrix} -\lambda & 0 & 0 \\ 0 & -\lambda & [(C - A)/B]\,\omega_{x0} \\ 0 & [(A - B)/C]\,\omega_{x0} & -\lambda \end{vmatrix} = 0,$$

$$\lambda\left[\lambda^2 + \frac{(A - C)(A - B)}{BC}\,\omega_{x0}^2\right] = 0.$$

The roots of this equation are

$$\lambda_1 = 0, \qquad \lambda_{2,3} = \pm \, \omega_{x0}[(C - A)(A - B)/BC]^{1/2}.$$

If $C < A < B$ or $C > A > B$, then one of the roots λ_i has a positive real part. According to Theorem 5 of Section 1.5.3, the instability of the stationary solution can be predicted with certainty from the linearized equations (2.2.26). This stationary motion consists of rotation about the intermediate principal axis of inertia.

Two imaginary roots are obtained for $A < B < C$ or $A > B > C$, which are rotations around the axis corresponding to the smallest or largest principal moment of inertia, respectively. Since $\lambda_1 = 0$ always holds, this is the *critical case*. It is now impossible to determine stability or instability from the linearized equations (2.2.26), and we must return to the non-linear variational equations (2.2.25).

The following positive definite *Lyapunov function* is used in the case $A < B < C$ (rotation about the axis corresponding to the *smallest* principal momentum of inertia):

$$V = B(A - B)\xi_2^2 + C(A - C)\xi_3^2 + [B\xi_2^2 + C\xi_3^2 + A(\xi^2 + 2\xi_1\omega_{x0})]^2.$$

It is easy to see that $\dot V \equiv 0$, so that the stationary solution is stable according to Theorem 1 of Section 1.5.3.

The following positive definite Lyapunov function is used for the rotation around the axis corresponding to the largest principal moment of inertia for $A > B > C$:

$$V = B(A - B)\xi_2^2 + C(A - C)\xi_3^2 + [B\xi_2^2 + C\xi_3^2 + A(\xi_1^2 + 2\xi_1\omega_{z0})]^2.$$

$\dot V \equiv 0$ also holds, so that the stationary motion is stable.

Example 2: Let us now consider a rocket controlled by an internal rotor. This system can be considered as a *gyroscope with rotation rate-dependent self-excitation.*[1]

The equations of motion are

$$A\dot\omega_x + (C - B)\omega_y\omega_z = 0,$$

$$B\dot\omega_y + (A - C)\omega_z\omega_x = -D\omega_z,$$

$$C\dot\omega_z + (B - A)\omega_x\omega_y = D\omega_y, \qquad (2.2.27)$$

where A, B, C, and D are constant parameters of the mechanical system.

One *stationary solution* of (2.2.27) is given by

$$\omega_{x0} = [D/(C - A)], \quad \omega_{y0} = 0, \quad \omega_{z0} = 0.$$

Let us investigate the stability of this solution. The equations $\omega_x = \omega_{x0} + \xi_1$ $\omega_y = \xi_2$, $\omega_z = \xi_3$ are substituted into (2.1.54) and the following non-linear varia-

---

[1]Leipholz, H., A contribution to the problem of a gyroscope with rotation rate dependent self-excitation, *Ing. Arch.* **32**, 255–285, 1963.

tional equations are obtained:

$$A\dot{\xi}_1 + (C - B)\xi_2\xi_3 = 0,$$

$$B\dot{\xi}_2 + (A - C)(\omega_{x0} + \xi_1)\xi_3 = -D\xi_3,$$

$$C\dot{\xi}_3 + (B - A)(\omega_{x0} + \xi_1)\xi_2 = D\xi_2. \qquad (2.2.28)$$

There is no point in linearizing these equations because it is not possible to determine stability with certainty from the linearized equations, as is easily derived. Therefore, the system (2.2.28) is used, which has the following *integrals:*

$$B\xi_2^2 + C\xi_3^2 + A(\omega_{x0} + \xi_1)^2 = \text{const},$$

$$B^2\xi_2^2 + C^2\xi_3^2 + A^2[(\omega_{x0} + \xi_1) + D/A]^2 = \text{const}.$$

By using combinations of these integrals, it is possible to *obtain a Lyapunov function.* Let us select

$$V = B(K - B)\xi_2^2 + C(K - C)\xi_3^2 + A(K - A)\xi_1^2 + 2A[\omega_{x0}(K - A) - D]\xi_1.$$

$\dot{V} = 0$ holds, because V is made up of integrals of (2.2.28). V becomes positive definite if

$$K > B, \qquad K > C, \qquad K > A, \qquad K = (A\omega_{x0} + D)/\omega_{x0}$$

holds simultaneously. These conditions are equivalent to

$$D > 0, \qquad \omega_{x0} < D/(C - A), \qquad \omega_{x0} < D/(B - A). \qquad (2.2.29)$$

These are the stability conditions for the stationary motion given above. If (2.2.29) are satisfied, V is positive definite and the motion is stable according to Theorem 1 of Section 1.5.3.

Example 3: Let us now investigate the stability of a *monorail.* The *Scherl* system contains a vertical gyroscopic pendulum. Fig. 51 shows a diagram of the system. θ_W, M_W are the moments of inertia and mass of the carriage, θ_K, M_K are the moments of inertia and mass of the the pendulum including the frame, and D is the constant angular momentum of the gyroscope. In addition, it is assumed that the rotation angles and their derivatives α, β, $\dot{\alpha}$, $\dot{\beta}$ are small quantities. Under these assumptions, the angular momentum vector

$$D = (\theta_W\dot{\alpha} + D\beta, \theta_K\dot{\beta} - D\alpha, D)$$

and the torque vector

$$M = (M_W g s_1\alpha, M_K g s_2\beta, 0)$$

and $dD/dt = M$ lead to the following scalar equations:

$$\theta_W\ddot{\alpha} + D\dot{\beta} = M_W g s_1\alpha,$$

$$\theta_K\ddot{\beta} - D\dot{\alpha} = M_K g s_2\beta. \qquad (2.2.30)$$

Fig. 51. Monorail

The characteristic equation of the system is

$$\begin{vmatrix} \theta_W \lambda^2 - M_W g s_1 & D\lambda \\ -D\lambda & \theta_K \lambda^2 - M_K g s_2 \end{vmatrix} = 0,$$

or

$$\theta_W \theta_K \lambda^4 + (D^2 - \theta_W M_K g s_2 - \theta_K M_W g s_1) \lambda^2 + M_W M_K g^2 s_1 s_2 = 0.$$

All the Hurwitz determinants of this fourth-order equation are identically zero, as can easily be calculated. Therefore, the necessary condition for asymptotic stability and for the applicability of the linearized equations (2.2.30), which would imply stability (that is, roots with negative real parts only) can never be satisfied. Therefore a moment generator must be added to the system, which introduces a moment around the y axis proportional to α. It is also necessary to have damping about the x axis proportional to the angular rate $\dot{\alpha}$. Differential equations (2.2.30) then become

$$\theta_W \ddot{\alpha} + D\dot{\beta} + K\dot{\alpha} = M_W g s_1 a,$$

$$\theta_K \ddot{\beta} - D\dot{\alpha} = M_K g s_2 \beta + R\alpha, \qquad (2.2.31)$$

with the characteristic equation

$$\begin{vmatrix} \theta_W \lambda^2 + K\lambda - M_W g s_1 & D\lambda \\ -R - D\lambda & \theta_K \lambda^2 - M_k g s_2 \end{vmatrix} = 0.$$

They can also be written

$$\theta_M \theta_K \lambda^4 + K\theta_K \lambda^3 + (D^2 - \theta_W M_K g s_2 - \theta_K M_W g s_1)\lambda^2$$
$$+ (DR - M_K g s_2 K)\lambda + M_W M_K g^2 s_1 s_2 = 0.$$

The following conditions must be satisfied for asymptotic stability according to the *Hurwitz criterion:*

$$D^2 > M_W g s_1 \theta_K + M_K g s_2 \theta_W,$$

$$DR > M_K g s_2 K,$$

$$K\theta_K (D^2 - M_W g s_1 \theta_K - M_K g s_2 \theta_W)(DR - M_K g s_2 K)$$
$$- \theta_W \theta_K (DR - M_K g s_2 K)^2 - K^2 \theta_k^2 M_W M_K g^2 s_1 s_2 > 0.$$

As can be seen, all three conditions can be simultaneously satisfied by a sufficiently large angular momentum D of the gyroscope after all other parameters of the system have been specified. The characteristic equation then has only roots with negative real parts. Stability prevails and the stability can be determined with the linearized equations (2.2.31).

Example 4: Finally, let us consider a *gyroscopic pendulum with an erection motor* (Fig. 52). There are friction moments about the x and y axes proportional to the angular rates $\dot{\alpha}$, $\dot{\beta}$. The supporting moment is about the y axis and is a non-linear function of the angle α. Weight forces are neglected. The quantities α, β, $\dot{\alpha}$, $\dot{\beta}$ are again assumed to be small, and the rotor has the angular momentum D. The equations of motion are obtained in exactly the same way as in the previous example from

$$\boldsymbol{D} = (\theta_A \dot{\alpha} + D\beta, \ \theta_J \dot{\beta} - D\alpha, \ D),$$

$$\boldsymbol{M} = (-K_x \dot{\alpha}, \ F(\alpha) - K_y \dot{\beta}, \ 0),$$

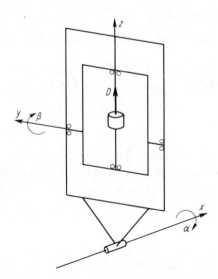

Fig. 52. Gyroscopic pendulum with erection motor

and $dD/dt = M$. They are given by

$$\theta_A\ddot{\alpha} + D\dot{\beta} + K_x\dot{\alpha} = 0,$$

$$\theta_J\ddot{\beta} - D\dot{\alpha} + K_y\dot{\beta} = F(\alpha).$$

θ_A is the moment of inertia of the entire system with respect to the x axis, θ_J the moment of inertia of the system consisting of the rotor and the inner frame with respect to the y axis, and K_x, K_y are the damping coefficients.

Equation (2.2.32) can be transformed into a third-order equation with respect to α

$$\theta_A\theta_J\dddot{\alpha} + (\theta_JK_x + \theta_AK_y)\ddot{\alpha} + (D^2 + K_xK_y)\dot{\alpha} + DF(\alpha) = 0.$$

The abbreviations

$$(1/\theta_A\theta_J)(\theta_JK_x + \theta_AK_y) = a, \qquad (1/\theta_A\theta_J)(D^2 + K_xK_y) = b, \qquad (1/\theta_A\theta_J)D = e$$

and substitutions $x = \alpha$ and $eF(\alpha) = f(\alpha)$ lead to

$$\dddot{x} + a\ddot{x} + b\dot{x} + f(x) = 0,$$

which is exactly the differential equation (2.2.12) of Section 2.2.2. It is not necessary to discuss this equation again. The existence of stable oscillations of a supported gyroscopic pendulum can be determined according to the discussion in Section 2.2.2.

2.2.4 Stability of Aircraft, Rockets, Satellites

Example 1: Let us first consider the center-of-gravity motion of an aircraft moving in the x, z plane as shown in Fig. 53. The aircraft is symmetrical with respect to this plane. The velocity of the aircraft center of gravity is v, the weight of the aircraft is $G = mg$, and the angle between the trajectory and the x axis is β.

The equations of motion of the center of gravity with respect to the axis parallel and perpendicular to the trajectory are given by

$$m\dot{v} = -W + T + mg \sin \beta,$$

$$mv\dot{\beta} \cong -A + \mu g \cos \beta. \qquad (2.2.33)$$

W is the drag, T the component of the propulsion force in the flight direction, and A is the lift, which includes the component of the propulsion force perpendicular to the trajectory. It is assumed that $T = W$. The first equation in (2.2.33) can be integrated by first multiplying by v and noting that $v \sin \beta = \dot{z}$. After integration the result is

$$v^2 = 2gz + c, \qquad (2.2.34)$$

where c is a constant of integration.

In addition, let us assume that the lift A is proportional to the square velocity, $A = kv^2$. For $T = W$, the system $\beta = 0$, $v = v_0 = $ const, $A = mg$ is a

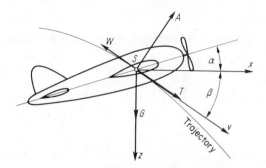

Fig. 53. Forces on the aircraft

solution. Therefore, $mg = kv_0^2$ must hold and the final result is

$$A = mg(v^2/v_0^2). \tag{2.2.35}$$

Furthermore

$$\frac{d\beta}{dt} = \frac{d\beta}{dz}\frac{dz}{ds}\frac{ds}{dt} = v\frac{d\beta}{dz}\sin\beta. \tag{2.2.36}$$

Substitution of (2.2.34), (2.2.35), and (2.2.36) into the second equation of (2.2.33) results in

$$(2gz + c)(d\beta/dz)\sin\beta = -(g/v_0^2)(2gz + c) + g\cos\beta.$$

It follows from $q_1 = \beta$, $q_2 = 2gz + c$, $dz = dq_2/2g$ that

$$\frac{dq_1}{dq_2} = \frac{\cos q_1 - 1/v_0^2 \cdot q_2}{2q_2\sin q_1} \tag{2.2.37}$$

or

$$\dot{q}_1 = (1/\alpha)[\cos q_1 - (1/v_0^2)q_2], \qquad \dot{q}_2 = (1/\alpha)2q_2\sin q_1,$$

where α is a proportionality factor.

There are stationary solutions to Eq. (2.2.37) for $\dot{q}_1 = \dot{q}_2 = 0$, which are obtained from the equations

$$\cos q_1 - (1/v_0^2)q_2 = 0, \qquad 2q_2\sin q_1 = 0.$$

As was mentioned previously, one of the solutions is

$$q_{1s} = \beta = 0, \qquad q_{2s} = 2gz_s + c = v_0^2. \tag{2.2.38}$$

This corresponds to constant velocity at a constant altitude $z_s = (v_0^2 - c)/2g$.

Let us carry out the discussion in the *phase plane*. According to Section 1.4, the following abbreviated linear variational equations may be used in the immediate vicinity of the stationary point:

$$\dot{\xi}_1 = (\partial F_i/\partial q_k)_{qs} \cdot \xi_k.$$

This is a simple stationary point.

In our case,

$$F_1(q_1, q_2) = (1/\alpha)(\cos q_1 - 1/v_0^2 \cdot q_2), \quad F_2(q_1, q_2) = (1/\alpha)2q_2 \sin q_1$$

and therefore

$$\left(\frac{\partial F_1}{\partial q_1}\right)_{qs} = \frac{1}{\alpha}(-\sin q_1)_{\substack{q_1=0 \\ q_2=v_0^2}} = 0, \qquad \left(\frac{\partial F_1}{\partial q_2}\right)_{qs} = -\left(\frac{1}{\alpha v^2}\right)_{\substack{q_1=0 \\ q_2=v_0^2}} = -\frac{1}{\alpha v_0^2}$$

$$\left(\frac{\partial F_2}{\partial q_1}\right)_{qs} = \frac{1}{\alpha}(2q_2 \cos q_1)_{\substack{q_1=0 \\ q_2=v_0^2}} = \frac{2v_0^2}{\alpha}, \qquad \left(\frac{\partial F_1}{\partial q_2}\right)_{qs} = \frac{1}{\alpha}(2 \sin q_1)_{\substack{q_1=0 \\ q_2=v_0^2}} = 0,$$

which means that it is indeed such a point, because not all the derivatives vanish simultaneously. Therefore

$$\dot{\xi}_1 = -(1/\alpha v_0^2), \qquad \dot{\xi}_2 = (2v_0^2/\alpha)\xi_1$$

can be used. The integration of this equation results in

$$\xi_1^2 + (1/2v_0^4)\xi_2^2 = \text{const},$$

which is the equation for the phase curves surrounding the stationary point. Since they are ellipses, the stationary point is a stable center. It follows that stationary flight at constant altitude is stable for the assumptions made above.

Example 2: Let us now consider the motion of a *rocket* without propulsion. The rocket is symmetrical with respect to the ζ axis and rotates at the angular rate ω (Fig. 54). As soon as the axis of rotation ζ deviates from the direction z of the velocity vector v by the angle θ, the drag moment W affects the rocket. Let us determine whether the special motion of the rocket given by $\theta = 0$ is stable and determine whether it is insensitive to small disturbances which result from the moment caused by W.

An orthogonal axis system is used to describe the rocket motion and the

Fig. 54. Rocket stabilized by angular momentum

Fig. 55. Axis angles and angular rates

deflection of the rocket axis by the angle θ. This system is comprised of the ζ axis, the node axis ξ and the orthogonal axis η (Fig. 55). The rotation of this system is given by the angular rates

$$\omega_\xi = \dot{\theta}, \qquad \omega_\eta = \dot{\psi}\sin\theta, \qquad \omega_\zeta = \dot{\psi}\cos\theta. \qquad (2.2.39)$$

The angular momentum vector has the components

$$D_\xi = A\dot{\theta}, \qquad D_\eta = A\dot{\psi}\sin\theta, \qquad D_\zeta = C(\omega + \dot{\psi}\cos\theta), \qquad (2.2.40)$$

where A, A, C are the corresponding moments of inertia.

The equations of motion are obtained from the scalar components of the vector equation

$$M = \partial D/\partial t + \omega \times D$$

and from $M = (Wl\sin\theta, 0, 0)$. This result is

$$Wl\sin\theta = A\ddot{\theta} + C(\omega + \dot{\psi}\cos\theta)\dot{\psi}\sin\theta - A\dot{\psi}^2\sin\theta\cos\theta,$$

$$0 = A(d/dt)(\dot{\psi}\sin\theta) + A\dot{\psi}\cos\theta - C\dot{\theta}(\omega + \dot{\psi}\cos\theta),$$

$$0 = C(d/dt)(\omega + \dot{\psi}\cos\theta). \qquad (2.2.41)$$

Two integrals are used to reduce the system. The first integral is obtained from the third line of (2.2.41):

$$\omega + \dot{\psi}\cos\theta = n = \text{const.} \qquad (2.2.42)$$

The second one is the energy integral

$$E = T + U = \tfrac{1}{2}Cn^2 + \tfrac{1}{2}A(\dot{\theta}^2 + \dot{\psi}^2\sin^2\theta) + Wl\cos\theta, \qquad (2.2.43)$$

which is transformed in the following way. We have $M_z = 0$, and the corresponding angular momentum component satisfies

$$D_z = D_\zeta\cos\theta + D_\eta\sin\theta = Cn\cos\theta + A\dot{\psi}\sin^2\theta = \text{const.} \qquad (2.2.44)$$

It follows that

$$\dot{\psi} = (D_z - Cn\cos\theta)/A\sin^2\theta,$$

which is substituted into Eq. (2.2.43). The result is

$$E - \frac{Cn^2}{2} = \frac{A\dot{\theta}^2}{2} + \frac{(D_z - Cn\cos\theta)^2}{2A\sin^2\theta} + Wl\cos\theta. \qquad (2.2.45)$$

By means of the substitutions

$$(2/A)(E - (Cn^2/2)) = \alpha, \qquad 2Wl/A = \beta,$$
$$D_z/A = \gamma, \qquad Cn/A = D, \qquad \cos\theta = u, \qquad (2.2.46)$$

Eq. (2.2.45)) becomes

$$\dot{u}^2 = (1 - u^2)(\alpha - \beta u) - (\gamma - Du)^2 = f(u). \qquad (2.2.47)$$

This last relationship makes it possible to describe the rocket motion in the $\dot{u}$, u plane by using the phase curves $\dot{u} = [f(u)]^{1/2}$. The characteristics of the 'potential function' $f(u)$ determine the course of any phase curve. A motion which is physically possible only results if $f(u)$ is positive, because $\dot{u}$ must be real (Fig. 56). The stationary solution for $\dot{u} \equiv 0$ is obtained from (2.2.47). It is given by $u = 1$ because this makes the first term on the right-hand side of (2.2.47) equal to zero. However, for $u = 1$ we also have $\theta = 0$ and therefore $D_z = C_n$ according to (2.2.44), which means that $\gamma = D$ according to (2.2.46). This means that the second term on the right-hand side of (2.2.47) is also equal to zero and therefore $f(u) = 0$, which must be the case for the stationary solution. The motion of the rocket corresponding to $u = 1$, $\theta = 0$, the stability of which is being investigated, is exactly the stationary solution of (2.2.47). The *phase diagram* (Fig. 56) shows that $f(u)$ must have *a maximum* at $u = 1.0$ for stability to prevail. This guarantees that a disturbed motion will occur with u values (θ values) that remain sufficiently close to the stationary values $u = 1$ ($\theta = 0$).

$f(u)$ has a maximum at $u = 1$ if the conditions $f'(1) = 0$ and $f''(1) < 0$ are satisfied.

We have $\mathrm{d}f/\mathrm{d}u = -2u(\alpha - \beta u) - \beta(1 - u^2) + 2D(\gamma - Du)$. This expression

Fig. 56. Potential function and phase diagram

is zero for $u = 1$ provided that $\alpha = \beta$. This is because we can make use of the fact that $\gamma = D$ holds for $u = 1$.

The other condition at $u = 1$ is

$$d^2 f/du^2 = -2(\alpha - \beta u) + 2\beta u + 2\beta u - 2D^2 < 0.$$

We make use of the relationship $\alpha = \beta$ so that

$$f''(1) = 4\beta - 2D^2 < 0.$$

It follows that

$$\beta < \tfrac{1}{2}D^2.$$

Because of (2.2.46), this inequality becomes

$$n > (2/C)(WlA)^{1/2}. \tag{2.2.48}$$

The condition (2.2.48) causes the function $f(u)$ to be stable — that is, to have a maximum at $u = 1(\theta = 0)$. If the rocket rotates around the z axis, which coincides with the ζ axis in this case, with an angular rate n satisfying (2.2.48), the rocket motion is stable.

Example 3: The *flutter of airfoils* is an important problem in flight dynamics. Flutter is an instability phenomenon which can occur at certain critical flight velocities. This problem really belongs in the field of aero-elasticity. However, it is possible to discuss a simplified version of the problem in this section without losing sight of its characteristic features.

Let us consider the problem shown in Fig. 57. The wing carries out torsional oscillations around the axis F whose amplitude is the angle q_1. The rudder is deflected by the angle q_2 about the axis R. The centers of gravity S_F and S_R are located along the axis. J_F is the moment of inertia of the wing with respect to the axis F and J_R the moment of inertia of the rudder with respect to the

Fig. 57. Flutter of an airfoil

axis R, m_R is the mass of the rudder, and l is the distance between the axis F and R. The expression for the kinetic energy of the system is

$$T = \tfrac{1}{2}(J_F \dot{q}_1^2 + J_R \dot{q}_2^2 + m_R l^2 \dot{q}_1^2). \tag{2.2.49}$$

The generalized forces applied to the system are

$$Q_1 = -(k - c_{11}v^2)q_1 - c_{12}v^2 q_2, \qquad Q_2 = c_{21}v^2 q_1 - c_{22}v^2 q_2. \tag{2.2.50}$$

These forces are restoring moments, aerodynamic moments, and rotational spring moments about the axis R. All coefficients in (2.2.50) are constant and positive. v is the velocity of the wing.

Substitution of (2.2.49) and (2.2.50) into the Lagrange equations

$$\frac{\mathrm{d}}{\mathrm{d}t}\left(\frac{\partial T}{\partial \dot{q}_k}\right) - \frac{\partial T}{\partial \dot{q}_k} = Q_k$$

results in the system of differential equations

$$(J_F + m_R l^2)\ddot{q}_1 = (c_{11}v^2 - k)q_1 - c_{12}v^2 q_2,$$
$$J_R \ddot{q}_2 = c_{21}v^2 q_1 - c_{22}v^2 q_2. \tag{2.2.51}$$

One stationary solution is given by $q_{10} = q_{20} = 0$, and the stability of this solution will now be investigated. The following system of variational equations is obtained from (2.2.51) by means of the trial solution $q_i = q_{i0} + \xi_i$:

$$(J_F + m_R l^2)\ddot{\xi}_1 = (c_{11}v^2 - k)\xi_1 - c_{12}v^2 \xi_2,$$
$$J_R \ddot{\xi}_2 = c_{21}v^2 \xi_1 - c_{22}v^2 \xi_2.$$

The characteristic equation is

$$\begin{vmatrix} (J_F + m_R l^2)\lambda^2 + (k - c_{11}v^2) & c_{12}v^2 \\ -c_{21}v^2 & J_R\lambda^2 + c_{22}v^2 \end{vmatrix} = 0.$$

It is given by

$$(J_F + m_R l^2)J_R \lambda^4 + [(J_F + m_R l^2)c_{22}v^2 + J_R(k - c_{11}v^2)]\lambda^2$$
$$+ (k - c_{11}v^2)c_{22}v^2 + c_{12}c_{21}v^4 = 0.$$

The roots of this equation are

$$\lambda_{1,2,3,4} = \pm\left\{-\frac{(J_F + m_R l^2)c_{22}v^2 + J_R(k - c_{11}v^2)}{2(J_F + m_R l^2)J_R}\right.$$
$$\pm\left[\frac{[(J_F + m_R l^2)c_{22}v^2 + J_R(k - c_{11}v^2)]^2}{4(J_F + m_R l^2)^2 J_R^2}\right.$$
$$\left.\left.-\frac{(k - c_{11}v^2)c_{22}v^2 + c_{12}c_{21}v^4}{(J_F + m_R l^2)J_R}\right]^{1/2}\right\}^{1/2}. \tag{2.2.52}$$

By means of abbreviations

$$\frac{(J_F + m_R l^2)c_{22}v^2 + J_R(k - c_{11}v^2)}{(J_F + m_R l^2)J_R} = a,$$

$$\frac{(k - c_{11}v^2)c_{22}v^2 + c_{12}c_{21}v^4}{(J_F + m_R l^2)J_R} = b, \qquad (2.2.53)$$

Eq. (2.2.52) becomes

$$\lambda_{1,2,3,4} = \pm \left[\tfrac{1}{2} a \pm (\tfrac{1}{4} a^2 - b)^{\frac{1}{2}} \right]^{\frac{1}{2}}. \qquad (2.2.54)$$

The characteristic exponents are all purely imaginary if the radicand of the outer square root is negative and real. In this case, there are four different and simple roots which have zero real parts, which means that stability prevails according to Theorem 3 of Section 1.3.1. On the other hand, if the radicant of the outer square root is positive or complex, roots λ having a positive real part exist. The stationary solution is then unstable according to Theorem 2 of Section 1.3.1.

Therefore, the following two conditions

$$\tfrac{1}{4} a^2 - b > 0, \qquad \tfrac{1}{2} a > (\tfrac{1}{4} a^2 - b)^{\frac{1}{2}} \qquad (2.2.55)$$

must be satisified for stability.

Let us consider the first condition. According to (2.2.53)

$$[(L_F + m_R l^2)c_{22}v^2 + J_R(k - c_{11}v^2)]^2$$
$$- 4(J_F + m_R l^2)J_R[(k - c_{11}v^2)c_{22}v^2 + c_{12}c_{21}v^4] > 0.$$

This can also be written in the form

$$[(J_F + m_R l^2)c_{22}v^2 - J_R(k - c_{11}v^2)]^2 > 4(J_F + m_R l^2)J_R c_{12}c_{21}v^4.$$

This inequality is satisfied by

$$(J_F + m_R l^2)c_{22}v^2 - J_R(k - c_{11}v^2) > 2v^2[(J_F + m_R l^2)J_R c_{21}c_{21}]^{\frac{1}{2}} \qquad (2.2.56)$$

as well as by

$$J_R(k - c_{11}v^2) - (J_F + m_R l^2)c_{22}v^2 > 2v^2[(J_F + m_R l^2)J_R c_{12}c_{21}]^{\frac{1}{2}}. \qquad (2.2.57)$$

If the first condition in (2.2.55) is satisfied, the second one is also satisfied provided that b is positive. This is true for $k > c_{11}v^2$ according to (2.2.53), which is always satisfied for the velocity region under consideration. Therefore, the question of stability is determined solely from (2.2.56) and (2.2.57). Therefore, according to (2.2.56), flutter cannot occur for

$$v^2 > \frac{J_R k}{(J_F + m_R l^2)c_{22} + J_R c_{11} - 2[(J_F + m_R l^2)J_R c_{12}c_{21}]^{\frac{1}{2}}} = v^2_{\text{crit}, 0}$$

and according to (2.2.57), cannot occur for

$$v^2 < \frac{J_R k}{(J_F + m_R l^2)c_{22} + J_R c_{11} + 2[(J_F + m_R l^2)J_R c_{12}c_{21}]^{\frac{1}{2}}} = v^2_{\text{crit}, u}.$$

188

On the other hand, flutter can be expected for velocities in the range

$$v_{\text{crit}, u} < v < v_{\text{crit}, 0}.$$

This velocity range must be avoided.

Example 4: As a final example, let us consider the stability of motion of a *satellite.* We will follow a method developed by Magnus.[1]

Let 0 be the common center of mass of the earth and the satellite, 0_E the center of mass of the earth, 0_S the center of mass of the satellite, dm a mass element of the satellite, m_E the mass of the earth, m_S the mass of the satellite, $m = m_E + m_S$ the total mass. We will employ the space-fixed orthogonal coordinate system ξ, η, ζ and the principal axis coordinate system x, y, z fixed on the satellite. The vectors r, r_E, r_S, r' and $r_a = r_S - r_E = r - r'$ are defined in Fig. 58 and are used to formulate Newton's law

$$\ddot{r}_S + \gamma \frac{m_E}{m_S} \int \frac{r}{r^3} \, dm = 0. \tag{2.2.58}$$

γ is the gravitational constant. The integration must be carried out over the satellite. The angular momentum theorem with respect to 0_S and the moving orthogonal coordinate system x, y, z is

$$\frac{\partial D_S}{\partial t} + \omega \times D_S = - \gamma m_E \frac{r' \times r}{r^3} \, dm. \tag{2.2.59}$$

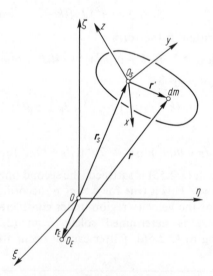

Fig. 58. Satellite motion

[1]Magnus, K., The rod satellite in radially symmetric field of gravitation. *Z. Flugwiss.* pp. 233–241, 1963; Contributions to the investigation of the rotational motion of rigid-body satellites along circular orbits. *Jahrb. Wiss. Ges. Luft Raumfahrt* pp. 174–180, 1963.

$D_S = \int [r' \times (\omega + r')] \, dm$ is the angular momentum of the satellite and ω is the rotation vector of the satellite with respect to the space-fixed system.

We have

$$r' = e_x x + e_y y + e_z z$$

$$r_a = r_a(e_x \gamma_x + e_y \gamma_y + e_z \gamma_z). \qquad (2.2.60)$$

r_a is the vertical direction and γ_x, γ_y, γ_z are the direction cosines with respect to the body-fixed system. It is important to note that $|r'| \ll |r_a|$. The following series expansion holds:

$$\frac{1}{r^3} = \frac{1}{r_a^3} - 3(r_a r') \frac{1}{r_a^5} - \frac{3}{2} \frac{r'^2}{r_a^5} \cdots \qquad (2.2.61)$$

Only the first three terms are retained, and the following ones are suppressed because they are small. It follows that

$$\int \frac{r}{r^3} \, dm = \int \frac{r_a + r'}{r^3} \, dm = \int \frac{r_a}{r_a^3} \, dm - 3 \int \frac{r_a(r_a r')}{r_a^5} \, dm$$

$$- \frac{3}{2} \int \frac{r_a r'^2}{r_a^5} \, dm$$

$$+ \int \frac{r'}{r_a^3} \, dm - 3 \int \frac{r'(r_a r')}{r_a^5} \, dm$$

$$- \frac{3}{2} \int \frac{r' r'^2}{r_a^5} \, dm \cdots$$

We have $\int r_a \, dm = m_S r_a$, $\int r' \, dm = 0$, and we will neglect all terms which contain r_a^4 and higher powers of r_a in the denominator. The expression

$$\int \frac{r}{r^3} \, dm \approx \frac{m_S r_a}{r_a^3} \qquad (2.2.62)$$

remains. Using the relationships

$$r_S = (m_E/m)r_a, \qquad r_S = (m_E/m)r_a, \qquad (2.2.63)$$

the following approximation for Newton's law (2.2.58) is obtained:

$$\ddot{r}_S + \gamma(m_E^3/m^2 r_S^3)r_S = 0. \qquad (2.2.64)$$

The angular momentum theorem can also be simplified. Because of (2.2.61) we have

$$\int \frac{r' \times r}{r^3} \, dm = \int \frac{r' \times (r_a + r')}{r^3} \, dm$$

$$= -r_a \times \int \frac{r'}{r_a^3} \, dm - 3 \int \frac{(r' \times r_a)(r_a r')}{r_a^5} \, dm \cdots,$$

if small terms are again suppressed. The first is zero so that

$$\int \frac{r' \times r}{r^3} \, dm \approx -\frac{3}{r_a^5} \int (r' \times r_a)(r_a r') \, dm$$

$$= -\frac{3}{r_a^5} \left[\int r'(r_a r') \, dm \right] \times r_a \qquad (2.2.65)$$

remains. The following expression is calculated from (2.2.60)

$$\int r'(r_a r') \, dm = P r_a - r_a (e_x \gamma_x A + e_y \gamma_y B + e_z \gamma_z C). \qquad (2.2.66)$$

A, B, C are the principal moments of inertia:

$$A = \int (y^2 + z^2) \, dm, \qquad B = \int (z^2 + x^2) \, dm, \qquad C = \int (x^2 + y^2) \, dm,$$

and $P = \int (x^2 + y^2 + z^2) \, dm$ is the polar moment of inertia. Making use of (2.2.60) and (2.2.66), we can rewrite Eq. (2.2.65) as follows:

$$\int \frac{r' \times r}{r^3} \approx -\frac{3}{r_a^5} [e_x \gamma_y \gamma_z (C - B) + e_y \gamma_z \gamma_x (A - C) + e_z \gamma_x \gamma_y (B - A)]. \qquad (2.2.67)$$

Substitution of (2.2.67) into the theorem of angular momentum (2.2.59), together with (2.2.63), results in

$$\partial D_S / \partial t + \omega \times D_S$$
$$= (3 \gamma m_E^4 / m^3 r_S^3) [e_x \gamma_y \gamma_z (C - B) + e_y \gamma_z \gamma_x (A - C) + e_z \gamma_x \gamma_y (B - A)].$$

Using the abbreviation

$$K = 3 \gamma m_E^4 / m^3 r_S^3 \approx 3 \gamma m_E^3 / m^2 r_S^3 \qquad (2.2.68)$$

and with $\omega = (\omega_x, \omega_y, \omega_z)$, the scalar form of this equation is

$$A \dot{\omega}_x + (C - B) \omega_y \omega_z = K \gamma_y \gamma_z (C - B),$$
$$B \dot{\omega}_y + (A - C) \omega_z \gamma_x = K \gamma_z \zeta_x (A - C),$$
$$C \dot{\omega}_z + (B - A) \omega_x \omega_y = K \gamma_x \gamma_y (B - A).$$

If the satellite is assumed to have symmetry so that $A = B$, and introducing the notation

$$a = (C/A) - 1,$$

the following equations are obtained from the theorem of angular momentum:

$$\left. \begin{aligned} \dot{\omega}_x + a \omega_y \omega_z &= a K \gamma_y \gamma_z, \\ \dot{\omega}_y - a \omega_z \omega_x &= -a K \gamma_z \gamma_x, \\ \dot{\omega}_z &= 0. \end{aligned} \right\} \qquad (2.2.69)$$

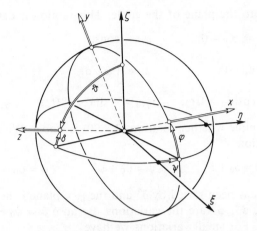

Fig. 59. Reference axes and angles

Let us not determine the particular solutions of (2.2.64) and (2.2.69) for the motion of the satellite. Such a solution can be found for (2.2.64), for example, i.e., for $\ddot{r}_S + \frac{1}{3}Kr_S = 0$, using the trial solution $|r_S| = \text{const}$. In this case, $K = \text{const}$ so that the vector r_S rotates in a plane with the angular velocity $\Omega = (K/3)^{1/2}$. This means that the center of gravity of the satellite moves along a plane circular trajectory. A particular solution of (2.2.69) for the rotation of the satellite about its center of gravity is given by

$$\omega_{z0} = \text{const}, \qquad \omega_{x0} = \omega_{y0} = 0, \qquad \gamma_{z0} = 0.$$

It corresponds to the rotation of the satellite around the z axis which is always perpendicular to the circular trajectory plane for this solution.

The three angles ϕ, ψ, δ are used to uniquely determine the position of the satellite in space, assuming that $|\delta| < \pi/2$ (Fig. 59). Let us select the space-fixed orthogonal coordinate system in such a way that the ξ, ζ plane is the plane of motion of the satellite. Because $r_S \approx r_a$, the vertical vector r_a, which is a radius of the orbit, rotates uniformly around the origin of the fixed axes at the angular rate Ω. Therefore,

$$\gamma_{x0} = \cos(\phi_0 - \Omega t), \quad \gamma_{y0} = -\sin(\phi_0 - \Omega t).$$

If the satellite rotates around the z axis at the rate ω_{z0}, the body-fixed axis system x, y, z will also rotate. Fig. 59 shows that $\phi_0 = \int \omega_{z0} \, dt$ must hold.

The particular solution in question, which satisfied the equations of motion and kinematic equations, can be summarized as follows. The satellite's center of gravity describes a circular motion around the earth's center at the angular rate $\Omega = (K/3)^{1/2}$. The radius r_a of this orbit is constant. In addition, the satellite rotates about an axis through the center of gravity z, which remains perpendicular to the plane of the circle. This motion is described by

$$\omega_{x0} = = \omega_{y0} = 0, \qquad \omega_{z0} = \text{const},$$

$$\delta_0 = \psi_0 = 0, \qquad \phi_0 = \int \omega_{z0}\, dt,$$

$$\gamma_{x0} = \cos(\phi_0 - \Omega t), \qquad \gamma_{y0} = -\sin(\phi_0 - \Omega t), \qquad \gamma_{z0} = 0. \qquad (2.2.70)$$

Let us determine whether solution (2.2.70) is stable. We shall consider the perturbed motion

$$\delta = \delta_0 + \delta^*, \qquad \psi = \psi_0 + \psi^*, \qquad \phi = \phi_0 + \phi^*,$$

i.e. the z axis has been raised by δ^* and the ξ, ζ plane is no longer the plane of motion. δ^*, ψ^*, ϕ^* are the variations. Because $\delta_0 = \psi_0 = 0$, we may write $\delta = \delta^*$, $\psi = \psi^*$. For small variations we have

$$\omega_x = \dot{\psi} \sin \phi - \dot{\delta} \cos \phi, \qquad \gamma_x = \cos(\phi - \Omega t),$$

$$\omega_y = \dot{\psi} \cos \phi + \dot{\delta} \sin \phi, \qquad \psi_y = -\sin(\phi - \Omega t),$$

$$\omega_z = \dot{\phi}, \qquad \gamma_z = \psi \cos \Omega t + \delta \sin \Omega t. \qquad (2.2.71)$$

If (2.2.71) is substituted into the first equation of (2.2.69) (because the second equation gives the same result and the third equation gives $\dot{\phi} = \text{const}$), we obtain the variational equation

$$\ddot{\psi} \sin \phi - \ddot{\delta} \cos \phi + \dot{\phi}(1 + \alpha)(\dot{\psi} \cos \phi + \dot{\delta} \sin \phi)$$
$$= - aK \sin(\phi - \Omega t)(\psi \cos \Omega + \delta \sin \Omega t).$$

Introducing the new variables

$$\alpha = \psi \cos \Omega t + \delta \sin \Omega t, \qquad \beta = \psi \sin \Omega t - \delta \cos \Omega t,$$

we obtain

$$\cos(\phi - \Omega t)\{\dot{\alpha}[\dot{\phi}(1 + a)2\Omega] + \ddot{\beta} + \beta[\Omega\dot{\phi}(1 + \alpha) - \Omega^2]\}$$
$$+ \sin(\phi - \Omega t)\{\ddot{\alpha} + \alpha[\Omega\dot{\phi}(1 + a) - \Omega^2 + aK] - \dot{\beta}[\dot{\phi}(1 + a) - 2\Omega]\} = 0.$$

Since this equation must always hold, the following two differential equations are obtained for α and β:

$$\ddot{\alpha} + c_{11}\alpha - c_{12}\dot{\beta} = 0, \qquad (2.2.72)$$
$$\ddot{\beta} + c_{22}\beta + c_{12}\dot{\alpha} = 0,$$

with the constant coefficients

$$c_{11} = \Omega\dot{\phi}(1 + a) - \Omega^2 + aK,$$
$$c_{12} = \dot{\phi}(1 + a) - 2\Omega,$$
$$c_{22} = \Omega\dot{\phi}(1 + a) - \Omega^2.$$

The quantity $\dot{\phi}$ is constant because of the third equation in (2.2.71) and (2.2.69). The characteristic equation of (2.2.72) follows from

$$\begin{vmatrix} \lambda^2 + c_{11} & -\lambda c_{12} \\ \lambda c_{12} & \lambda^2 + c_{22} \end{vmatrix} = 0$$

and is given by

$$\lambda^4 + \lambda^2 (c_{22} + c_{11} + c_{12}^2) + c_{11} c_{22} = 0. \tag{2.2.73}$$

The existence of four imaginary roots (2.2.73) is a necessary condition for stability. This condition is satisfied if λ^2, calculated from (2.2.73), is negative and real for all cases. This corresponds to conditions

$$c_{22} + c_{11} + c_{12}^2 > 0, \qquad c_{11} c_{22} > 0, \qquad (c_{22} + c_{11} + c_{12}^2)^2 - 4 c_{11} c_{22} > 0.$$
$$\tag{2.2.74}$$

The stability conditions (2.2.74) can be written in the following form, making use of $K = 3\Omega^2$ and the new quantities $p = C/A = (a + 1)$ and $\nu = \dot{\phi}\Omega$:

$$\nu^2 p^2 - 2\nu p + 3a + 2 > 0,$$

$$\nu^2 p^2 + \nu p(3a - 2) + (1 - 3a) > 0,$$

$$\nu^4 p^4 - 4\nu^3 p^3 + (4 + 6a)\nu^2 p^2 - 24a\nu p + 9a^2 + 24a > 0.$$

They have been evaluated in the form of a *stability diagram* in Fig. 60. This diagram shows that for $C = 0$ or $p = 0$, the rod satellite is always unstable with respect to the particular solution (2.2.76). On the other hand, a satellite having a flattened moment of inertia ellipsoid ($p > 1$) is always stable for $\nu > 1$. Satellites with a long moment of inertia ellipsoid ($p < 1$) may be stabilized with respect to the motion (2.2.70) for sufficiently large ν, i.e., for sufficiently large angular momentum with respect to the symmetry axis (z axis) perpendicular to the orbit.

Fig. 60. Stability diagram for satellites

194

Exercises 2.2

1) Given the non-linear differential equation

$$\ddot{x} + x = \varepsilon F\left(x, \frac{dx}{dt}\right)$$

where ε is a small parameter and F is a non-linear function of its arguments.

Assume that the differential equation has a periodic solution $x \approx A \cos t$ with the period $T = 2\pi$ representing a limit in the phase space. Show that the stability condition for the limit cycle is

$$\varepsilon \int_0^2 \frac{\partial F(A \cos t, -A \sin t)}{\partial y} dt < 0.$$

Apply this result to van der Pol's equation

$$\ddot{x} + x = \varepsilon(1 - x^2)\frac{dx}{dt}, \qquad 0 < \varepsilon \ll 1;$$

show that

$$\varepsilon \int_0^{2\pi} \frac{\partial F}{\partial y} dt = -2\pi\varepsilon < 0$$

in this case, which proves stability of the limit cycle.

2) For the control shown in Fig. 48 assume that the non-linear function f is given by

$$f = \begin{cases} c = \text{const for } x > 0, \\ -mc = \text{const for } x < 0. \end{cases}$$

Which condition for m must be satisfied for a stable oscillation to exist in the non-linear control system, assuming that m could take on not only positive but also negative values?

3) A ship is equipped with a stabilizing gyroscope. Let ϕ be the angle by which the ship rolls about its longitudinal axis and θ the angle by which the gyroscope's frame rotates about the trunnion axis. Then, the differential equations of motion of the ship and the gyroscope read

$$I\ddot{\phi} + K\dot{\phi} - C\omega\dot{\theta} + Wh\phi = M \sin bt,$$

$$A\ddot{\theta} + k\dot{\theta} + C\omega\dot{\phi} + wl\theta = 0,$$

where K and k are coefficients of friction, $M \sin bt$ is the moment of wave action about axis of roll, I is the moment of inertia of ship about axis of roll, A is the moment of inertia of gyro and frame about the trunnion axis, W is the weight of the ship, w is a pendulous weight attached to the gyro frame, h is the metacentric height of the ship, l is the moment arm of w. Find the condition under which the motion is stable.

4) The equations of motion of an airplane guided by an autopilot are

$$x_1 = x_2, \qquad \dot{x}_2 = -Mx_2 - Nx_3, \; \dot{x}_3 = \Phi\left(ax_1 + px_2 - \frac{1}{k}x_3\right),$$

where M, N are positive constants, a, E, G, l are control parameters, $\Phi(\sigma)$ is a function of σ with the properties $\Phi(0) = 0$, $\sigma\Phi(\sigma) > 0$ for $\sigma \neq 0$, $p = E - MG^2$, $k = l[1 + lNG^2]^{-1}$.

Investigate the stability of the trivial soluton $x_i^* = 0$, $i = 1, 2, 3$, using Lyapunov's direct method.

Hint: use

$$V = A(MN^{-1}x_2 + x_3)^2 + Bx_3^2 + C \int_0^\sigma \Phi(\sigma)\, d\sigma$$

as a Lyapunov function, where A, B, C are (as yet undetermined) positive constants.

5) The equations of motion of a bridge exposed to wind forces and modelled as a system with two degrees of freedom read

$$\ddot{q}_1 + bq_1 + cq_2 = 0,$$

$$\ddot{q}_2 + dq_1 + eq_2 = 0.$$

The generalized coordinates are q_1 and q_2. Quantities, b, c, d, and e are structural and load parameters. Find conditions for these parameters which rule out flutter of the bridge.

2.3
Problems of Elastomechanics

As outlined in Section 1.1, we will now discuss stability problems in elasto-mechanics in connection with the fundamental concepts contained in the stability theory discussed previously. There exists a well-developed and inde-pendent theory for elastomechanics, which can be found in the literature. In addition to Refs. (1) and (2), the reader is also referred to several important works (8), which by no means exhaust the list. It is not necessary to repeat the methods and examples of these works. Instead we will critically question their contents. Modern studies on the stability of elastic bodies indeed frequently question the validity and applicability of the old concepts and methods, and the theory of elastic-body stability has, therefore, often been reformulated and classified within the overall mechanical theory of stability.

The modern theory of elastic-body stability is as follows. The behaviour of a special state of the elastic body (for example, an equilibrium state considered as the stationary solution of the fundamental equations of elastostatics) is investigated as a function of the *system parameter p* (for example, load, temperature, etc.). This behaviour is expressed by means of a *characteristic quantity* ω which determines whether stability or instability prevails. Let us assume that the functional dependence on the parameter is given by $\omega = \omega(p)$. By plotting this law, we obtain a diagram from which it is possible to deter-mine whether the corresponding state is stable or unstable. Let us assume that the curve $\omega = \omega(p)$ passes into the instability region of ω for certain critical values of p. It follows that the state of the elastic body is no longer stable for such parameter values. The term *eigenvalue curve* $\omega = \omega(p)$ is often used, because the parameter is frequently an eigenvalue in the mathematical treat-ment of the problem. Actually, one is dealing with *structural stability* accord-ing to the concepts introduced in Section 1.1.2, because the stability of a special state is often influenced by a perturbation of structural parameters.

In practice, the *lower stability boundary* is usually of interest, which is the smallest value $p_{crit,min}$ of the parameter p, usually assumed to be positive, for which instability first occurs. The behaviour of the elastic body after this lower bound has been exceeded is called its 'postcritical' behaviour. Euler in-vestigated the postcritical behaviour of elastic rods. The works by Tricomi (9),

196

Wang (10), or Langhaar (8.5) contain modern investigations of this problem. Non-linear equations and the corresponding special methods must be used. At present, a great deal of work is being done on the postcritical behaviour of plates and shells. Koiter (11) has reviewed this subject.

The stability behaviour of bodies in the non-elastic range, as well as disturbances caused by undesirable inaccuracies of elastic constructions, are also of great interest. Since such inaccuracies are of a stochastic nature, one is lead to problems of stochastic stability.

We shall not deal further with these general problems. Instead, we will discuss a linearized theory of buckling of *ideal and elastic* bodies. We will attempt to determine the lowest critical value of the system parameter (the so-called Euler value). Let us first consider elastostatic problems. The stability problems of elastokinetics will be considered subsequently.

2.3.1 Stability Criteria of Elastostatics

Two criteria have been used up to the present to determine stability regions and stability boundaries: These are the *energy criterion* and the *static criterion*. The application of these criteria leads to the so-called *energy method* and *equilibrium method*. The *energy method* uses the fact that the principle of virtual displacements can be extended to elastic bodies with infinite degrees of freedom. The relationship $\delta\Pi = 0$ holds in the equilibrium state (and in statics this is always the special state for which the stability is investigated) provided that the system is conservative or, more generally, monogenic according to Lanczos (6). The potential is $\Pi = \Pi_i + \Pi_a$ (where Π_i is the deformation energy and Π_a the potential of external loads). This relationship follows immediately from the principle of virtual work. The total potential Π takes on a stationary value in the equilibrium position. Let us assume that $\Pi = \Pi(h)$, i.e., the potential depends on a system parameter that specifies the position. Therefore, $\delta\Pi(h_0) = 0$ holds for the parameter value h_0 corresponding to equilibrium. In order to investigate the effect of h on Π, h_0 is varied and

$$\Pi(h_0 + \delta h) = \Pi(h_0) + (\partial\Pi/\partial h)_{h0}\,\delta h + \tfrac{1}{2}(\partial^2\Pi/\partial h^2)_{h0}\,\delta h^2 + \cdots$$
$$= \Pi(h_0) + \delta\Pi(h_0) + \tfrac{1}{2}\delta^2\Pi(h_0) + \cdots$$

is assumed. This can immediately be transformed to the relationship

$$\Pi(h_0 + \delta h) = \Pi(h_0) + \tfrac{1}{2}\delta^2\Pi(h_0) + \cdots \qquad .$$

In the vicinity of the equilibrium position, i.e., for small variations δh the behaviour of $\Pi(h_0 + \delta h)$ is given by

$$(\partial^2\Pi/\partial h^2)_{h0}\,\delta h^2 = \delta^2\Pi(h_0).$$

The following results are obtained analogous to *Dirichlet's* theorem, which is valid for a point system, and from a consideration of the energy relationships.

If $\delta^2\Pi < 0$, energy is released when h_0 is varied. The body can move away from the equilibrium position that is unstable. On the other hand, if $\delta^2\Pi > 0$, energy is consumed when h_0 is varied. The body cannot leave the equilibrium position by itself. The equilibrium is stable. The stability boundary is given by $\delta^2\Pi = 0$ and the equilibrium is neutral. The stability criterion therefore amounts to the requirement $\delta^2\Pi > 0$. For *stability*, $\delta^2\Pi$ must be *positive definite* and Π must have a *minimum* in the equilibrium position.

Often stability is determined from a condition $\Delta\Pi_i = \Delta A_a$, where $\Delta\Pi_i$ and ΔA_a are the increases of the potential of the internal forces and the work of the external forces for variation of the equilibrium position. For example, Timoshenko (1.6) used this criterion successfully. Because $\Delta Aa = -\Delta\Pi a$, this is identical to

$$\Delta(\Pi_i + \Pi_a) = \Delta\Pi = 0,$$

and because $\Delta\Pi = \delta\Pi + \frac{1}{2}\delta^2\Pi \equiv \frac{1}{2}\delta^2\Pi$, this again amounts to $\delta^2\Pi = 0$.

There is another variation of the energy criterion advanced by E. Trefftz. He postulated that $\delta^2\Pi_a \equiv 0$ for loads that only depend linearly on h. Therefore, the discussion can be restricted to $\delta^2\Pi_i$. What is true for $\delta^2\Pi$ can then be stated for $\delta^2\Pi_i$.

In the *equilibrium method* it is assumed that because of the neutral equilibrium state on the stability boundary, *neighbouring* equilibrium positions can exist for the same load conditions. If we assume the trivial equilibrium positions in our discussion, for which displacements are zero for the chosen coordinate system, the application of the equilibrium method postulates that *non-trivial equilibrium positions* exist. In this case, this may be called the *static stability criterion,* because the critical parameter value $p_{crit, min}$, follows from the proof that non-trivial equilibrium positions exist. Instability would occur after this parameter value is exceeded.

In the case of *conservative or monogenic* problems, which are usually treated without being explicitly mentioned as such, the energy and static methods lead to the *same* basic criterion. This may be shown as follows:

The energy method specifies $\delta^2\Pi = 0$ at the stability boundary. The static method assumes that in addition to the trivial equilibrium position h_0 there is a non-trivial equilibrium position $h_1 = h_0 + \Delta h$. It must satisfy $\delta\Pi(h_1) = 0$ because it is an equilibrium position, and $\delta\Pi(h_0) = 0$ holds for the trivial equilibrium position.

We have

$$\Pi(h_1) = \Pi(h_0 + \Delta h) = \Pi(h_0) + \delta\Pi(h_0) + \frac{1}{2}\delta^2\Pi(h_0) + \cdots$$

and therefore because of $\delta\Pi(h_0) = 0$,

$$\Pi(h_1) = \Pi(h_0) + \frac{1}{2}\delta^2\Pi(h_0) + \cdots \equiv \Pi(h_0) + \frac{1}{2}(\partial^2\Pi/\partial h^2)_{h0}(\Delta h)^2 + \cdots$$

Furthermore

$$\delta\Pi(h_1) = \delta\Pi(h_0 + \Delta h) = [\partial\Pi(h_1)/\partial(\Delta h)]\,\delta(\Delta h)$$

is calculated and by substitution of $\Pi(h_1)$ we have

$$\delta\Pi(h_1) = (\partial^2\Pi/\partial h^2)_{h0}\,\Delta h\,\delta(\Delta h).$$

The condition $\delta\Pi(h_1) = 0$ which follows from the existence of the non-trivial solution implies

$$(\partial^2\Pi/\partial h^2)_{h0}\, \Delta h = 0$$

and therefore the condition

$$\delta^2\Pi(h_0) \equiv (\partial^2\Pi/\partial h^2)_{h0}(\Delta h)^2 = 0,$$

which correspond to the energy criterion. It has thus been shown that the same conclusions may be reached for both criteria.

Finally, it should be mentioned that it is not necessary to be restricted to only *one* load parameter p or one position parameter h. In more complicated cases it can become necessary to calculate with several parameters. Multiple parameter systems have for example been dealt with by K. Huseyin.[1]

The first serious objections to the universal validity of the two classical stability criteria mentioned above were made by H. Ziegler. He showed that the static criterion can only be safely used for conservative problems. This is because there are non-conservative problems (for example, the Beck problem[2]), for which there are no non-trivial equilibrium positions. The assumptions for the criterion are therefore not satisfied. This result makes it necessary to distinguish two classes of problems in the stability theory of elastomechanics. One class encompasses the conservative problems that are monogenic in the broader sense of the word. These are the problems for which the internal as well as the external forces can be derived from a potential. Only for these problems can the principle of virtual displacements be transformed into the simple form $\delta\Pi = 0$, which is the foundation for the static as well as for the classical energy criteria, as we have seen. Since the vast majority of problems in practice are of this type, it was always assumed that these conditions were satisfied until incorrect results made it necessary to reconsider the problems with greater attention. The other class encompasses the non-conservative problems, which are polygenic in the broad sense. These are the problems for which the external forces cannot be derived from a potential (the internal forces of elastic bodies always have a potential). The 'follower' forces are examples for this, which occur in problems in aeroelasticity and which occur more often in modern technology.

There are also objections to the classical energy criterion even if applied to conservative systems. This is because the positive definite property of the expression $\delta^2\Pi$ according to the *Dirichlet* theorem does not necessarily imply stability of equilibrium according to the rigorous definitions of Lyapunov for two- or three-dimensional continua. Lyapunov's definition of stability stipulates that the displacements are sufficiently and arbitrarily small at all

[1]Huseyin, K., *Vibrations and Stability of Multiple Parameter Systems*, Sijthoff and Noordhoff, Alphen aan den Rijn, The Netherlands, 1978.
[2]Beck, M., The buckling load of a tangentially loaded beam clamped on one side, *Z. Angew, Math. Phys.* 3, 255–228, 476–477, 1952.

times and at every point of the continuum. These doubts were already formulated by E. Hellinger and G. Hamel.[1] Shield and Green[2] showed that the objections to the energy criterion are justified.

Ziegler recommends a third stability criterion in cases where the classical energy criterion can definitely not be applied and where the static criterion may also not be applicable because of the absence of a potential for the external forces. This is the *kinetic criterion,* which consists of the following: small displacements from the equilibrium position of the elastic body under consideration are carried out. Then the disturbed motion which occurs is discussed. If it remains bounded for all times, stability of the equilibrium position is implied.

The introduction of this third criterion and the class of non-conservative problems immediately leads to questions that will subsequently be discussed. (1) What mathematical methods must be applied in the non-conservative case (the Ritz method does not qualify because $\delta\Pi = 0$ no longer holds)? (2) Can the static criterion continue to hold for non-conservative problems (the classical energy criterion certainly does not hold)? (3) Does the mass distribution of the elastic body and its damping influence the stability, because there is now a motion to be considered? (4) Under which condition is it legitimate to use linearized systems for stability considerations?

2.3.2 Application of General Methods of Elastic Systems with Time-independent Parameters and Forces

Stability problems are by nature kinetic problems. This is because, according to the general theory developed in Part 1 of the book, they consist of observing the motion about the equilibrium position of interest, which is caused by a perturbation. Therefore, we will use this kinetic point of view in the following. In spite of this, we will essentially discuss elastostatic stability-problems when we are dealing with stability of equilibrium positions. On the other hand, later on we will discuss elastokinetic stability problems that involve the stability of stationary motions.

Making use of the ideas of Ziegler, we will use kinetic methods and kinetic stability criteria in the investigation of elastostatic problems, which follow the fundamental principles of a general theory. In this fashion, let us investigate the *stability of an equilibrium position,* which will be called the *original state.* For simplicity, a reference system is assumed for which all displacements $w(x, y, z, t)$ are zero in the equilibrium position. The original state is disturbed so that a small motion of the body occurs. According to the general theory, the corresponding equations of motion are the variational equations. Their solution leads to a decision regarding the stability of the original state. Even

---

[1] Hamel, G., Theoretische Mechanik, p. 270, Springer, Berlin, 1949.
[2] Shield, R. T., and Green, A. E., On certain methods in the stability theory of continuous systems, *Arch. Ratl. Mech. Anal.* **12**, 354–360, 1963.

though apparently the displacements themselves are substituted in the variational equations in the following, it should be remembered that these are actually variations of the displacement of the original state. This fact is not so obvious because of the selection of the reference system.

In order to obtain variational equations, the principle of virtual displacements is used. According to Hellinger, the latter is at least a permissible axiom in the field of continuum mechanics. The perturbation consists of an infinitesimal displacement which is compatible with the constraints of the body. The work equations which follow from the principle are then formulated, and the equations of motion and the boundary conditions can be obtained from them.

Example: Let us consider the *buckling of a column* clamped at one end with a tangential (follower) point load P_t and a vertical (constant direction) point load P_v (Fig. 61). The disturbance consists of a lateral deflection of the column from its trivial equilibrium position. The column will then carry out small oscillations around the original state (vertical column axis). The points of the column axis undergo the transverse displacements $w(x, t)$. For these displacements, the principle of virtual displacements yields

$$\int_0^l \mu w_{tt}\, \delta w\, \mathrm{d}x + \delta\Pi_i + (P_v + P_t)\, \delta u(l) + P_t w_x(l)\, \delta w(l) = 0. \qquad (2.3.1)$$

μ is the linear mass density, l the length, Π_i the potential energy of the internal forces of the column, w and u are the horizontal and vertical displacements of the column axis. Subscripts correspond to the partial derivative with respect to the independent variable t (time) and x (abscissa of the column axis). If it is assumed that the axis of the column is incompressible, the extension is given by

$$\varepsilon \approx u_x + (w_x^2/2) = 0,$$

Fig. 61. Buckling rod with undirectional force P_v and follower force P_t

so that

$$u(l) = \int_0^l u_x \, \mathrm{d}x = -\frac{1}{2} \int_0^l w_x^2 \, \mathrm{d}x, \qquad \delta u(l) = - \int_0^l w_x \, \delta w_x \, \mathrm{d}x,$$

and therefore Eq. (2.3.1) becomes

$$\int_0^l \mu w_{tt} \, \delta w \, \mathrm{d}x + \delta \Pi_i - (P_v + P_t) \int_0^l w_x \, \delta w_x \, \mathrm{d}x + P_t w_x(l) \, \delta w(l) = 0.$$

We also have

$$(P_v + P_t) \int_0^l w_x \, \delta w_x \, \mathrm{d}x = \delta \left[\tfrac{1}{2} (P_v + P_t) \int_0^l w_x^2 \, \mathrm{d}x \right] = - \delta \Pi_a,$$

where Π_a is the potential of the external forces. Equation (2.3.1) can be written in the form

$$\int_0^l \mu w_{tt} \, \delta w \, \mathrm{d}x + \delta(\Pi_i + \Pi_a) + P_t w_x(l) \, \delta w(l) = 0. \qquad (2.3.2)$$

Moreover,

$$\Pi_i = \tfrac{1}{2} \int_0^l \alpha \, w_{xx}^2 \, \mathrm{d}x,$$

where α is the bending stiffness of the column, assumed to be constant. After the variation is carried out and partial integration is used, we obtain

$$\delta(\Pi_i + \Pi_a) = \delta \left[\frac{1}{2} \int_0^l [\alpha w_{xx}^2 - (P_v + P_t) w_x^2] \, \mathrm{d}x \right]$$

$$= \int_0^l [\alpha w_{xxxx} + (P_v + P_t) w_{xx}] \, \delta w \, \mathrm{d}x$$

$$+ [-(P_v + P_t) w_x \, \delta w + \alpha w_{xx} \, \delta w_x - \alpha w_{xxx} \, \delta w]_0^l.$$

Substitution into (2.3.2) results in

$$\int_0^l [\mu w_{tt} + \alpha w_{xxxx} + (P_v + P_t) w_{xx}] \, \delta w \, \mathrm{d}x$$

$$+ [-(P_v + P_t) w_x \, \delta w + \alpha w_{xx} \, \delta w_x - \alpha w_{xxx} \, \delta w]_0^l + P_t w_x(l) \, \delta w(l) = 0.$$

Because of the prescribed geometric boundary conditions

$$w(0, t) = w_x(0, t) = 0,$$

the requirement that all boundary terms vanish, and because of the arbitrary nature of the variation δw, the differential equation

$$\mu w_{tt} + \alpha w_{xxxx} + (P_v + P_t) w_{xx} = 0 \qquad (2.3.3)$$

and the dynamic boundary conditions

$$w_{xx}(l, t) = 0, \qquad w_{xxx}(l, t) = -(P_v / \alpha) w_x(l, t) \qquad (2.3.4)$$

are obtained.

Equation (2.3.3) is the desired *variational equation,* which describes the perturbed motion around the original state.

The original state is called stable if the displacements $w(x, y, z, t)$ remain sufficiently small for sufficiently small initial perturbations. In the example given above, this would have to be satisfied by the lateral displacements $w(x, t)$ if the column axis is subjected to a small transverse deflection from the vertical original state. This stability definition coincides with the definition of Lyapunov given in Section 1.1.2: If the initial perturbation of the original state is selected small enough, for stability the displacements w should remain within arbitrary limits at all times. The variational equations determine whether or not this is possible. In our example above, this is determined by Eq. (2.3.3) and their boundary conditions.

The following two *remarks* are in order.

In the example, we used the linearized equations, which we shall also do in the following. As stated above, this should be justified by showing that the displacements as well as their derivatives remain small when the original state is disturbed. Only when this is true should the classical theory of elasticity, which is based on linear constitutive equations, hold.

Also, in the example we did not vary all possible displacement components. Instead, only a special variation was carried out, the variation of the horizontal component w. This is certainly permissible in certain cases, because inspection of the problem will show which component and variation is really decisive for the behaviour of the body. In principle, it is important to realize that, if not all variations are carried out, only *conditional* stability conditions with respect to special perturbations are obtained. This will not always be stated in the following. We will accept the restriction that our stability results are only conditional because of the special variations of the displacements considered. Usually this restriction is unimportant in practice.

2.3.3 Conservative Stability Problems and the Application of the Classical Energy Method

Conservative problems are especially important in particle mechanics and in the mechanics of rigid bodies, because there is an *energy integral* that can be used as the *Lyapunov function* to determine the stability of equilibrium positions. This is mentioned in Section 1.5.3 in connection with the theorem of *Lagrange–Dirichlet.* It is natural to proceed in the same way for conservative problems of elastomechanics. The total energy of the elastic body is given by

$$H = \int_V (\tfrac{1}{2}\mu \dot{w}^2 + \Pi^*)\, \mathrm{d}V = \int_V \mathscr{H}\, \mathrm{d}V \qquad (2.3.5)$$

where V is the volume, μ the mass density, w the displacement vector, and Π^* the specific potential of the internal and external forces of the body. It can be shown that the variational equations for the motion about the equilibrium position are given by

$$\delta H/\delta p_k = \dot{w}_k, \qquad \delta H/\delta w_k = -\dot{p}_k, \qquad p_k = \mu \dot{w}_k, \qquad (2.3.6)$$

where $\mathscr{H} = \mu w^2/2 + \Pi^*$ is the Hamiltonian density appearing in Eq. (2.3.5).

Example: Let us consider a flat *plate* of thickness h which carries out small oscillations about the equilibrium position. If F is the area of the plate, we have

$$H = \int_F \left\{ \frac{1}{2\mu^*} p^2 + \frac{D}{2} [w_{xx}^2 + 2(1 - v)w_{xy}^2 + 2vw_{xx}w_{yy} + w_{yy}^2] \right\} dF$$

$$= \int_F \mathscr{H} \, dF,$$

where

$$p = \mu^* \dot{w}, \qquad \mu^* = h\mu, \qquad \Pi^* = w_{xx}^2 + 2(1 - v)w_{xy}^2 + 2vw_{xx}w_{yy} + w_{yy}^2, \quad w(x, y, t)$$

is the vertical deflection and D is the plate stiffness. According to the first equation of (2.3.6)

$$\frac{\delta H}{\delta p} \equiv \frac{\partial \mathscr{H}}{\partial p} = \frac{p}{\mu^*} = \dot{w},$$

and therefore

$$p = \mu^* \dot{w}.$$

The second equation of (2.3.6) yields

$$\frac{\delta H}{\delta w} = \frac{\partial \mathscr{H}}{\partial w} - \frac{\partial}{\partial x} \left(\frac{\partial \mathscr{H}}{\partial w_x} \right) - \frac{\partial}{\partial y} \left(\frac{\partial \mathscr{H}}{\partial w_y} \right) + \frac{\partial^2}{\partial x^2} \left(\frac{d \mathscr{H}}{\partial w_{xx}} \right)$$

$$+ \frac{\partial^2}{\partial x \, \partial y} \left(\frac{\partial \mathscr{H}}{\partial w_{xy}} \right) + \frac{\partial^2}{\partial y^2} \left(\frac{\partial \mathscr{H}}{\partial w_{yy}} \right) = -\dot{p}.$$

Differentiation of this expression and the relationship $\dot{p} = \mu^* \ddot{w}$ results in

$$D(w_{xxxx} + 2w_{xxyy} + w_{yyyy}) = -\mu^* \ddot{w} \qquad \text{or} \qquad \Delta\Delta w = -(\mu^*/D)\ddot{w},$$

respectively, which is the equation of motion.

If $\partial \mathscr{H}/\partial t = 0$, which is always assumed to hold, H is an integral of the variational equation because of (2.3.6). Making use of (2.3.6), we obtain

$$\frac{dH}{dt} = \int_V \left[\frac{\delta H}{\delta w_k} \frac{dw_k}{dt} + \frac{\delta H}{\delta p_k} \frac{dp_k}{dt} \right] dV$$

$$= \int_V \left[-\dot{p}_k \frac{dw_k}{dt} + \dot{w}_k \frac{dp_k}{dt} \right] dV \equiv 0,$$

which must be true for an integral.

Just as is done in particle mechanics and mechanics of rigid bodies, an attempt will be made to use the functional H as the Lyapunov functional assuming that it is positive definite. This is always appropriate if the bounded nature of w were to follow from the fact that w always lies on the phase surface (and this is the case because $dH/dt = 0$). Unfortunately, this proof cannot always be given easily. Therefore, there is a *restricted applicability of the classical energy criterion*. This can be shown as follows. Let H be positive definite. Equation (2.3.5) shows that this is the case if $\Pi = \int \Pi^* \, dV$ is positive definite, which means that the potential of the internal and external forces is positive definite. Now Π is the potential of the varied state. Therefore, we have

$$\Pi = \Pi_g + \delta\Pi_g + \tfrac{1}{2}\,\delta^2\Pi_g.$$

Where the subscript g refers to the original state.

Since the original state is an equilibrium state, $\delta\Pi_g = 0$ must hold according to the principle of virtual displacements. Therefore

$$\Pi = \Pi_g + \tfrac{1}{2}\,\delta^2\Pi_g.$$

Since it is always possible to have $\Pi_g \geqslant 0$, it follows that for a positive definite H and consequently positive definite Π, the quantity $\delta^2\Pi_g$ is also positive definite, which is the condition for the classical energy stability criterion. Yet for positive definite H, w is not necessarily always sufficiently bounded and therefore the equilibrium state cannot be shown to be stable inspite of the fact that H and $\delta^2\Pi_g$ are positive definite. Therefore, the energy stability criterion is not reliable in all cases.

Let us first assume the case in which the energy criterion does not fail for the selected definition of stability.

Example: Let us consider an elastic *buckling column with a conservative load* (Fig. 62). The origin of the coordinate system is located at a fixed support. Then $w(0, t) = 0$ holds for this point and all the other 'Euler cases' shown in Fig. 63. The internal and external forces have the potential

$$\Pi = \frac{1}{2} \int_0^l (\alpha w_{xx}^2 - P w_x^2) \, dx. \tag{2.3.7}$$

Let us assume that it is positive definite and has a small magnitude, because the deflections w are the result of a small initial disturbance. Because $w(0, t) = 0$, and if

$$\| w_x \| = \left(\int_0^l w_x^2 \, dx \right)^{1/2}$$

is the norm of w_x, using the Cauchy–Schwarz inequality results in the estimation

$$w^2 \leqslant x \int_0^x w_x^2 \, dx \leqslant l \int_0^l w_x^2 \, dx = l \| w_x \|^2. \tag{2.3.8}$$

206

Fig. 62. Buckling column
with conservative load

Fig. 63. Euler cases of the buckling
column

Furthermore, the same inequality gives

$$[w_x(x,t) - w_x(0,t)]^2 = \left[\int_0^x w_{xx}\,dx\right]^2 \leq x \int_0^x w_{xx}^2\,dx$$

$$\leq x \int_0^l w_{xx}^2\,dx \leq x \|w_{xx}\|^2,$$

where $\|w_{xx}\|$ is the norm of w_{xx}. Therefore

$$\int_0^l [w_x(x,t) - w_x(0,t)]^2\,dx \leq \frac{l^2}{2}\|w_{xx}\|^2$$

is obtained. After the integration on the left-hand side of the inequality is carried out, we obtain

$$\|w_x\|^2 - 2w_x(0,t)w(l,t) + w_x^2(0,t)l \leq (l^2/2)\|w_{xx}\|^2.$$

In all cases shown in Figs. 62 and 63 $w_x(0,t) = 0$ or $w(l,t) = 0$ holds. Therefore, we always have

$$\|w_x\|^2 \leq (l^2/2)\|w_{xx}\|^2, \tag{2.3.9}$$

and finally, because of (2.3.8),

$$w^2 \leq (l^3/2)\|w_{xx}\|^2. \tag{2.3.10}$$

Because the potential was assumed to be positive definite, we have

$$\Pi = (\alpha/2)\|w_{xx}\|^2 - (P/2)\|w_x\|^2 > 0. \tag{2.3.11}$$

With (2.3.9), Π becomes

$$\Pi \geq \tfrac{1}{2}\|w_{xx}\|^2(\alpha - (Pl^2/2)). \tag{2.3.12}$$

We also impose the requirement

$$\tfrac{1}{2} \| w_{xx} \|^2 (\alpha - Pl^2/2) > 0. \tag{2.3.13}$$

This is the case for

$$P < 2\alpha/l^2. \tag{2.3.14}$$

P then remains below the critical Euler value P_E, which is equal to $2.46\,\alpha/l^2$ for the example shown in Fig. 62.

If (2.3.14) holds, then because of (2.3.12) we have

$$\| w_{xx} \|^2 \leqslant 4\pi/(2\alpha - Pl^2),$$

and it follows from (2.3.10) that

$$w^2 \leqslant 2l^3\Pi/(2\alpha - Pl^2). \tag{2.3.15}$$

Because of the assumption (2.3.11) and (2.3.14) it is therefore possible to obtain an upper bound for the *magnitude* of the displacements w that guarantees that these magnitudes are small enough for stability to prevail. The smaller P the more the above statement is valid, i.e., the more it will remain below the stability bound given by (2.3.14).

In the above it was shown that a positive definite Π and a sufficiently small load P guarantees stability of the elastic column according to Lyapunov's stability definition. It is necessary to show that the derivative of w and therefore the *strain* remains *small* for the column under these assumptions.

We have

$$[w_x(x,t) - w_x(0,t)]^2 \leqslant x \| w_{xx} \|^2 \leqslant l \| w_{xx} \|^2 \leqslant 4l\Pi/(2\alpha - Pl^2),$$

or

$$| w_x(x,t) - w_x(0,t) | \leqslant 2(l\Pi)^{\frac{1}{2}}/(2\alpha - Pl^2)^{\frac{1}{2}}.$$

We have the boundary condition $w_x(0,t) = 0$ for columns built in at the bottom. Therefore, the relationships

$$w_x^2(x,t) \leqslant 4l\Pi/(2\alpha - Pl^2) \qquad \text{or} \qquad | w_x x, t | \leqslant 2[l\Pi/(2\alpha - Pl^2)]^{\frac{1}{2}}$$

are immediately obtained. These are the desired bounds for w_x.

For columns hinged at the bottom, there exists a certain $x = \xi$ for which $w_x(\xi, t) = 0$ holds. Since the inequality given above holds for any x, we must therefore have

$$w_x^2(0,t) \leqslant 4l\Pi/(2\alpha - Pl^2) \qquad \text{or} \qquad | w_x(0,t) | \leqslant 2[l\Pi/(2\alpha - Pl^2)]^{\frac{1}{2}}.$$

Since

$$| w_x(x,t) - w_x(0,t) | \leq 2[l\Pi/(2\alpha - Pl^2)]^{\frac{1}{2}},$$

and owing to the upper bound for the magnitude of $w_x(0,t)$ just established, the extreme case is

$$| w_x(x,t) | = 4[l\Pi/(2\alpha - Pl^2)]^{\frac{1}{2}}.$$

This again shows the upper bound for w_x.

We have therefore shown that w and w_x are sufficiently small, which means it is possible to apply the linearized theory of elasticity and the classical energy criterion.

As we already know, similar results do not necessarily hold for *multidimensional bodies*. In the case of the elastic plate we will see that for positive definite Π the classical energy criterion may yield a buckling load that is too high as to exclude the occurrence of wrinkles in the plate's surface. But then, the plate cannot be considered to be stable in the sense of Lyapunov's definition which requires that $|w|$ be small at any point on the plate's surface and at any time. Hence, it is evident in such a case that the classical energy criterion does not provide a dependable stability limit.

In order to illustrate this situation, consider the thin, elastic Kirchhoff plate shown in Fig. 64. If w is the vertical deflection of the plate, D the plate's bending stiffness, and v Poisson's ratio, then condition $\Pi > 0$, positive definite, reads explicitly

$$\Pi = \int\int_S \left\{ D\left[w_{xx}^2 + 2w_{xx}w_{yy} + w_{yy}^2 + 2(1-v)(w_{xy}^2 - w_{xx}w_{yy}) - \frac{q}{D}(a-x)w_x^2 \right] \right\} \, dS \geqslant 0. \qquad (2.3.16)$$

In this inequality,

$$w_x = \frac{\partial w}{\partial x}, \qquad w_{xx} = \frac{\partial^2 w}{\partial x^2}, \qquad w_{yy} = \frac{\partial^2 w}{\partial y^2}, \qquad w_{xy} = \frac{\partial^2 w}{\partial x \partial y},$$

and S denotes the surface of the plate. Later on, also the quantity

$$w_y = \partial w / \partial y$$

will be used.

Let the plate problem be given in the form of the boundary-eigenvalue problem

$$\left. \begin{array}{l} D\nabla^4 w + q[(a-x)w_x]_x = 0, \\ \text{plus boundary conditions,} \end{array} \right\} \qquad (P)$$

and let $q_c^{(0)}$ be the smallest eigenvalue of this problem. The specific value $q_c^{(0)}$ of q is usually considered to be the buckling load of the plate in the sense of the classical energy method.

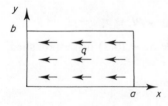

Fig. 64. Thin, elastic plate subjected to uniformly distributed, compressive forces

Now, use (P) to set up the identity

$$\int\int_S \{D\nabla^4 w + q\,[(a-x)w_x]_x\}\,w\,\mathrm{d}S = 0,$$

which by means of Gauss' integral theory and an application of the boundary conditions can be transformed into

$$\int\int_S \left\{ D\left[(\nabla^2 w)^2 + 2(1-v)(w_{xy}^2 - w_{xx}w_{yy}) - \frac{q}{D}(a-x)w_x^2\right]\right\}\,\mathrm{d}S = 0.$$

The equal sign holds specifically for $q \equiv q_c^{(0)}$. Hence

$$\int\int_S \{D[(\nabla^2 w)^2 + 2(1-v)(w_{xy} - w_{xx}w_{yy})]\,\}\,\mathrm{d}S = \frac{q_c^{(0)}}{D}\int\int_S (a-x)w_x^2\,\mathrm{d}S.$$

$$(2.3.17)$$

Using (2.3.17) in (2.3.16) yields the inequality

$$\Pi = (q_c^{(0)} - q)\frac{1}{D}\int\int_S (a-x)w_x^2\,\mathrm{d}S \geqslant 0$$

as the stability criterion. This inequality is obviously satisfied under the condition

$$q \leqslant q_c^{(0)} \qquad\qquad (2.3.18)$$

for the compressive load q. Therefore, the lowest eigenvalue $q_c^{(0)}$ of problem (P) seems indeed to be the limit load, and any load q below $q_c^{(0)}$ seems not to impair the stability of the plate. But what kind of stability is one implying? Orienting oneself according to Lyapunov's stability condition, one would expect that condition (2.3.18), which makes the total potential energy Π positive definite, would at least grant smallness of the plate's deflections, i.e.

$$\sup_{x,y\in S} |w| < \varepsilon$$

in the case of stability. Yet, is this expectation justified? That question cannot be answered off-hand in the positive. As a glance at (2.3.16) shows, Π does not even include the quantity w. Thus, it is therefore insufficient to conclude the stability investigation of the plate with the calculation of the lowest eigenvalue $q_c^{(0)}$ of problem (P) as it is commonly done when applying the classical energy method. One must in fact search further, and one must try to detect the very nature of stability following from a positive definite Π which is a consequence of (2.3.18).

Conclusions to be drawn from (2.3.16) require the application of adequate inequalities which are frequently obtained through Sobolev's embedding theorems, like Friedrich's or Poincaré's inequalities[1]. However, the form in which, for example, Friedrich's inequality is usually found in the literature is

---

[1]Smirnov, V. I., *A course of Higher Mathematics*, Vol. V, Addison-Wesley, Reading, MA., 1964.

often too weak for an application. Therefore, a more general form of this ine-quality has been developed[1], which already holds when the plate is clamped only along the edge $x = 0$ (see Fig. 64), while not satisfying any prescribed boundary conditions elsewhere. In this book, an even more general inequality substituting the one by Friedrich will be derived in accordance with a communication received from G. Fichera.

Consider the boundary value problem

$$\Delta u + \lambda u = 0, \qquad u_n = 0 \text{ on } B_1, \qquad u = 0 \text{ on } B_2, \qquad \text{(a)}$$

where u is a sufficiently smooth function, $u_n \equiv \partial u / \partial n$, λ an eigenvalue, $B_1 + B_2 = B$ where B is the boundary of the rectangle corresponding to the plate's surface in Fig. 64 and finally, B_1 is that part of B along which no edge conditions are prescribed for the plate, while B_2 is the remainder of B. It may specifically be pointed out that the possibility: $B_2 \equiv 0$, $B_1 \equiv B$, is not ruled out.

Green's identity yields

$$\iint_S u \, \Delta u \, dS = - \iint_S (u_x^2 + u_y^2) \, dS - \int_B u \, \frac{\partial u}{\partial n} \, dB. \qquad \text{(b)}$$

Hence, due to (b) and the boundary conditions

$$\iint_S u \, \Delta u \, dS + \lambda \iint_S u^2 \, dS = - \iint_S (u_x^2 + u_y^2) \, dS + \lambda \iint_S u^2 \, dS = 0. \text{ (c)}$$

From (c) it follows that

$$\lambda = \iint_S (u_x^2 + u_y^2) \, dS / \iint_S u^2 \, dS. \qquad \text{(d)}$$

Now, introduce the function

$$v = u + \eta, \qquad \text{(e)}$$

where η is a variation of u, satisfying the boundary conditions

$$\eta = \text{arbitrary along } B_1, \qquad \eta = 0 \text{ along } B_2, \qquad \text{(f, 1)}$$

$$v = \text{arbitrary along } B_1, \qquad v \equiv u = 0 \text{ along } B_2. \qquad \text{(f, 2)}$$

Consider

$$\delta\lambda = \frac{\left[\iint_S u^2 \, dS\right] \cdot \left[\iint_S 2(u_x \, \delta u_x + u_y \, \delta u_y) \, dS\right] - 2\left[\iint_S u \, \delta u \, dS\right]\left[\iint_S (u_x^2 + u_y^2) \, dS\right]}{\left[\iint_S u^2 \, dS\right]^2} \qquad \text{(g)}$$

under the assumption that $\delta u \equiv \eta$, which follows from (e). Then

$$\delta\lambda \left[\iint_S u^2 \, dS\right] = 2 \iint_S (u_x \eta_x + u_y \eta_y) \, dS - 2\lambda \iint_S u\eta \, dS$$

[1]Leipholz, H. H. E., On a special case of Friedrich's inequality, *Mech. Res. Comm.* 7, 329–333, 1980.

and

$$\frac{1}{2}\left[\iint_S u^2 \, \mathrm{d}S\right]\delta\lambda = \iint_S (u_x\eta_x + u_y\eta_y - \lambda u\eta) \, \mathrm{d}S. \qquad (\mathrm{h})$$

Yet,

$$\iint_S (u_x\eta_x + u_y\eta_y) \, \mathrm{d}S = -\iint_S \eta \, \Delta u \, \mathrm{d}S - \int_B \eta \frac{\partial u}{\partial n} \, \mathrm{d}B. \qquad (\mathrm{i})$$

Using (i) in (h) yields

$$\frac{1}{2}\left[\iint_S u^2 \, \mathrm{d}S\right]\delta\lambda = -\iint_S (\Delta u - \lambda u)\eta \, \mathrm{d}S - \int_B \eta \frac{\partial u}{\partial n} \, \mathrm{d}B. \qquad (\mathrm{k})$$

By virtue of (a) and (f, 1) the terms on the right-hand side of (k) vanish. Hence,

$$\delta\lambda = 0 \qquad \text{and} \qquad \lambda = \min \qquad \text{for } u \qquad (\mathrm{l})$$

in the space of admissible functions v defined by (e) and (f, 2). Hence,

$$\lambda \leqslant \iint_S (v_x^2 + v_y^2) \, \mathrm{d}S \Big/ \iint_S v^2 \, \mathrm{d}S \qquad (\mathrm{m})$$

where λ is the smallest eigenvalue of a rectangular membrane with a free edge along B_1, but satisfying there the 'natural condition', and with a supported edge along B_2.

Consider a plate with no prescribed boundary conditions along B_1 and clamped along B_2. Then the deflection w of the plate satisfies the conditions

$$w = 0, \qquad w_x = 0, \qquad w_y = 0 \qquad \text{along } B_2. \qquad (\mathrm{n})$$

Note again, that $B_2 \equiv 0$, $B_1 \equiv B$ is not excluded, as mentioned earlier.

First, choose $v \equiv w$ which is possible by virtue of (f, 2) and (n). Then (m) yields

$$\iint_S w^2 \, \mathrm{d}S \leqslant \frac{1}{\lambda} \iint_S (w_x^2 + w_y^2) \, \mathrm{d}S. \qquad (\mathrm{o})$$

Secondly, choose consecutively $v \equiv w_x$ and $v \equiv w_y$, which again is possible due to (f, 2) and (n). In these cases, (m) leads to

$$\iint_S w_x^2 \, \mathrm{d}S \leqslant \frac{1}{\lambda} \iint_S (w_{xx}^2 + w_{xy}^2) \, \mathrm{d}S \quad \text{and} \quad \iint_S w_y^2 \, \mathrm{d}S \leqslant \frac{1}{\lambda} \int_S (w_{xy}^2 + w_{yy}^2) \, \mathrm{d}S.$$

Combining the two inequalities results in

$$\iint_S (w_x^2 + w_y^2) \, \mathrm{d}S \leqslant \frac{1}{\lambda} \iint_S (w_{xx}^2 + 2w_{xy}^2 + w_{yy}^2) \, \mathrm{d}S. \qquad (\mathrm{p})$$

Inequalities (o) and (p) are the desired relations which will be used in the following calculations.

Obviously,

$$w_{xx}^2 + w_{yy}^2 + 2w_{xx}w_{yy} + 2(1 - v)(w_{xy}^2 - w_{xx}w_{yy}) > (1 - v)(w_{xx}^2 + 2\,w_{xy}^2 + w_{yy}^2),$$

since $v < 1$. Hence, inequality (2.3.16) can be changed into

$$\Pi > \iint_S \left\{ D\left[(1 - v)(w_{xx}^2 + 2w_{xy}^2 + w_{yy}^2) - \frac{q}{D}(a - x)w_x^2 \right] \right\} dS \geq 0.$$

Let $0 < \eta < 1$. Then the preceding inequality can be replaced by

$$\Pi > \iint_S D(1 - \eta)(1 - v)(w_{xx}^2 + 2w_{xy}^2 + w_{yy}^2) \, dS$$

$$+ \iint_S D\left[\eta(1 - v)(w_{xx}^2 + 2w_{xy}^2 + w_{yy}^2) - \frac{q}{D}(a - x)w_x^2 \right] dS \geq 0. \quad (2.3.19)$$

The first integral in (2.3.19) is never negative. The second integral in (2.3.19) has to be evaluated further. According to (p)

$$\iint_S D[\eta(1 - v)(w_{xx}^2 + 2w_{xy}^2 + w_{yy}^2)] \, dS > \lambda D\eta(1 - v) \iint w_x^2 \, dS.$$

Hence,

$$\iint_S D\left[\eta(1 - v)(w_{xx}^2 + 2w_{xy}^2 + w_{yy}^2) - \frac{q}{D}(a - x)w_x^2 \right] dS >$$

$$> \iint_S (\lambda D(1 - v)\eta - qa)w_x^2 \, dS.$$

From this inequality follows that the second integral in (2.3.19) is never negative if

$$q < (\lambda\eta(1 - v)D)/a. \quad (2.3.20)$$

Assuming that condition (2.3.20) is satisfied,

$$\Pi > \iint_S D(1 - \eta)(1 - v)(w_{xx}^2 + 2w_{xy}^2 + w_{yy}^2) \, dS \geq 0. \quad (2.3.21)$$

By means of (o) and (p) one finds that

$$\iint_S (w^2 + w_x^2 + w_y^2 + w_{xx}^2 + w_{xy}^2 + w_{yy}^2) \, dS < \left(\frac{1}{\lambda^2} + \frac{1}{\lambda} + 1 \right)$$

$$\iint_S (w_{xx}^2 + 2w_{xy}^2 + w_{yy}^2) \, dS.$$

Since $\lambda > 0$, and since

$$\left[\left[\int_S (w^2 + w_x^2 + w_y^2 + w_{xy}^2 + w_{yy}^2) \, dS \right]^{1/2} \equiv \| w \|_{w_2^2} \right]$$

is the Sobolev norm in the w_2^2-space, the previous relation can be rewritten to yield

$$(\| w \| w_2^2)^2 < \left(\frac{\lambda + 1}{\lambda} \right)^2 \iint_S (w_{xx}^2 + 2w_{xy}^2 + w_{yy}^2) \, dS. \quad (2.3.22)$$

213

Comparing (2.3.22) with (2.3.21) yields

$$\Pi > \frac{D(1-\eta)(1-\upsilon)\lambda^2}{(1+\lambda)^2}\,(\|\,w\,\|_{w_2^2})^2.$$

Moreover, a Sobolev embedding theorem[1] states that for the case of a two-dimensional continuum like a plate,

$$\sup_{w\in S}|\,w\,| \leqslant C\,\|\,w\,\|_{w_2^2},$$

where C is an appropriate constant. Therefore,

$$\sup_{w\in S}|\,w\,|^2 \leqslant \frac{C^2}{D(1-\eta)(1-\upsilon)}\left(1+\frac{1}{\lambda}\right)^2\Pi. \qquad (2.3.23)$$

On the basis of (2.3.23) one can conclude as follows: bounded and positive definite total energy Π of the plate guarantees smallness of the deflections w throughout the plate for all possible boundary conditions. This smallness is slightly affected by the boundary conditions through the various values which λ assumes for the different boundary conditions.

The preceding statement seems to be satisfactory with regard to the plate's stability, as it apparently confirms the conclusions of the classical theory concerning the effect of the positive definiteness of Π on the smallness of the deflections of a stable plate. Yet, this is not completely true: while the classical theory grants stability for loads satisfying inequality (2.3.18), the stability and smallness of deflections granting inequality (2.3.23) is based on the stricter condition (2.3.20), which prescribes an upper level for safe loading definitely below the classical buckling load $q_c^{(0)}$. It is also interesting to note that when trying on the one hand to push up the upper bound of the permissible load in condition (2.3.20) by making η as close as possible to 'one', one lowers on the other hand the quantity $(1-\eta)$, thus, increasing the upper bound of the maximum deflections according to (2.3.23). But, that means a potential deterioration of stability.

The preceding deliberations have shown a pronounced difference between the column and the plate problem. In the first case, the classical energy criterion yielded a critical load value which limited a safe range of loads within which the column behaved stable in the rigorous sense of Lyapunov's stability condition. In the second case, the case of the plate, the classical energy criterion seems to be unreliable as it yields a critical load limit which does not guarantee smallness of the plate's deflections at all points of the plate's surface. In order to satisfy Lyapunov's stability of smallness of the deflections, one must restrict oneself to the load condition (2.3.20), which prescribes a stability limit which is below the limit given by $q_c^{(0)}$

One may conclude in the light of these findings that indiscriminate application of the classical energy criterion is dangerous and can lead to an

[1]Smirnov, V. I., *A Course of Higher Mathematics*, Vol. V, Addison-Wesley, Reading, MA, p. 340, 1964.

overestimation of stability especially in the case of systems with dimensions larger than one.

2.3.4 Outline of an Algebraic Theory

The principle of virtual work gives the following equation for the perturbed motion of an elastic body about the original state:

$$\int_V \mu w_{tt} \, \delta w \, \mathrm{d}V + \delta(\Pi_i + \Pi_a) - \int_0 k_a \, \delta w \, \mathrm{d}O - \sum_i K_{ai} \, \delta w_i = 0. \qquad (2.3.24)$$

The following notation is used in equation (2.3.24): μ is the mass density, w the displacement vector, Π_i the potential of the internal forces, Π_a the potential of the external surface forces (if this is present), k_a are the non-conservative or polygenic external surface forces, K_{ai} the non-conservative or polygenic external concentrated forces, δw_i the displacement vectors of their points of application, V is the volume, O the surface area of the elastic body, and the subscripts tt to the variable w indicate partial differentiation.

Example: As is already known from Section 2.3.2, it is possible to obtain the differential equations and the boundary conditions of the stability problems under discussion from (2.3.24). For simplicity, let us demonstrate this for the column shown in Fig. 65 with the conservative load P_v and the non-conservative loads P_t and q. Equation (2.3.24) gives

$$\int_0^l \mu \ddot{w} \, \delta w \, \mathrm{d}x + \delta \left\{ \frac{1}{2} \int_0^l \left[\alpha w_{xx}^2 - \left(P_v + P_t + \int_0^l q(\xi) \, \mathrm{d}\xi \right) w_x^2 \right] \mathrm{d}x \right\}$$

$$+ \int_0^l q w_x \, \mathrm{d}x + P_t w_x(l) \, \delta w(l) = 0. \qquad (2.3.25)$$

The second term of (2.3.25) results in the following expression after the varia-

Fig. 65. Buckling column with non-conservative load

tion is carried out:

$$\delta\left\{\frac{1}{2}\int_0^l\left[\alpha w_{xx}^2 - \left(P_v + P_t + \int_x^l q(\xi)\,d\xi\right) w_x^2\right] dx\right\}$$

$$= \int_0^l (\alpha w_{xxxx} + Q w_{xx} - q w_x)\,\delta w\,dx + [\alpha w_{xx}\,\delta w_x - \alpha w_{xxx}\,\delta w - Q w_x\,\delta w]_0^l,$$
$$(2.3.26)$$

with

$$Q = P_v + P_t + \int_x^l q(\xi)\,d\xi.$$

If this is substituted into (2.3.25), the result is

$$\int_0^l (\mu\ddot{w} + \alpha w_{xxxx} + Q w_{xx})\,\delta w\,dx + P_t w_x(l)\,\delta w(l)$$

$$+ \left[\alpha w_{xx}\,\delta w_x - \alpha w_{xxx}\,\delta w - P_t w_x\,\delta w - \left(P_v + \int_x^l q(\xi)\,d\xi\right) w_x\,\delta w\right]_0^l = 0,$$

or

$$\int_0^l [\mu\ddot{w} + L(w)]\,\delta w\,dx + R(w) = 0, \qquad (2.3.27)$$

where L is the linear differential operator $\alpha\partial^4/\partial x^4 + Q\partial^2/\partial x^2$ and $R(w)$ is the boundary value expression

$$R(w) = P_t w_x(0, t)\,\delta w(0, t)$$

$$+ \left\{\alpha w_{xx}\,\delta w_x - \alpha w_{xxx}\,\delta w - \left[P_v + \int_0^l q(\xi)\,d\xi\right] w_x\,\delta w\right\}_{x=0}^{x=l}.$$

We require $R(w) = 0$, which results in the boundary values for the problem. The column shown in Fig. 65 is clamped at the bottom, which implies the geometric conditions $w_x(0, t) = w(0, t) = 0$. If $R(w) = 0$ holds, the dynamic boundary conditions

$$w_{xx}(l, t) = 0, \qquad w_{xxx}(l, t) = -(P_v/\alpha)w_x(l, t)$$

must be satisfied also. If the column is supported in a different way, the geometric boundary conditions are different and dynamical boundary conditions change as well. Because the variation δw is arbitrary, we must have

$$\mu\ddot{w} + L(w) = 0, \qquad (2.3.28)$$

which leads to the differential equations for the perturbed motion. Equation (2.3.27) therefore describes all the features of the column for the selected loading conditions.

The next step in solving the stability problem at hand consists of solving the partial differential equation obtained from the principle of virtual work. In

216

our example this is Eq. (2.3.28). Since an exact solution is usually not possible, the question of a *sufficient mathematical approximation method* is raised, which was already pointed out at the end of Section 2.3.1. Since we are considering the presence of non-conservative or polygenic loads, the differential expression $L(w)$ in (2.3.28) can be completely general and, in particular, it can be non-self-adjoint. This means that the Ritz method cannot be used. Yet, according to Section 1.7.5, the Galerkin method can still be used and converges. We have thus obtained the following result: The *Galerkin method* becomes the most general basis of the stability theory of elastomechanics. This method was investigated in Section 1.7.5, and the conditions of applicability of the method can be obtained from there.

We will go further than the mathematical discussion of Section 1.7.5 by showing that the *Galerkin* method also converges for *non-conservative column problems*.

The discussion will be based on known theorems of functional analysis which can be found in the work of Michlin (12). The theorems imply the following.

Let the equation $u + Tu = 0$ be satisfied by the elements u of a Hilbert space H. The approximate solution obtained according to the Galerkin method converges, based on the norm of H, towards the exact solution u_0, if the equation has a unique solution and the operator T is completely continuous in H.

According to (2.3.28), the following differential equation holds for *column problems*:

$$\mu \ddot{w} + \alpha w_{xxxx} + Q w_{xx} = 0,$$

which is a special case of the general type

$$w_{xxxx} + a_1 w_{xx} + a_2 w_{tt} + a_3 w_t = 0. \tag{2.3.29}$$

The product trial solution $w = e^{\omega t}\phi(x)$ applied to (2.3.29), and its boundary conditions results in the ordinary boundary value problem

$$\phi^{\text{iv}} + a_1 \phi'' + b \phi = 0, \qquad U[\phi]_R = 0, \qquad b = a_2\omega^2 + a_3\omega. \tag{2.3.30}$$

In operator notation, the differential equation (2.3.30) is

$$A\phi + K\phi = 0,$$

where

$$A = \mathrm{d}^4/\mathrm{d}x^4, \qquad K = a_1(\mathrm{d}^2/\mathrm{d}x^2) + b,$$

which can be transformed into

$$\phi + A^{-1}K\phi = 0.$$

According to the theorem above, in order to apply the Galerkin method it is necessary to show that the operator $T = A^{-1}K$ is *completely continuous,* because Eq. (2.3.30) will certainly have a unique solution for the prescribed boundary conditions.

As Michlin shows, $T = A^{-1}K$ is completely continuous if A^{-1} is completely continuous and K is bounded.

We have $A^{-1}\phi(x) = \int_0^l G(x, \xi)\phi(\xi)\,d\xi$, where l is the column length and $G(x, \xi)$ is the Green function corresponding to the operator $A = d^4/dx^4$ for the boundary conditions $U[\phi]_R = 0$.

The following

$$U[\phi]_R = \begin{cases} \phi(0) = \phi(l) = \phi''(0) = \phi''(l) & = 0 \text{ (column hinged at both ends)} \\ \phi(0) = \phi(l) = \phi'(0) = \phi'(l) & = 0 \text{ (column clamped at both} \\ & \quad \text{ends)} \\ \phi(0) = \phi(l) = \phi'(0) = \phi''(l) & = 0 \text{ (column clamped at one end} \\ & \quad \text{and hinged at the other} \\ & \quad \text{end)} \\ \phi(0) = \phi'(0) = \phi''(l) = 0, \phi'''(l) = 0 & \text{ (column clamped at one end} \\ & \quad \text{and free at the other end} \\ & \quad \text{without a conservative} \\ & \quad \text{load } P_v \equiv 0) \end{cases}$$

are boundary conditions for which a Green's function[1] can easily be found. This function is continuous and bounded, and it can be shown that A^{-1} is completely continuous because $G(x, \xi)$ is quadratically integrable.

It remains to be shown that K is bounded. We have

$$\| K\phi \| \leqslant |a_1|_{max} \| \phi'' \| + |b|_{max} |\phi\|, \tag{2.3.31}$$

where

$$\| u \| = \left(\int_0^l u^2\,dx \right)^{1/2}$$

is the ordinary norm of an element u. The scalar product

$$(A\phi, \phi) = \int_0^l (A\phi) \cdot \phi\,dx$$

can be changed into the following form by partial integration and by consideration of the boundary conditions:

$$(A\phi, \phi) = \int_0^l \phi''^2\,dx = \| \phi'' \|^2. \tag{2.3.32}$$

Applying the Cauchy–Schwarz inequality, we obtain

$$\phi^2 \leqslant x \int_0^x \phi'^2\,dx \leqslant x \int_0^l \phi'^2\,dx = x \| \phi' \|^2$$

and therefore

$$\| \phi \|^2 \leqslant (l^2/2) \| \phi' \|^2. \tag{2.3.33}$$

[1]Collatz, L., *Eigenvalue Problems and their Numerical Treatment*, Akademische Verlagsgesellschaft Geest and Portig K. -G., Leipzig, 1945.

From (2.3.9) we have

$$\| \phi' \|^2 \leqslant (l^2/2) \| \phi'' \|^2, \qquad (2.3.34)$$

and making use of (2.3.33), we obtain

$$\| \phi \|^2 \leqslant (l^2/4) \| \phi'' \|^2. \qquad (2.3.35)$$

If the calculation is carried out in a function space that is bounded by

$$(A\phi, \phi) \leqslant N^2,$$

then, according to (2.3.32) and (2.3.35), we have

$$\| \phi'' \| \leqslant N, \qquad \| \phi \| \leqslant (l^2/2)N.$$

Substituting into (2.3.31) results in

$$\| K\phi \| \leqslant CN$$

with the appropriately chosen constants C and N. Consequently, K is bounded. This means that all conditions for completely continuous T, and therefore for the convergence of the Galerkin method, are satisfied. It should be pointed out again that the convergence *is with respect to the norm of H*. This norm is defined as follows for an element ϕ of H:

$$\| \phi \|_{(H)} = (A\phi, \phi)^{\frac{1}{2}}.$$

Now that the convergence of the Galerkin method has been established, the results of this method will be examined.

Let us use the trial solution

$$w(x, t) = a_i w_i(x, t) = a_i f(t)\phi_i(x) = f_i(t)\phi_i(x). \qquad (2.3.36)$$

A complete and orthonormal function system $\phi_i(x)$ has been selected, which satisfies all the boundary conditions so that

$$R(w) = R[f_i(t)\phi_i(x)] = 0.$$

As already discussed in Section 1.7.5, the *eigenfunctions $\phi_i(x)$ of the 'abbreviated' problem* are selected. They are obtained from (2.3.28) with the product trial solution $w = e^{\omega t}\phi(x)$ if all terms that disturb the self-adjoint property are crossed out.

Substitution of (2.3.36) into (2.3.27) results in

$$\int_0^l \{\mu[\ddot{f}_k\phi_k(x)] + L[f_k(t)\phi_k(x)]\} [f(t)\phi_r(x)\,\delta a_r]\,\mathrm{d}x = 0. \quad (2.3.37)$$

Since the variations δa_r are arbitrary, (2.3.37) becomes the following system of equations:

$$\int_0^l \mu f(t)\phi_i(x)[\ddot{f}_k(t)\,\phi_k(x)]\,\mathrm{d}x + \int_0^l f(t)L[f_k(t)\phi_k(x)]\phi_i(x)\,\mathrm{d}x = 0$$

$$i = 1, 2, 3, \ldots. \qquad (2.3.28)$$

Let us now consider the orthonormal property of the coordinates $\phi_k(x)$ and let us set $\mu = \text{const}$. It is assumed that $f(t)$ is a function that does not vanish. Equation (2.3.38) then becomes

$$\mu \ddot{f}_i + f_k \int_0^l L(\phi_k)\phi_i \, dx = 0, \qquad i = 1, 2, 3, \ldots . \qquad (2.3.39)$$

Making use of

$$a_{ik} = (1/\mu) \int_0^l L(\phi_k)\phi_i \, dx,$$

(2.3.39) can be written in the form of the linear, homogeneous system of differential equations

$$\ddot{f}_i + a_{ik}f_k = 0, \qquad i, k = 1, 2, 3, \ldots . \qquad (2.3.40)$$

The trial solution $f_i = a_i e^{\omega t}$ is used so that $f(t) = e^{\omega t}$, which justifies the assumptions made above regarding $f(t)$.

From (2.3.40) it can be seen that the Galerkin method makes it possible to treat the stability problem of elastic-body equilibrium the same way as the problem of small oscillations that a discrete system of point masses with an infinite number of degrees of freedom carries out around its original position.

Introducing

$$f_k(t) = a_k e^{\omega t}; \qquad (2.3.41)$$

Eq. (2.3.40) becomes

$$(\mathbf{A} + \omega^2 \mathbf{E})\mathbf{X} = 0, \qquad (2.3.42)$$

where $\mathbf{A} = (a_{ik})$, $\mathbf{X} = (a_i)$ and $\mathbf{E}$ is the unit matrix. The stability problem is reduced to a discussion of the algebraic equation (2.2.42) for the components a_i of the vector $\mathbf{X}$ that are the coefficients of the trial solution (2.3.41). Since all further calculations are carried out using algebra and because stability is expressed in terms of algebraic theorems, we are justified in calling this theory the *algebraic stability theory,* which is the title of the following section.

Algebraic stability theory

The following equation is obtained from (2.3.42):

$$\det(\mathbf{A} + \omega^2 \mathbf{E}) = 0. \qquad (2.3.43)$$

If

$$\omega = \pm(-\lambda_A)^{1/2} \qquad (2.3.44)$$

and if λ_A is the eigenvalue of the matrix $\mathbf{A}$, Eq. (2.3.43) is equivalent to the eigenvalue problem

$$\det(\mathbf{A} - \lambda_A \mathbf{E}) = 0. \qquad (2.3.45)$$

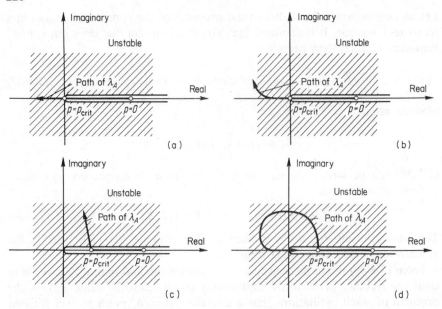

Fig. 66. Path of the smallest eigenvalue in the complex plane. (a) and (b) Behaviour of λ_A for a symmetric matrix **A**, for a matrix **A** which can be made symmetric, and for a slightly asymmetric matrix **A**. (c) and (d) Behaviour of λ_A for a strongly asymmetric matrix **A**

According to (2.3.36) and (2.3.41), $w(x, t) = a_k\, e^{\omega t} \phi_k(x)$, so that the behaviour of the disturbed motion as well as stability is decided by the term $e^{\omega t}$. If ω is *imaginary*, there is an oscillating motion that can be called *stable*. Whether or not this is the case depends, according to (2.3.44), on the behaviour of the eigenvalues λ_A. In the final analysis, stability is decided by λ_A and the structure of the matrix **A**. The stability investigation is carried out by tracing the path of λ_A in the complex plane (Fig.66). The positive real axis is the stability region (only if λ_A is positive and real can ω be imaginary according to (2.3.44), which is sufficient for stability). The path of λ_A results because λ_A is a function of a parameter p (for example, the load). If p is varied, the image point of λ_A changes position in the complex plane. Stability is lost when the image point leaves the positive real axis. This can occur if either λ_A passes through the origin or if λ_A becomes complex. The parameter value for which this occurs is the *critical value* P_{crit}, which is the buckling value. We shall restrict ourselves to considering only the *smallest* critical value, which corresponds to the case where one of the eigenvalues λ_A leaves the stability region for the first time.

Matrix structure investigations

The behaviour of λ_A depends on the *structure of the matrix A*.

Two cases must be distinguished: **A** is symmetric or Hermitian or **A** is asymmetric or non-Hermitian.

Let us consider the first case and assume that $\mathbf{A}$ is *symmetric or Hermitian*. Then $\mathbf{A}$ has only *real eigenvalues* and the image point λ_A must always move along the real axis. As shown in Fig. 66a, stability can only be lost if λ_A passes through the origin and becomes negative. ω then has a positive real part according to (2.3.44), and instability occurs according to Theorem 2 of Section 1.3.1. The stability boundary is given by $\lambda_A = 0$.

λ_A is a root of the characteristic equation (2.3.45) and can only be zero if the absolute term of this equation, — that is, det $\mathbf{A}$ — vanishes. In the case of a symmetric matrix $\mathbf{A}$, instead of $\lambda_A = 0$, the criterion

$$\det \mathbf{A} = 0 \qquad (2.3.46)$$

can also be used, which is nothing other than the *static stability criterion,* as we shall now see. If a static calculation is carried out under the assumption that a non-trivial equilibrium position exists, then according to the principle of virtual work the following is obtained in place of (2.3.24):

$$\delta(\Pi_i + \Pi_a) - \int_0^l k_a \, \delta w \, dO - \sum_i K_{ai} \, \delta w_i = 0. \qquad (2.3.24a)$$

Instead of (2.3.27), we obtain

$$\int_0^l L(w) \, \delta w \, dx + R(w) = 0. \qquad (2.3.27a)$$

Instead of using (2.3.36), we now use the trial solution which is independent of time t:

$$w(x) = \rho_i \phi_i(x), \qquad (2.3.36a)$$

and (2.3.27a) becomes

$$\int_0^l L(\rho_k \phi_k)(\phi_r \, \delta \rho_r) \, dx = 0. \qquad (2.3.37a)$$

From (2.3.37a) it finally follows that

$$a_{ik}\rho_k = 0, \qquad a_{ik} = \int_0^l L(\phi_k)\phi_i \, dx, \qquad i = 1, 2, 3, \ldots . \qquad (2.3.40a)$$

and a non-trivial equilibrium position exists if the homogeneous system of equations (2.3.40a) results in a non-trivial solution for the coefficients ρ_k. The necessary condition for this is det $\mathbf{A} = 0$. But this is exactly (2.3.46), which shows that this condition is indeed the static stability criterion expressed in an algebraic way, because, if non-trivial equilibrium positions exist, the stability disappears.

Based on these results, we may now attempt to answer the second question of Section 2.3.1, which is concerned with the validity of the static stability criterion. It can at least be used if the matrix $\mathbf{A}$ is symmetric!

It will be shown below that $\mathbf{A}$ is always *symmetric* for *conservative or monogenic* problems, so that the static criterion is acceptable for this class of problems.

In the conservative or monogenic case, Eq. (2.3.24a) reduces to the following: because of $k_a = 0$, $K_{ai} = 0$

$$\delta(\Pi_i + \Pi_a) = \delta\Pi = 0. \qquad (2.3.24b)$$

According to the discussion in Section 2.3.3,

$$\Pi = \Pi_g + \tfrac{1}{2}\,\delta^2\Pi_g.$$

Since $\delta\Pi_g = 0$ holds for the original state according to the definition, it follows that

$$\delta\Pi = \tfrac{1}{2}\,\delta(\delta^2\Pi_g)$$

so that (2.3.24b) becomes $\tfrac{1}{2}\,\delta(\delta^2\Pi_g) = 0$.

The following relationship can be found from (2.3.26):

$$\tfrac{1}{2}\,\delta(\delta^2\Pi_g) = \int_0^l L(w)\,\delta w\,\mathrm{d}x + R(w),$$

and in addition this must now be made zero. With the trial solution $w = \rho_i\phi_i(x)$, which satisfies the boundary conditions, we find $R(w) = 0$, and the relationship

$$\tfrac{1}{2}\,\delta(\delta^2\Pi) = \int_0^l L(w)\,\delta w\,\mathrm{d}x = 0$$

is obtained. Instead of using (2.3.27a), we may substitute (2.3.36a) into the completely equivalent relationship $\tfrac{1}{2}\,\delta(\delta^2\Pi_g) = 0$. $\delta^2\Pi_g$ consists only of second-order terms, and, in particular, for the example of a column problem, it consists of *quadratic terms* only. For the elastic column, it has the form

$$\delta^2\Pi_g = \int_0^l \sum_i c_i D_i^2(w)\,\mathrm{d}x, \qquad (2.3.47)$$

where c_i are coefficients and $D_i(w)$ are various differential quotients. Because of (2.3.47) we have

$$\delta(\delta^2\Pi_g) = 2\int_0^l \sum_i c_i D_i(w)\,\delta D_i(w)\,\mathrm{d}x = 0,$$

which according to (2.3.36a), becomes

$$\int_0^l \sum_i (c_i D_i)\left[\sum_k \rho_k\phi_k\right] D_i\left[\sum_k \phi_k\,\delta\rho_k\right]\mathrm{d}x = 0.$$

Since the variations are independent, the system of equations

$$\sum \rho_k \int_0^l \sum_i \{c_i D_i(\phi_k)D_i(\phi_j)\}\,\mathrm{d}x = 0, \qquad j = 1, 2, 3, \ldots,$$

follows. It can be written in the abbreviated form

$$a_{jk}\rho_k = 0, \qquad j = 1, 2, 3, \ldots.$$

where

$$a_{jk} = \int_0^l \sum_i c_i D_i(\phi_k) D_i(\phi_j)\, \mathrm{d}x. \tag{2.3.48}$$

The static stability criterion det $\mathbf{A}$ = det a_{jk} = 0 makes use of a symmetric matrix $\mathbf{A} = (a_{jk})$, because $a_{jk} = a_{kj}$ follows from (2.3.48). If follows from this discussion that the matrix $\mathbf{A}$ is always symmetric for conservative or monogenic problems and the static stability criterion can be used.

The *self-adjoint property* of the differential expression $L(\phi)$ follows from the symmetry of the a_{jk}. This differential expression plays a role in the differential equation of the problem. If a_{jk} (2.3.48) is symmetric, the element a_{ik} (2.3.40a) that is equivalent to it must also be symmetric. This means that

$$\int_0^l L(\phi_k)\phi_i\, \mathrm{d}x = \int_0^l L(\phi_i)\phi_k\, \mathrm{d}x,$$

which means that $L(\phi)$ is self-adjoint. Therefore, conservative or monogenic stability problems have self-adjoint differential equations so that the *Ritz method* can be applied to them, as was shown in Section 1.7.5.

Example: Let us use the buckling column with conservative load as an example. We have $P_t = 0, q = 0, P_v \neq 0$

$$\delta^2\Pi_g = \tfrac{1}{2} \int_0^l (\alpha w_{xx}^2 - P_v w_x^2)\, \mathrm{d}x.$$

$\delta^2\Pi_g$ has quadratic terms as was postulated. From $\delta(\delta^2\Pi_g) = 0$ it follows that

$$\int_0^l (\alpha w_{xx}\, \delta w_{xx} - P_v w_x\, \delta w_x)\, \mathrm{d}x = 0.$$

Therefore, $c_1 = \alpha$, $c_2 = -P_v$, $D_1 = \partial^2/\partial x^2$, $D_2 = \partial/\partial x$ and the trial solution $w = \rho_i\phi_i(x)$ results in

$$a_{jk} = \int_0^l (\alpha\phi_k''\phi_j'' - P_v\phi_k'\phi_j')\, \mathrm{d}x,$$

which is symmetric. The differential equation of the problem is

$$L(w) = \alpha w_{xxxx} + P_v w_{xx} = 0,$$

or, after making use of $w = \rho_i\phi_i(x)$,

$$L(\phi_i) = \alpha\phi_i^{iv} + P_v\phi_i'' = 0.$$

For the boundary conditions

$$U[\phi]_R = \begin{cases} \phi(0) = \phi(l) = \phi''(0) = \phi''(l) = 0, \text{ (column hinged at both ends)} \\ \phi(0) = \phi(l) = \phi'(0) = \phi'(l) = 0, \text{ (column clamped at both ends)} \\ \phi(0) = \phi(l) = \phi'(0) = \phi''(l) = 0, \text{ (column clamped at one end and} \\ \qquad\qquad\qquad\qquad\qquad\qquad\qquad \text{hinged at the other end)} \\ \phi(0) = \phi'(0) = \phi''(l) = 0, \phi'''(l) = -(P_v/\alpha)\phi'(l), \text{ (column} \\ \qquad\qquad\qquad\qquad\qquad\qquad \text{clamped at one end and} \\ \qquad\qquad\qquad\qquad\qquad\qquad \text{free at the other end)} \end{cases}$$

224

The differential expression $L(\phi)$ is self-adjoint, as can easily be established. This is because for two functions u, v, which satisfy these boundary conditions, the relationship

$$\int_0^l L(v)\, u\, dx = \int_0^l L(u)v\, dx$$

always holds.
The discussion above can be summarized as follows.

Theorem: *For conservative or monogenic stability problems the matrix* **A** *is symmetric, the differential equation is self-adjoint, and the static stability criterion is applicable.*

It is interesting to note that the static criterion is not restricted to conservative or monogenic problems. It can remain valid for *non-conservative* problems, as will now be shown.

The nature of a stability problem can be inferred from the load frequency curve (eigenvalue curve) of the buckling column hinged at both ends with a concentrated load having constant direction. The buckling value P_{crit} and the stability boundary is the location where this curve intersects the load axis (P axis). Since $\omega \equiv 0$ for the intersection point, P_{crit} is the load for which a nontrivial equilibrium position is possible. It may therefore be concluded that only when the eigenvalue curves interesect the load axis can non-trivial equilibrium positions exist. The eigenvalue curves of other conservative problems have a similar course.

Example: For the column clamped at one end and for a concentrated load load with fixed direction at the free end, if

$$a = \mu/\alpha, \qquad p = P/\alpha,$$
$$\lambda_{1,2} = [\tfrac{1}{2}(4a\omega^2 + p^2)^{1/2} \mp p]^{1/2}$$

(where μ is the specific mass, α the bending stiffness, P the load on the column), the eigenvalue equation is

$$2a\omega^2 - a^{1/2}p\omega \sinh(\lambda_1 l)\sin(\lambda_2 l) + (2a\omega^2 + p^2)\cosh(\lambda_1 l)\cos(\lambda_2 l) = 0.$$

The important branch that results in the buckling load is shown in Fig.67.
It is possible for non-conservative problems to have completely similar eigenvalue curves. The approximate eigenvalue equation for a beam pinned at both ends with a uniformly distributed tangential load (Fig. 68) is given by

$$q = 56.7\frac{\alpha}{l^3} - 0.15\,\mu l\omega^2 \pm \left(1424.6\frac{\alpha^2}{l^6} + 2.5\frac{\alpha\mu\omega^2}{l^2} + 0.011\,\mu^2 l^2\omega^4\right)^{1/2}.$$

Fig. 69 shows two of the branches. The eigenvalue curve intersects the load axis in this case as well. Therefore, non-trivial equilibrium positions exist. Even though the problem is non-conservative, the stability can only disappear, as can be seen from the course of the eigenvalue curve, because of the fact that

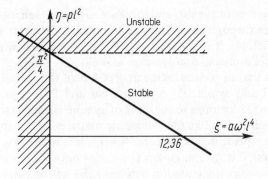

$$2\xi - \xi^{1/2}\nu|\eta|\sinh\lambda_1\sin\lambda_2 + (2\xi + \eta^2)\cosh\lambda_1\cos\lambda_2 = 0$$

$$\lambda_{1,2} = \sqrt{\frac{(4\xi + \eta^2)^{1/2} \mp \eta}{2}}$$

Fig. 67. Eigenvalue curve

Fig. 68. Buckling column with non-conservative load

Fig. 69. Eigenvalue curve

226

these intersection points exist and because ω^2 as a function of the load parameter passes through zero. The stability limit is therefore given by $\omega = 0$. According to (2.3.44), it is given by $\lambda_A = 0$, i.e., the static criterion det $\mathbf{A} = 0$ is decisive in this non-conservative case as well.

It can be seen that in such cases the matrix $\mathbf{A}$ is of the special form $\mathbf{A} = \mathbf{HD}$, where $\mathbf{H}$ is real and symmetric or Hermitian and $\mathbf{D}$ is a positive diagonal matrix. Therefore, it belongs to the class of symmetrizable matrices that have only real eigenvalues just like the symmetric matrices. If $\mathbf{A}$ can be made symmetric, it can be represented in the form $\mathbf{A} = \mathbf{T}^{-1}\tilde{\mathbf{H}}\mathbf{T}$, where $\tilde{\mathbf{H}}$ is real symmetric or Hermitian and $\mathbf{T}$ is non-singular. The eigenvalues $\lambda_A = -\omega^2$ of $\mathbf{A}$ are the roots of the characteristic equation $\det(\mathbf{A} - \lambda_A \mathbf{E}) = 0$. It can be transformed into

$$\det(\mathbf{A} - \lambda_A\mathbf{E}) = \det(\mathbf{T}^{-1}\tilde{\mathbf{H}}\mathbf{T} - \lambda_A\mathbf{E}) = \det[\mathbf{T}^{-1}(\tilde{\mathbf{H}} - \lambda_A\mathbf{E})\mathbf{T}]$$
$$= \det\mathbf{T}^{-1}\det\mathbf{T}\det(\tilde{\mathbf{H}} - \lambda_A\mathbf{E}) = 0,$$

or

$$\det(\mathbf{A} - \lambda_A\mathbf{E}) = \det(\tilde{\mathbf{H}} - \lambda_A\mathbf{E}) = 0.$$

From this it is seen that λ_A is also an eigenvalue of the symmetric or Hermitian matrix $\tilde{\mathbf{H}}$ and therefore is always real. This means that the image point of $\lambda_A = -\omega^2$ can only move along the real axis when the load parameter is varied, just as was the case in Fig.66a for conservative problems. Stability can only be lost when λ_A passes through the origin, and this means that the static criterion is applicable and decisive.

The 'Pflüger' buckling column shown in Fig.68 is an example of *non-conservative problems with a symmetrizable matrix*. Another example is the beam hinged at both ends with the axial torsion T (Fig. 70). An investigation

Fig. 70. Column with axial torsion moment

when a column may have a symmetrizable matrix and is therefore a so-called divergence type system, has been carried out in [1].

The final results can be expressed as follows.

Theorem: *The static stability criterion is applicable and decisive in the case of non-conservative or polygenic stability problems if the matrix* **A** *of the problem can be made symmetrical, i.e., if* **A** *is symmetrizable.*

Even if the problem is non-conservative or polygenic and the matrix cannot be made symmetrical, i.e., there is a *truly asymmetrical* (or non-Hermitian) matrix, the static criterion sometimes can be applicable and decisive.

Example: Let us assume that the structure of the matrix **A** is governed by a parameter. The column discussed in Section 1.7.5 with a subtangential compressive load (Fig. 35) is an example of this. The parameter α, which specifies the inclination of the column, determines the structure of the matrix **A**. The reader is referred to the calculation presented in Section 1.7.5. According to this calculation and to the degree of approximation considered, the eigenvalue problem is given by

$$\det[\mathbf{A} + \omega^2 \mathbf{C}] = 0,$$

where

$$\mathbf{A} = \begin{pmatrix} c_1 & d_1 \\ d_2 & c_2 \end{pmatrix}, \qquad \mathbf{C} = \begin{pmatrix} a_1 & b \\ b & a_2 \end{pmatrix},$$

according to (1.7.76). The meaning of the elements $a_1, a_2, b, c_1, c_2, d_1, d_2$ can be found in Section 1.7.5. Apparently the matrix **C** is a function of the parameter α, because it is a function of c_2 and d_1. The matrix **C** takes the place of the matrix **E** of the special eigenvalue problem $\det[\mathbf{A} + \omega^2 \mathbf{E}] = 0$. Since **C** is symmetric and always positive definite, the nature of the problem is not changed.

It is assumed that the elements of the matrix **A** are entire algebraic and continuous functions of the structural parameter α and the load parameter p, as is the case in the example ($p \equiv P$, where P is the concentrated load applied to the column). Let us assume that **A** becomes symmetric for $\alpha = 0$ and is also positive definite for $p = 0$. In other words: The non-conservative problem becomes a conservative one for $\alpha = 0$, and for $p = 0$ it becomes the problem of free oscillations of the elastic body.

It can easily be seen that this is the case for the example mentioned above. Assuming that only positive values of p and the smallest critical load is of interest, we must still determine whether and how long the static criterion remains in effect. For this purpose, let us consider the *stability diagram* plotted in the p, α plane. It is shown in Fig. 36 for example, and Fig. 71 shows a more

[1]Leipholz, H., *Direct Variational Methods and Eigenvalue Problems in Engineering*, pp. 62–74, Noordhoff Int. Publ., Leyden, 1977.

228

general case. The boundaries of the instability regions are partly given by the static criterion det $\mathbf{A} = 0$ and partly by the kinetic criterion, which will now be discussed. Which of the criteria is applicable depends on the different α intervals.

It is certain that at least *one* interval must exist, the interval $0 \leqslant \alpha \leqslant \alpha_{g,min}$, in which the static criterion holds in spite of the non-conservative nature of the problem. This means that the static criterion remains in force for sufficiently small α, even through for $\alpha > 0$ the matrix $\mathbf{A}$ is non-symmetric. This can be expressed in another way. The static criterion remains in force for a 'weakly asymmetric' matrix $\mathbf{A}$!

Proof: $\mathbf{A}$ is symmetric for $\alpha = 0$. Consequently, it only has real eigenvalues λ_A. This property of $\mathbf{A}$ will remain in spite of the asymmetry of $\mathbf{A}$ which occurs for $\alpha > 0$, because of the continuous dependence of the matrix elements on α in the region $0 \leqslant \alpha \leqslant \alpha_{g,min}$. In this region the stability can only disappear for increasing p if one of the real eigenvalues λ_A changes from positive to negative. The stability boundary is given by $\lambda_A = 0$, which implies that the static criterion holds in this α interval.

Let us summarize these facts as follows.

Theorem: *In non-conservative or polygenic stability problems, in which the matrix $\mathbf{A}$ depends on a structural parameter α, the static stability criterion applies and is decisive if $\mathbf{A}$ is slightly asymmetric (or slightly non-Hermitian) for $0 \leqslant \alpha \leqslant \alpha_{g,min} \cdot \alpha_{g,min}$ is a limiting value which depends on the problem under discussion.*

We have now completely answered the question of the validity of the static stability criterion. The criterion is applicable and decisive for all conservative or monogenic stability problems, as well as for those non-conservative or polygenic stability problems having a matrix $\mathbf{A}$ that can be made symmetrical or is slightly asymmetric. Let us give a broader definition of the term *slightly asymmetric* matrix. As Fig. 66b shows, the static criterion remains applicable and decisive if the image point of the eigenvalue λ_A of an symmetric matrix $\mathbf{A}$ only enters the complex plane for sufficiently large load parameter p, if it has *first* passed through the origin, which is the first and lowest stability boundary. Therefore, matrices for which λ_A has this property can also be called slightly asymmetric, even though their eigenvalues do not always remain real, as is the case for symmetric matrices or matrices which can be made symmetrical.

Let us consider the second case where $\mathbf{A}$ is asymmetric (or non-Hermitian). Just as in the first case, for stability, λ_A must be real and $\lambda_A \geqslant 0$, because then according to (2.3.44) ω is imaginary and the perturbed motion consists of small oscillations around the equilibrium position. Let us assume that $\mathbf{A}$ is '*strongly asymmetric*' (or '*strongly non-Hermitian*'), which means that none of the special cases discussed above is being considered such as the case of matrices $\mathbf{A}$ which can be made symmetrical or slightly asymmetrical matrices. This means that the matrix can have complex eigenvalues for sufficiently large values of the parameter p without any of the eigenvalues first becoming zero. As soon as λ_4 becomes complex, one of the ω values must have a positive real part according to (2.3.44), which means that stability is lost. This process can be visualized as follows. The fact that one eigenvalue λ_4 becomes complex for p_{crit} means that the image point of λ_4 moves from the real positive axis into the complex plane, the region of instability, as Figs. 66c and 66d show, *before* it has passed through the origin, which is a stability boundary. This first (or only) stability boundary is now the edge of the groove in the complex plane outside of the origin.

Under these conditions is can be said that the equilibrium position of the body at rest has become *kinetically unstable,* because instability occurs owing to induced oscillations. It is not caused by the fact that the body passes through non-trivial equilibrium positions, as would be the case in the classical conservative case. These relationships can be followed by means of the corresponding eigenvalue curve.

Example: Let us consider the *'Beck' buckling column* shown in Fig. 72. It is clamped at one end, and there is a tangential compression force P_t at the free end. Figure 73 shows the corresponding eigenvalue curve, which is the dependence of P_t on $\omega^2 \equiv -\lambda_4$. The eigenvalue curve never intersects the load axis. No non-trivial equilibrium positions exist, and the static stability criterion cannot be used. Instability occurs because there are no real but only complex values of ω^2 and therefore of λ_4 for $P_t > P_{t,crit}$.

A new set of circumstances has arisen that cannot be treated with the classical criteria, the static and the classical energy criteria. This type of instability can only be discussed in terms of the third criterion, the *kinetic* criterion.

Kinetic instability cannot exist if it is specified that all roots λ_4 of the characteristic equation $\det(\mathbf{A} - \lambda_4 \mathbf{E}) = 0$ remain real. For this, the matrix which will be called the 'Borchardt Matrix,'[1] must be *positive definite.* The elements S_k of this matrix are sums of powers of the roots of the characteristic equation.

$$\mathbf{B} = \begin{bmatrix} S_0 & S_1 \ldots S_{n-1} \\ S_1 & S \ldots S_n \\ \cdots\cdots\cdots\cdots\cdots \\ S_{n-1} & S_n \ldots S_{2n-2} \end{bmatrix} \qquad (2.3.49)$$

The criterion can therefore be formulated as follows.

Kinetic stability criterion: If the stability problem is non-conservative and if its matrix is strongly asymmetric or strongly non-Hermitian, the stability disappears if one of the eigenvalues λ_4 becomes complex. This happens when the Borchardt matrix is no longer positive definite. Therefore, the kinetic criterion may also be called the 'Borchardt' criterion.

---

[1]Borchardt, A., *J. Math.* pp. 50–67, 1847.

Fig. 71. Stability diagram

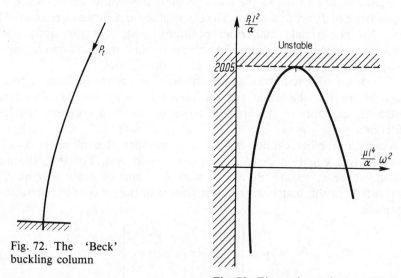

Fig. 72. The 'Beck'
buckling column

Fig. 73. Eigenvalue of the 'Beck'
buckling column

It should be pointed out that if the calculation is restricted to matrices **A** of second and third order, which depends on the degree of approximations used, the kinetic stability criterion is greatly simplified. This is because in this case it is sufficient to set the *discriminant* of the characteristic equation equal to zero in order to calculate the stability boundary.

Example: Let us consider the column shown in Fig. 72. It is a special case of the column shown in Fig. 65 ($P_v = 0$, $q = 0$, $P \neq 0$). We must set $Q \equiv P_t$ and

$$L(w) = \alpha \partial^4 w/\partial x^4 + P_t \partial^2 w/\partial x^2,$$

$$w(0, t) = w_x(0, t) = w_{xx}(l, t) = w_{xxx}(l, t) = 0.$$

In order to form the matrix $\mathbf{A} = (a_{ik})$, let us form

$$a_{ik} = (1/\mu) \int_0^l L(\phi_k)\phi_i \, dx. \tag{2.3.50}$$

The ϕ_k are the eigenfunctions of the 'abbreviated' problem

$$\mu\omega^2\phi(x) + \alpha\phi^{iv}(x) = 0,$$
$$\phi(0) = \phi'(0) = \phi''(l) = \phi'''(l) = 0, \tag{2.3.51}$$

which is obtained from $\mu\ddot{w} + L(w) = 0$ and the corresponding boundary conditions after $w = e^{\omega t}\phi(x)$ is substituted. The term containing P_t in $L(w)$ is crossed out. The boundary value problem (2.3.51) is self-adjoint. Two of its eigenfunctions are used, i.e., the orthonormal functions

$$\phi_k(x) = \varkappa_k\left[c_{1k}\left(\cos\frac{m_k x}{l} - \cosh\frac{m_k x}{l}\right) - c_{2k}\left(\sin\frac{m_k x}{l} - \sinh\frac{m_k x}{l}\right)\right], \quad k = 1, 2$$

$$m_1 = 1.875, \qquad m_2 = 4.694,$$

$$c_{11} = 4.148, \qquad c_{21} = 3.037, \qquad c_{12} = 53.640, \qquad c_{22} = 54.631,$$

$$\varkappa_1 = (17.259 \, l)^{-1/2}, \qquad \varkappa_2 = (2868.283 \, l)^{-1/2}.$$

The elements of $\mathbf{A}$ are obtained from (2.3.50) as follows:

$$a_{11} = \frac{1}{17.259 \, \mu l}\left(\frac{\alpha}{l^3} \cdot 1.875^4 \cdot 17.259 + \frac{P_t}{l} 1.875^2 \cdot 4.277\right),$$

$$a_{12} = \frac{P_t}{l}\frac{1.875^2 \cdot 118.707}{(17.259 \cdot 2868.283)^{1/2}\mu l},$$

$$a_{21} = -\frac{P_t}{l}\frac{4.694^2 \cdot 118.664}{(17.259 \cdot 2868.283)^{1/2}\mu l},$$

$$a_{22} = \frac{1}{2868.283\mu l}\left(\frac{\alpha}{l^3} \cdot 4.694^4 \cdot 2868.283 - \frac{P_t}{l} \cdot 4.694^2 \cdot 1733.631\right). \tag{2.3.52}$$

The characteristic equation $\det(\mathbf{A} - \lambda_A\mathbf{E}) = 0$ results in

$$\lambda_A^2 - (a_{11} + a_{22})\lambda_A + a_{11}a_{22} - a_{12}a_{21} = 0.$$

The kinetic stability criterion consists of setting the discriminant D of this quadratic equation equal to zero. For $D < 0$, λ_A can take on complex values, and the stability can become questionable. Therefore, the stability boundary follows from $D = 0$. If the expressions (2.3.52) are substituted into

$$D = (a_{11} - a_{22})^2 + 4a_{12}a_{21} = 0,$$

an equation is obtained for the critical value P_t which follows from this equation as

$$P_{t,\text{crit}} = 20.07(\alpha/l^2).$$

The exact value calculated by *Beck* is $P_{t,\text{crit}} = 20.05\alpha/l^2$. The accuracy of our approximate calculation is remarkable considering that we only used two eigenfunctions. We have therefore shown how effective the Galerkin method is.

The realization that non-conservative or polygenic problems exist for which only the kinetic and not static stability criterion holds — thus making it necessary to carry out the kinetic calculation — leads us to the third question posed in Section 2.3.1.

Influence of the mass distribution

When it is permissible to use the static criterion, there are no quantities which contain the mass distribution, because of det $\mathbf{A} = \det(a_{ik}) = 0$. The mass distribution is of no consequence in this case. This is not the case if it is necessary to carry out the kinetic calculation, as mentioned above. Then the mass distribution is important and must be taken into account. The calculations are then based on $\det(\mathbf{A} + \omega^2\mathbf{E}) = 0$, where $\mathbf{A} = m\mathbf{A}^*$. The matrix $\mathbf{A}^*$ depends on geometric, elastic and load parameters while m is a function of the mass distribution. In the non-self-adjoint case, one is led from $\det(\mathbf{A} + \omega^2\mathbf{E}) = 0$ to the characteristic equation $\lambda^n + b_1\lambda^{n-1} + \cdots + b_{n-1}\lambda + b_n = 0$, $\lambda = \omega^2$, where the coefficients b_i, $i = 1, 2, \ldots, n$ are functions of m. Thus, solving the characteristic equation for ω^2, the frequency ω of the small vibrations of the elastic system about its trivial equilibrium position becomes a function of m. Consequently, the stability of the system depends on m too. A paper by Pflüger[1] is devoted to the subject that considers a concentrated mass M at the end of the column (Fig. 74). If ml is the column mass, there is a pronounced dependence of the load parameter $p = P_t l^2/\alpha$ on the mass distribution parameters $\mu_1 = M/ml$ or $\mu_2 = ml/M$ shown in Fig. 75, according to this calculation.

Another investigation has been made concerning the effect of the mass distribution on the stability of Beck's rod (Fig. 72)[2]. The result is the following: If M_s is the total mass of the rod and if a concentrated mass $M = xM_s$ (x is a numerical factor) having the weight W is added at $x = \xi$ to the rod, the maximal follower force P that the rod can carry at its end without buckling it obtained for $\xi = [0.70 - 0.04\,x]\,l$, $x \leqslant 5$. This result is certainly quite different from the situation encountered in the case of the Euler-buckling of the rod. In that case, M should be located at $\xi = 0$ for a maximum value of P to occur.

[1]Pflüger, A., The stability of a tangentially loaded compression column.
[2]Leipholz, H. and Lindner, G., Über den Einfluss der Massenverteilung auf das nichtkonservative Knicken von Stäben, *Ing. Arch.* 39, 187–194, 1970.

Fig. 74. Influence of mass distribution for non-conservative
stability problems

Fig. 75. Influence of mass
distribution on the magnitude of
the 'follower' buckling force

Because of this difference in results, mass distribution should for example be taken into account for rocket mechanics. The maximum thrust the rocket can endure without buckling, can be obtained by placing a payload at a position ξ calculated according to the given equation above, since the thrust is a non-conservative follower load.

Further investigations on the influence of the mass distribution are due to B. Buchser and H. Leipholz[1], R. N. Kapoor and H. Leipholz[2/3], H. Leipholz[4] and R. Piche.[5] It is of specific interest to note that according to these investigations the mass distribution can decisively determine the mode of instability: buckling or flutter. It is also of interest to learn that a certain mass distribution, tensile loading, which is commonly considered to be harmless as far as instability is concerned, can lead to flutter instability.

The influence of damping

Consideration of this problem also touches on the fourth question of justifying the linearization. A detailed answer to it will be given in the following.

Theorem 5 of Section 1.5.3, given by Lyapunov, states that it is possible to carry out the stability calculation with linearized equations as long as all exponents of the characteristic equation have negative real parts. If only one characteristic exponent has a vanishing real part, the critical case occurs and the calculations with the linearized equations become questionable.

[1]Buchser, B. and Leipholz, H. ZAMP 24 255–269, 1973.
[2/3]Kapoor, R. N. and Leipholz, H., Ing. Arch. 43, 233–239, 1974 and ZAMP 54, 205–208, 1974
[4]Leipholz, H., Ing. Arch. 48, 185–195, 1979.
[5]Piche, R. Optimization of dlan Distribution of Elastic rods subjected to Follower Forces, MASc Thesis, Univ. of Waterloo, Ontario, 1982.

If we assume that there is no damping and if we use the system (2.3.40), we will use the theory of small oscillations and we will find that the equilibrium position of the elastic body is stable because the roots $\omega = \pm(-\lambda_4)^{\frac{1}{2}}$ of the characteristic equation are all purely imaginary. This is exactly the critical case according to the definition of Lyapunov, so that the stability results obtained from (2.3.40) will appear uncertain at first.

In order to satisfy Theorem 5 of Lyapunov, it would be better to consider damping. Every time the damping would make the roots ω of the characteristic equation have negative real parts, there would be asymptotic stability and the stability investigation could be carried out with the linearized equations.

In non-conservative stability problems it can also happen that the stability becomes worse, in contrast to conservative systems, if small damping is taken into account. In the transition from the damped to the undamped system, there can be a jump in the stability condition. H. Ziegler was probably the first to point this out.

All the following reasons make it necessary to investigate the influence of damping: occurrence of the critical case according to the Lyapunov definition when damping is absent, possible degradation of stability by damping, possible jumps in stability conditions in the transition from the damped to the undamped process.

When damping is considered, the term

$$\int_0^l \beta w_t \, \delta w \, dx$$

must be added to the relationship (2.3.25) for the buckling column, for example, β is the damping coefficient. Instead of (2.3.27), one obtains

$$\int_0^l [\mu \ddot{w} + L^*(w)] \, \delta w \, dx + R(w) = 0 \qquad (2.3.53)$$

where

$$L^*(w) = \alpha w_{xxxx} + Q w_{xx} + \beta w_t. \qquad (2.3.54)$$

The expression $R(w)$ remains unchanged. On the other hand, (2.3.37) now becomes

$$\int_0^l \{\mu[\ddot{f}_k \phi_k(x)] + \beta[\dot{f}_k \phi_k(x)] + L[f_k(t)\phi_k(x)]\}[f(t)\phi_r(x)\delta a_r] \, dx = 0$$

and in place of (2.3.39),

$$\mu \ddot{f}_i + f_k \int_0^l L(\phi_k)\phi_i \, dx + \beta \dot{f}_i = 0.$$

In place of (2.3.40), we have

$$\ddot{f}_i + a_{ik} f_k + b \dot{f}_i = 0, \qquad i, k = 1, 2, 3, \ldots. \qquad (2.3.55)$$

Equation (2.3.55) is equivalent to the equations of motion of a system of discrete point masses with an infinite number of degrees of freedom, which carry out *damped* small oscillations about its equilibrium position.

Until now, b and therefore $b = \beta/\mu$ were assumed to be constant. For purposes of greater generality, we will not exclude the case where the b's can be different, so that instead of (2.3.55) we have

$$\ddot{f}_i + a_{ik}f_k + b_{(i)}\dot{f}_i = 0 \qquad \text{(not to be summed over } i\text{)}. \qquad (2.3.55a)$$

The following eigenvalue equation is obtained from (2.3.55a) with $f_k = a_k \, e^{\omega t}$:

$$(\mathbf{A} + \omega\mathbf{B} + \omega^2\mathbf{E})\mathbf{X} = 0 \qquad (2.3.56)$$

in which there is the *diagonal* damping matrix $\mathbf{B} = (b_i)$, in contrast to (2.3.42).

Let $\mathbf{A}$ be *symmetric* and $\mathbf{B}$ be *an arbitrary positive diagonal matrix* so that the damping coefficients b_i, which are the elements of $\mathbf{B}$, are different from each other. This is the case of an initially conservative system that is disturbed by purely dissipative forces. To investigate this further, the eigenvalue problem

$$(\mathbf{A} + \omega\mathbf{B} + \omega^2\mathbf{E})\mathbf{X} = 0,$$

which contains the arbitrary vector $\mathbf{X}$, is multiplied on the left by the conjugate transposed vector $\mathbf{X}'$. The result is

$$A + \omega^*B + \omega^{*2} = 0,$$

where A, B are numbers that involve values of the matrices $\mathbf{A}$, $\mathbf{B}$. According to our assumption, A is always real, since $\mathbf{A}$ has only real elements because it is a symmetric matrix. B is real and positive, because $\mathbf{B}$ is symmetric and positive definite. The relationship

$$\omega^* = -\tfrac{1}{2}B \pm (\tfrac{1}{4}B^2 - A)^{\frac{1}{2}}$$

specifies the region of values in which the quantity ω must be located so that it corresponds to the original eigenvalue problem.

As long as $A > 0$, any possible ω^* and therefore the ω, which is of interest to us, has a negative real part and the equilibrium of the elastic body is asymptotically stable. This means that the linearized calculation is justified. The stability boundary is given by $A = 0$. Since we have $\lambda_{A,\min} \leqslant A$, $\lambda_{A,\min} = 0$ is sufficient for stability, which means that the static stability criterion is sufficient. This fact, established for symmetric $\mathbf{A}$ — that is, for conservative problems — agrees with the well-known theorem of Lord Kelvin stating that the stability of a conservative system is not disturbed by dissipative forces and that the originally stable equlibrium of the system even becomes asymptotically stable because of dissipative forces.

Let us assume that the problem is non-conservative but the matrix $\mathbf{A}$ is *symmetrizable or is weakly asymmetrical* and the damping matrix is positive and diagonal. All the elements are assumed to be the same, so that $B = bE$, i.e., all the damping coefficients are equal. Then from (2.3.56) we have

$$[A + (\omega b + \omega^2)E] X = 0,$$

which, because $\omega^2 + \omega b = -\lambda_A$, becomes

$$(A - \lambda_A E)X = 0.$$

Because of the assumption that A can be made symmetrical or is weakly asymmetrical, all the λ_A are real or only become complex if the λ_A, which was previously real, first passes through zero.

From the relationship $\omega^2 + \omega b = -\lambda_A$, it follows that

$$\omega = -\tfrac{1}{2}b \pm (\tfrac{1}{4}b^2 - \lambda_A)^{1/2} \tag{2.3.57}$$

will have a negative real part *only* when $\lambda_A > 0$, which means that the equilibrium position of the elastic body is asymptotically stable and the linearized calculation can be used. The stability boundary is given by $\lambda_A = 0$ because of the properties of λ_A. This means that it is also given by $\det A = 0$, which means that the static stability criterion is satisfied.

Let us assume that the stability problem is non-conservative, the matrix A is asymmetric and that $B = bE$, i.e., *all damping coefficients are equal*. The eigenvalue problem can again be written in the form

$$[A + (\omega b + \omega^2)E] X = 0,$$

which leads to the condition $\det [A + (\omega b + \omega^2)E] = 0$. If this is compared with the characteristic equation $\det [A - \lambda_A E] = 0$ corresponding to A, it can be seen that $\omega b + \omega^2 = -\lambda_A$ so that

$$\omega^2 + \omega b + \lambda_A = 0. \tag{2.3.58}$$

As before, we have

$$\omega = -\tfrac{1}{2}b \pm (\tfrac{1}{4}b^2 - \lambda_A)^{1/2}.$$

As long as the load parameter p remains sufficiently small, the λ_A are real, positive and move to the left along the real positive axis of the λ_A plane (Fig. 76) as p is increased. At a certain value p the image point of an eigenvalue λ_A passes through the point B into the complex plane and $\lambda_A = \rho + i\sigma$, so that it becomes complex. It follows from

$$\omega = \tfrac{1}{2}b \pm (\tfrac{1}{4}b^2 - \rho - i\sigma)^{1/2} \tag{2.3.59}$$

that ω has a negative real part if

$$\rho > \sigma^2/b^2, \tag{2.3.60}$$

which means that the equilibrium of the elastic body is asymptotically stable and the linearization can be used. The stability boundary is given by $\rho = (\sigma/b)^2$.

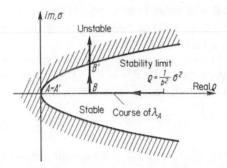

Fig. 76. Stability diagram taking
damping into account

A comparison of Fig. 76 with Figs. 66c and 66d shows that the instability region has become smaller because of damping. As b is increased, i.e., for stronger damping, the stability region becomes inflated and the damping has a favourable effect. The critical case of loss of stability occurs at B' where the image of λ_A crosses the stability boundary, and not at B. In particular it can be seen that the kinetic Borchardt criterion for the calculation without damping, which is represented by point B in Fig. 76, is sufficient for stability. In retrospect it can be seen that in a non-conservative case the corresponding criterion of the undamped case (static or kinetic criterion, according to Borchardt) remains at least sufficient in the case with damping, regardless of whether **A** is symmetrizable or asymmetric, if the damping matrix satisfies **B** = b**E**, i.e., all damping coefficients are equal. The damping cannot neutralize the stability. In addition, damping makes the stability become asymptotic and makes it possible to use the linearized calculation. Therefore, all presently known results regarding the stability of buckling columns for non-conservative loading would remain in effect, even though it was determined without damping, provided the damping is introduced in the same way as in (2.3.54). This means that $b_i = b$ for all i in (2.3.55), i.e., the same damping coefficients throughout. The formulation (2.3.54) contains all the cases discussed in the literature, if special values for $Q(x)$ and special boundary conditions $R(w)$ are assumed.

It is necessary to proceed with caution if **A** is asymmetric and the damping coefficients are different. As H. Ziegler pointed out, it can happen that small damping may have a *neutralizing* effect. The independent works of O. Bottema and V. V. Bolotin for second-order systems has shown that in the non-conservative case and for different damping coefficients the stability condition is discontinuous with respect to the undamped case.

This statement can be generalized as follows:

In the undamped case, the behaviour of the roots ω of the characteristic equation $\det(\mathbf{A} + \omega^2 \mathbf{E}) = 0$ must be established. Since $\omega = \pm(-\lambda_A)^{1/2}$ we must have $\lambda_A > 0$ and real for stability. This means that the roots of the algebraic equation $\det(\mathbf{A} - \lambda\mathbf{E}) = 0$ must remain real. A necessary and sufficient condition for this is that the matrix (2.3.49) is positive definite. This is the kinetic *Borchardt criterion for the undamped* case.

On the other hand, for the damping case the behaviour of the roots of the characteristic equation $\det(\mathbf{A} + \omega\mathbf{B} + \omega^2\mathbf{E}) = 0$ must be followed. If $\mathbf{A}$ is of order n, this leads to an algebraic equation of the type

$$\omega^{2n} + a_1\omega^{2n-1} + \cdots + a_{2n-1}\omega + a_{2n} = 0, \qquad (2.3.61)$$

where the coefficients $a_j(j = 1, 2, \ldots, 2n)$ are all different from zero and are formed in certain ways from the elements a_{ik}, b_i of the matrices $\mathbf{A}$ and $\mathbf{B}$.

Stability occurs when all roots ω have negative real parts. The necessary and sufficient conditions for this are $a_j > 0$ and the Hurwitz matrix must be positive definite. The Hurwitz matrix is not necessarily identical to (2.3.49), which is the Borchardt matrix. Therefore there will be *stability conditions* which are *different* from the undamped case.

In the special case $\mathbf{B} \equiv b\mathbf{E}$, the following discussion will show that for zero damping, the stability conditions which can follow from the Hurwitz criterion coincide exactly with the stability conditions which apply in the undamped case and are based on the Borchardt criterion. It is very fortunate that in this special case there is no jump in the stability conditions in the transition from the damped to the undamped system.

If, on the other hand, $\mathbf{B}$ is a general positive diagonal matrix, the coefficients of (2.3.61) will change as the damping is decreased. Even though the equation coefficients change, the difference in the b_i will cause a difference between the stability conditions applicable for (2.3.61) that follow from the Hurwitz criterion and the stability conditions for the undamped case that follow from the Borchardt criterion. This is true no matter how small the elements b_i of $\mathbf{B}$ are. In the limiting case, where there is still very small damping, there will be a jump in the stability conditions, which can be dangerous. This means that for non-conservative systems for which the diagonal damping matrix has elements that are all different, it is not possible to use the calculation method for the undamped system.

Let us give the method with the following *third-order system:*

$$\begin{vmatrix} \omega^2 + \omega b_1 + a_{11} & a_{12} & a_{13} \\ a_{21} & \omega^2 + \omega b_2 + a_{22} & a_{23} \\ a_{31} & a_{32} & \omega^2 + \omega b_3 + a_{33} \end{vmatrix} = 0.$$

The characteristic equation for the damped system is

$$\omega^6 + a_1\omega^5 + a_2\omega^4 + a_3\omega^3 + a_4\omega^2 + a_5\omega + a_6 = 0, \qquad (2.3.62)$$

where

$$a_1 = b_1 + b_2 + b_3,$$

$$a_2 = a_{11} + a_{22} + a_{33} + b_1b_2 + b_2b_3 + b_3b_1,$$

$$a_3 = a_{11}(b_2 + b_3) + a_{22}(b_3 + b_1) + a_{33}(b_1 + b_2) + b_1b_2b_3,$$

$$a_4 = (a_{11}a_{22} - a_{12}a_{21}) + (a_{22}a_{33} - a_{23}a_{32}) + (a_{33}a_{11} - a_{31}a_{13})$$
$$\qquad + a_{11}b_2b_3 + a_{22}b_3b_1 + a_{33}b_1b_2,$$

$$a_5 = b_1(a_{22}a_{33} - a_{23}a_{32}) + b_2(a_{33}a_{11} - a_{31}a_{13}) + b_3(a_{11}a_{23} - a_{12}a_{21}),$$

$$a_6 = a_{11}(a_{22}a_{33} - a_{23}a_{32}) + a_{12}(a_{23}a_{31} - a_{21}a_{33})$$
$$\qquad + a_{13}(a_{21}a_{32} - a_{31}a_{22}) = \det \mathbf{A}$$

(2.3.63)

In the undamped case, the characteristic equation is

$$\omega^6 + a_2^*\omega^4 + a_4^*\omega^2 + a_6^* = 0, \qquad (2.3.64)$$

where

$$a_2^* = a_{11} + a_{22} + a_{33},$$

$$a_4^* = (a_{11}a_{22} - a_{12}a_{21}) + (a_{22}a_{33} - a_{23}a_{32}) + (a_{33}a_{11} - a_{31}a_{13}), \qquad (2.3.65)$$

$$a_6^* = \det \mathbf{A}.$$

The stability of the *undamped* mechanical system exists if (2.3.64) has only pure imaginary roots ω. For this purpose, it is necessary that the auxiliary equation

$$(\omega^2)^3 + \alpha_2^*(\omega^2)^2 + a_4^*(\omega^2)^1 + a_6^* = 0,$$

considered as a cubic equation, have no complex but only real roots ω^2. This occurs when the discriminant D of this cubic equation is greater than zero. This condition results in the necessary stability condition in the form of the following inequality:

$$D = a_2^{*2}a_4^{*2} - 4a_2^{*3}a_6^* - 4a_4^{*3} + 18a_2^*a_4^*a_6^* - 27a_6^{*2} > 0. \qquad (\text{I})$$

Because of

$$D \equiv -4(a_4^* - \tfrac{1}{3}a_2^{*2})^3 - 27(a_6^* - \tfrac{1}{3}a_2^*a_4^* + (2/27)a_2^{*3})^2,$$

the condition $D > 0$ can only be satisfied if

$$a_4^* - \tfrac{1}{3}a_2^{*2} < 0,$$

or

$$a_2^{*2} > 3a_4^*. \qquad (\text{II})$$

This is another condition for stability of the undamped system. In the case *with damping*, let the b_i first all be equal to b. Since we are investigating the behaviour for small damping, let us also specify $|b| \ll 1$. This means that the higher powers of b can be ignored, and the following new equations are obtained for coefficients a_i of the characteristic equations (2.3.62), in place of Eqs. (2.3.63):

$$a_1^{**} = 3b, \qquad a_2^{**} = a_2^*, \qquad a_3^{**} = 2ba_2^*, \qquad a_4^{**} = a_4^* \qquad a_5^{**} = ba_4^*,$$
$$a_6^{**} = a_6^*. \qquad (2.3.66a)$$

The characteristic equation is

$$\omega^6 + a_1^{**}\omega^5 + a_2^{**}\omega^4 + a_3^{**}\omega^3 + a_4^{**}\omega^2 + a_5^{**}\omega + a_6^{**} = 0.$$

The Hurwitz criterion is used to investigate stability of the corresponding mechanical system. It follows from this that the damped mechanical system is stable if the two inequalities

$$(a_1^{**}a_2^{**} - a_3^{**})[a_5^{**}(a_4^{**}a_3^{**} - a_2^{**}a_5^{**}) + a_6^{**}(2a_1^{**}a_5^{**} - a_3^{**2})]$$
$$+ (a_1^{**}a_4^{**} - a_5^{**})[a_1^{**}a_3^{**}a_6^{**} - a_5^{**}(a_1^{**}a_4^{**} - a_5^{**})] - a_1^{**3}a_6^{**2} > 0$$

and

$$a_3^{**}(a_1^{**}a_2^{**} - a_3^{**}) - a_1^{**}(a_1^{**}a_4^{**} - a_5^{**}) > 0$$

are satisfied. If relationships (2.3.66a) are substituted for $a_i^{**}(i = 1, 2, \ldots, 5)$ in these inequalities, the result is

$$(3ba_2^* - 2ba_2^*)[ba_4^*(a_4^*2ba_2^* - a_2^*ba_4^*) + a_6^*(6b^2a_4^* - 4b^2a_2^{*2})]$$
$$+ (3ba_4^* - ba_4^*)[6b^2a_2^*a_6^* - ba_4^*(3ba_4^* - ba_4^*)] - 27b^3a_6^{*2} > 0$$

and

$$2ba_2^*(3ba_2^* - 2ba_2^*) - 3b(3ba_4^* - ba_4^*) > 0.$$

By cancellation of b^3 in the upper inequality and b^2 in the lower inequality, and after a simple calculation, the new inequalities

$$a_2^{*3}a_4^{*2} + 18a_2^*a_4^*a_6^* - 4a_2^{*3}a_6^* - 4a_4^{*3} - 27a_6^{*2} > 0 \quad \text{and} \quad 2a_2^{*2} - 6a_4^* > 0$$

are obtained. These are identical to the conditions (I) and (II) of the undamped case. It has been shown that the transition between the stability conditions for the undamped and damped case for zero damping occurs without a jump *only* if the damping coefficients $b_i = b$ are the *same*. On the other hand, if the $|b_i|$ differ for the damping case, and if they are again assumed to be small, $|b_i| \ll 1$, in the limit where the damping strives to zero, it is possible to ignore the products and powers of the b_i. If we set

$$b_2 = b_1(1 + e_2), \qquad b_3 = b_1(1 + e_3),$$

where e_2 and e_3 are suitable factors and $b_1 = b$, the following relationships are obtained for the coefficients a_i of (2.3.62) instead of (2.3.63) for these assumptions:

$$\bar{a}_1 = 3b + b(e_2 + e_3) = 3b + b\alpha_1,$$

$$\bar{a}_2 = a_2^*,$$

$$\bar{a}_3 = 2ba_2^* + b[a_{11}(e_2 + e_3) + a_{22}e_3 + a_{33}e_2] = 2ba_2^* + b\alpha_3,$$

$$\bar{a}_4 = a_4^*$$

$$\bar{a}_5 = ba_4^* + b[e_2(a_{33}a_{11} - a_{13}a_{31}) + e_3(a_{11}a_{22} - a_{12}a_{21})] = ba_4^* + b\alpha_5,$$

$$\bar{a}_6 = a_6^*. \tag{2.3.66b}$$

The meaning of the quantities $\alpha_1, \alpha_3, \alpha_5$ used as abbreviations is obvious. The characteristic equation is

$$\omega^6 + \bar{a}_1\omega^5 + \bar{a}_2\omega^4 + \bar{a}_3\omega^3 + \bar{a}_4\omega^2 + \bar{a}_5\omega + \bar{a}_6 = 0.$$

The Hurwitz stability criterion must again be used, which leads to the inequalities

$$(\bar{a}_1\bar{a}_2 - \bar{a}_3)[\bar{a}_5(\bar{a}_4\bar{a}_3 - \bar{a}_2\bar{a}_5) + \bar{a}_6(2\bar{a}_1\bar{a}_5 - \bar{a}_3^2)]$$
$$+ (\bar{a}_1\bar{a}_4 - \bar{a}_5)[\bar{a}_1\bar{a}_3\bar{a}_6 - \bar{a}_5(\bar{a}_1\bar{a}_4 - \bar{a}_5)] - \bar{a}_1^3\bar{a}_6^2 > 0$$

and

$$\bar{a}_3(\bar{a}_1\bar{a}_2 - \bar{a}_3) - \bar{a}_1(\bar{a}_1\bar{a}_4 - \bar{a}_5) > 0.$$

If (2.3.66b) is substituted, the result is

$$a_2^* a_4^{*2} + 18a_2^* a_6^* a_4^* - 4a_2^{*3} a_6^* - 4a_4^{*3} - 27a_6^{*2}$$
$$> (\alpha_3 - \alpha_1 a_2^*)\{(a_4^* + \alpha_5)[a_4^*(2a_2^* + \alpha_3) - a_2^*(a_4^* + \alpha_5)]$$
$$+ a_6^*[2(3 + \alpha_1)(a_4^* + \alpha_5) - (2a_2^* + \alpha_3)^2]\} - a_2^*\{a_4^*\alpha_3(a_4^* + \alpha_5)$$
$$- \alpha_5^2 a_2^* + a_6^*[\alpha_5(6 + 2\alpha_1) + 2a_4^*\alpha_1 - \alpha_3(4a_2^* + \alpha_3)]\}$$
$$+ (\alpha_5 - \alpha_1 a_4^*)\{(3 + \alpha_1)(2a_2^* + \alpha_3)a_6^* - (a_4^* + \alpha_5)[(3 + \alpha_1)a_4^*$$
$$- (a_4^* + \alpha_5)]\} - 2a_4^*\{\alpha_1[a_6^*(2a_2^* + \alpha_3) - a_4^{*2}] + 3\alpha_3 a_6^*$$
$$- \alpha_5[a_4^*(1 + \alpha_1) - \alpha_5]\} + (27\alpha_1 + 9\alpha_1^2 + \alpha_1^3)a_6^{*2}$$

and

$$2a_2^{*2} - 6a_4^* > a_2^*[\alpha_3 - \alpha_1(2a_2^* + \alpha_3)] + \alpha_3^2 + \alpha_1 a_4^*(5 + \alpha_1) - \alpha_5(3 + \alpha_1),$$

which differs from inequalities (I) (II) of the undamped case, because the right-hand sides of the inequalities are not zero, as was the case there. There is a *jump* in the stability conditions with respect to the undamped case. This jump disappears only when the quantities e_2 and e_3, which measure the difference in the damping coefficients, become zero. This makes the two conditions equal, because for $\alpha_1, \alpha_3, \alpha_5$ we have

$$\alpha_1 = e_2 + e_3,$$

$$\alpha_3 = a_{11}(e_2 + e_3) + a_{22}e_3 + a_{33}e_2,$$

$$\alpha_5 = e_2(a_{33}a_{11} - a_{13}a_{31}) + e_3(a_{11}a_{22} - a_{12}a_{21}),$$

242

and the right-hand side of the inequalities becomes zero for $e_2 = e_3 = 0$, as is easily established.

Let us now generalize our assumptions regarding the structure of the matrix **B** in another direction. The elements of **B** outside of the diagonal are assumed to be *skew symmetric*. This means that *gyroscopic forces* also affect the mechanical system. Yet, the restriction is made that the matrix **A** is *symmetric*. Problems of this type occur often in practice when the buckling of *pipes containing flows* is investigated.

The differential equation for a pipe hinged at both ends and having finite length is

$$\alpha w_{xxxx} + S w_{xx} + Q w_{xt} + \mu w_{tt} + \beta w_t + K w = 0,$$

where S and Q are load parameters and K is the modulus of the elastic foundation of the pipe. The boundary conditions for hinged supports are

$$w(0, t) = w(l, t) = w''(0, t) = w''(l, t) = 0,$$

so that

$$w = f_1 \sin(\pi/l)x + f_2 \sin(2\pi/l)x, \qquad f_i = a_i e^{\omega t} \qquad (i = 1, 2),$$

is a permissible, two-termed approximate trial solution. This leads to the system

$$
\begin{aligned}
\ddot{f}_1 + a_{11}f_1 + b_1\dot{f}_1 - b_{12}\dot{f}_2 = 0, \\
\ddot{f}_2 + a_{22}f_2 + b_{12}\dot{f}_1 + b_2\dot{f}_2 = 0,
\end{aligned}
\qquad (2.3.67)
$$

where

$$b_1 = b_2 = b = \beta/\mu, \qquad b_{12} = 8Q/3\mu l.$$

This is the case mentioned above:

$$\mathbf{A} = \begin{pmatrix} a_{11} & 0 \\ 0 & a_{22} \end{pmatrix}$$

is symmetric, whereas

$$\mathbf{B} = \begin{pmatrix} b_1 & -b_{12} \\ b_{12} & b_2 \end{pmatrix} \equiv \begin{pmatrix} b & -b_{12} \\ b_{12} & b \end{pmatrix}$$

satisfies the relationship $\mathbf{B} = b\mathbf{E} + \mathbf{G}$, where

$$\mathbf{G} = \begin{pmatrix} 0 & -b_{12} \\ b_{12} & 0 \end{pmatrix}$$

is skew symmetric.

After differentiation with respect to time, the system of Eqs. (2.3.67) can be written as follows:

$$(\mathbf{A} + \mathbf{B}\omega + \mathbf{E}\omega^2)\mathbf{X} \equiv (\mathbf{A} + b\mathbf{E}\omega + \mathbf{G}\omega + \mathbf{E}\omega^2)\mathbf{X} = 0, \qquad (2.3.68)$$

where **A** and **E** are symmetric, **G** is skew symmetric, and **E** is positive definite. In this case, Eq. (2.3.68) results in the relationship

$$A + b\omega^* + G\omega^* + \omega^{*2} = 0, \qquad (2.3.69)$$

where the numbers A and G are taken from values in the matrices **A** and **G**. For $\omega^* = -i\sigma$, Eq. (2.3.69) becomes

$$\sigma^2 + \sigma(ib + \widetilde{G}) - A = 0,$$

where A and $\widetilde{G} = iG$ are real. We have

$$\omega^* = -\tfrac{1}{2}b + \tfrac{1}{2}i\widetilde{G} \pm \tfrac{1}{2}[(b^2 - \widetilde{G}^2 - 4A) - i2b\widetilde{G}]^{1/2}.$$

It is easy to see that for $A < 0$, the quantity ω^* has a positive real part which causes the mechanical system to become unstable. Therefore, $A \geqslant 0$ must be required for stability. Since $\lambda_{A,\min}$ is the lower bound for the values that A can take on, the condition

$$\lambda_{A,\min} \geqslant 0$$

is also sufficient for stability. However, λ_A is the eigenvalue of the symmetric matrix **A**, which is always real. The condition $\lambda_{A,\min} \geqslant 0$ means that the static stability criterion obtained with the buckling determinant **A** is sufficient for the stability of the kinetic system.

Summary

The Galerkin method makes it possible to reduce the treatment of non-conservative or polygenic stability problems to the discussion of a discrete system with an infinite number of degrees of freedom. The latter can be approximately calculated by a reduction method using a finite number of degrees of freedom (with any degree of accuracy).

The presence of damping and asymptotic stability is required to justify the linearized calculation. This condition leads to the specification of the buckling load.

The static and kinetic criterion can be used to determine the buckling load. In the case *without damping,* the static criterion holds for conservative or monogenic problems with symmetric matrix **A** and for non-conservative or polygenic problems with matrices **A** which can be made symmetrical or which are slightly asymmetric. On the other hand, the kinetic Borchardt criterion holds for non-conservative or polygenic problems with a strongly asymmetrical matrix **A**.

When *damping is considered,* the static criterion remains sufficient if **A** is symmetric and the damping matrix **B** is an arbitrary positive diagonal matrix or a skew symmetric matrix. It also remains in force if **A** is symmetrizable or is slightly asymmetric and $\mathbf{B} = B\mathbf{E}$, $b > 0$.

If **A** is strongly asymmetric, the kinetic Borchardt criterion remains sufficient for the damping case, if $\mathbf{B} = b\mathbf{E}$, $b > 0$ again holds. On the other hand, if

B is a positive diagonal matrix with elements *that are different from each other,* there is a jump in the stability condition for asymmetric **A** with respect to the corresponding undamped case. Such a non-conservative or polygenic problem should therefore be calculated with the Hurwitz criterion and the calculation should not be made without damping.

The accompanying table summarizes the results given above.

Damping	Type of matrix	Sufficient stability criterion
Without	**A** symmetrical, symmetrizable, or slightly asymmetric	Static
	A strongly asymmetric	Kinetic (Borchardt)
With	**A** symmetric; **B** positive, diagonal with different elements or skew symmetric	Static
	A symmetrizable or slightly asymmetric; **B** = b**E**, $b > 0$	Static
	A strongly asymmetric; **B** = b**E**, $b > 0$	Kinetic (Borchardt)
	A asymmetric; **B** positive diagonal with different elements	Kinetic (Hurwitz)

For **A**, in the class of asymmetric matrices, a distinction was made between the symmetrizable and the true asymmetric matrices (matrices which cannot be made symmetrical). The true asymmetric matrices were divided into slightly asymmetric and strongly asymmetric matrices. The slightly asymmetric matrices are matrices whose eigenvalues always remain real when the load parameter p is varied, even though they are asymmetric, or at least they only become complex *after* the image point has passed through the origin as it moves along the real axis. The strongly asymmetric matrices have eigenvalues that, as the load parameter is varied, already have become complex *before* the image point has passed through the origin as it moves along the real axis.

The above discussion has shown the possible matrix structures and their influence on the type of stability problem. It is now possible to find the appropriate stability criterion and determine the method of calculation. The structure of the matrix **A** follows naturally from the nature of the given problem. On the other hand, it is especially difficult to make realistic assumptions for the damping matrix **B**. It would be necessary to use exact results from experiments. Since there are not yet enough such results available, it is very important that experiments be carried out to obtain the required data. The structure of **B** is discussed in the books by Bolotin (1.7, 1.8) already mentioned. According to this author, it is quite possible for the damping matrix to have elements which are different from each other. It must always be assumed that the internal resistance of the elastic column is proportional to the time variation of the deformation of the longitudinal fibers. Then Eqs. (2.3.53) and (2.3.54) must be supplemented as follows:

$$\int_0^l [\mu w_{tt} + \alpha w_{xxxx} + Q w_{xx} + \beta_1 w_{txxxx} + \beta_2 w_t]\, \delta w\, dx + R(w) = 0,$$

where β_1 is the coefficient of the internal damping and β_2 is the coefficient of the external damping.

Let us use the trial solution $w(x, t) = a_i f(t)\phi_i(x) = f_i(t)\phi_i(x)$, where $\phi_i(x)$ are the eigenfunctions of the free eigen-oscillation of the column.

Then

$$R(w) = 0, \qquad \phi_i^{IV} = k_{(i)}\phi_i, \qquad \int_0^l \phi_k^{IV}\phi_i\, dx = k_{(i)}\,\delta_{ik}, \qquad \int_0^l \phi_k\phi_i\, dx = \delta_{ik},$$

where $k_{(i)}$ is a constant which contains the square of the eigenfrequencies of the column as a factor. This trial solution then leads to

$$\int_0^l \{\mu(\ddot{f}_k\phi_k(x)) + [f_k(\alpha\phi_k^{IV}(x) + Q\phi_k''(x)]$$

$$+ [\dot{f}_k(\beta_1\phi_k^{IV}(x) + \beta_2\phi_k(x))]\}\,(f(t)\phi_r(x)\,\delta a_r)\, dx = 0.$$

Since the variations δa_r are independent, $f(t) \neq 0$ and due to the special properties of the ϕ_i, the following system of equations is obtained:

$$\mu\ddot{f}_i + f_i\alpha k_{(i)} + f_k \int_0^l Q\phi_k''\phi_i\, dx + \dot{f}_i(\beta_1 k_{(i)} + \beta_2) = 0, \qquad i, k = 1, 2, 3, \ldots.$$

If

$$a_{ik} = (1/\mu)\left(\alpha k_i\, \delta_{ik} + \int_0^l Q\phi_k''\phi_i\, dx\right), \qquad (1/\mu)(\beta_1 k_i + \beta_2) = b_i,$$

this can also be written in the form $\ddot{f}_i + a_{ik}f_k + b_{(i)}\dot{f}_i = 0$, $i, k = 1, 2, 3$. Therefore, $f_k = a_k\, e^{\omega t}$ leads to the eigenvalue equation

$$(\mathbf{A} + \omega\mathbf{B} + \omega^2\mathbf{E})\mathbf{X} = 0,$$

where $\mathbf{A} = (a_{ik})$, $\mathbf{X} = (a_i)$ and the $\mathbf{B} = (b_i)$ is now a positive diagonal matrix with elements that are all different!

Because of the assumptions made above regarding the internal damping of the elastic column, it would be advisable to reconsider the results of the stability calculation without damping for non-conservative beam problems, because the destabilizing effect of small damping must be taken into account, which is due to the structure of $\mathbf{B}$.

However, for sufficiently large values of β, damping re-assumes its stabilizing effect. Therefore all the results calculated for undamped beams remain valid in that case. See [1].

[1] Leipholz, H., *Stability of Elastic Systems*, Sijthoff and Noordhoff, Aalphen aan den Rijn, The Netherlands, pp. 378, 379, 1980.

Stability Discussion:

In order to determine the applicability of the algebraic theory discussed above, it is necessary to determine the conditions under which the bounded nature of the approximate solution

$$w_{(n)} = e^{\omega t}\phi_{(n)}(x), \qquad \phi_{(n)}(x) = \sum_{i=l}^{n} a_i\phi_i(x)$$

and its derivative guarantee that the true solution

$$w = e^{\omega t}\phi(x)$$

and its derivative will be bounded. The Galerkin method converges for the norm of the Hilbert space employed. For sufficiently large n we have

$$\| w - w_{(n)} \| = \varepsilon^{\frac{1}{2}},$$

where ε is arbitrarily small. Let us consider the *column problem* as an example. We have

$$\| w - w_{(n)} \|_{(H)}^2 = \int_0^l \{A(w - w_{(n)}, (w - w_{(n)}))\} \, dx + \varepsilon,$$

with $A = d^4/dx^4$. Therefore

$$\int_0^l (w^{IV} - w_{(n)}^{IV})(w - w_{(n)}) \, dx$$

$$= \int_0^l (w^{IV}w - w^{IV}w_{(n)} - w_{(n)}^{IV}w + w_{(n)}^{IV}w_{(n)}) \, dx = \varepsilon.$$

For two functions u and v, which satisfy the possible boundary conditions of the column, integration by parts yields

$$\int_0^l u^{IV}v \, dx = \int_0^l u''v'' \, dx.$$

w as well as $w_{(n)}$ satisfy the boundary conditions and therefore satisfy this condition. The result is

$$\int_0^l (w^{IV} - w_{(n)}^{IV})(w - w_{(n)}) \, dx = \int_0^l (w'' - w_{(n)}'')^2 \, dx = \varepsilon.$$

Since for the column problem we always have $w(0, t) = w_{(n)}(0, t) = 0$ because of the boundary conditions, we have

$$w - w_{(n)} = \int_0^x (w' - w_{(n)}') \, dx.$$

The Cauchy–Schwarz inequality results in

$$(w - w_{(n)})^2 \leqslant x \int_0^x (w' - w_{(n)}')^2 \, dx \leqslant x \int_0^l (w' - w_{(n)}')^2 \, dx$$

$$\leqslant l \int_0^l (w' - w_{(n)}')^2 \, dx.$$

Furthermore

$$w'(x, t) - w'(0, t) - w'_{(n)}(x, t) + w'_{(n)}(0, t) = \int_0^x (w'' - w''_{(n)}) \, dx.$$

Then, according to the Cauchy-Schwarz inequality

$$\{ [w'(x, t) - w'_{(n)}(x, t)] - [w'(0, t) - w'_{(n)}(0, t)] \}^2 \leqslant x \int_0^l (w'' - w''_{(n)})^2 \, dx$$

and therefore

$$\int_0^l \{ [w'(x, t) - w'_{(n)}(x, t)] - [w'(0, t) - w'_{(n)}(0, t)] \}^2 \, dx$$

$$\leqslant (l^2/2) \int_0^l (w'' - w''_{(n)})^2 \, dx = l^2 \varepsilon / 2.$$

We have

$$\int_0^l \{ [w'(x, t) - w'_{(n)}(x, t)] - [w'(0, t) - w'_{(n)}(0, t)] \}^2 \, dx$$

$$= \int_0^l [w'(x, t) - w'_{(n)}(x, t)]^2 \, dx - 2[w'(0, t) - w'_{(n)}(0, t)] \int_0^l [w'(x, t)$$

$$- w'_{(n)}(x, t)] \, dx + [w'(0, t) - w'_{(n)}(0, t)]^2 l$$

or, after carrying out the integration in the second term on the right-hand side and with $w(0, t) = w_{(n)}(0, t) = 0$, we obtain

$$\int_0^l \{ [w'(x, t) - w'_{(n)}(x, t)] - [w'(0, t) - w'_{(n)}(0, t)] \}^2 \, dx$$

$$= \int_0^l [w'(x, t) - w'_{(n)}(x, t)]^2 \, dx - 2[w'(0, t) - w'_{(n)}(0, t)] [w(l, t)$$

$$- w_{(n)}(l, t)] + [w'(0, t) - w'_{(n)}(0, t)]^2 l.$$

The possible boundary conditions of the column are either

$$w'(0, t) = w'_{(n)}(0, t) = 0 \qquad \text{or} \qquad w(l, t) = w_{(n)}(l, t) = 0$$

so that

$$\int_0^l \{ [w'(x, t) - w'_{(n)}(x, t)] - [w'(0, t)] - [w'_{(n)}(0, t)] \}^2 \, dx$$

$$= \int_0^l [w'(x, t) - w'_{(n)}(x, t)]^2 \, dx + [w'(0, t) - w'_{(n)}(0, t)]^2 l \leqslant l^2 \varepsilon / 2.$$

This results in

$$\int_0^l [w'(x, t) - w'_{(n)}(x, t)]^2 \, dx \leqslant l^2 \varepsilon / 2,$$

which finally leads to

$$(w - w_{(n)})^2 \leqslant l \int_0^l (w' - w'_{(n)})^2 \, dx \leqslant l^3 \varepsilon / 2.$$

Therefore

$$|w - w_{(n)}| \leqslant (l^3 \varepsilon / 2)^{1/2}$$

holds. *For elastic columns, convergence with respect to the norm of the Hilbert space means the same as convergence with respect to the magnitude.* This also means that the bounded nature of the approximate solution $w_{(n)}$ calculated for the stability case also implies that the true solution w will be bounded. Therefore, in the case of stability problems of elastic columns, the requirement of the Lyapunov stability definition, which requires w to be bounded everywhere, is satisfied.

Since we have based our calculation method on linear elasticity theory, which assumes that the strains are also small, it is necessary to show that not only w but also its *derivative* remain sufficiently bounded.

For this, we may use the relationship

$$\{[w'(x, t) - w'_{(n)}(x, t)] - [w'(0, t) - w'_{(n)}(0, t)]\}^2 \leqslant x \int_0^l (w'' - w''_{(n)})^2 \, dx.$$

It follows that

$$\{[w'(x, t) - w'_{(n)}(x, t)] - [w'(0, t) - w'_{(n)}(0, t)]\}^2$$

$$\leqslant l \int_0^l (w'' - w''_{(n)})^2 \, dx \leqslant l\varepsilon.$$

If the lower end of the column is clamped

$$w'(0, t) = w'_{(n)}(0, t) = 0.$$

Hence,

$$[w'(x, t) - w'_{(n)}(x, t)]^2 \leqslant l\varepsilon,$$
$$[w'(x, t) - w'_{(n)}(x, t)] \leqslant \sqrt{l\varepsilon},$$

respectively, i.e., w' converges with respect to the magnitude, like w.

If the ends of the column are hinged, we have

$$\int_0^l w'(x, t) \, dx = \int_0^l w'_{(n)}(x, t) \, dx = 0,$$

because the boundary conditions

$$w(l, t) = w_{(n)}(l, t) = w(0, t) = w_{(n)}(0, t) = 0$$

are satisfied. We may also set

$$\int_0^l [w'(x, t) - w'_{(n)}(x, t)] \, dx = 0.$$

The integrand is continuous, and therefore — according to the mean-value theorem — $w'(\xi, t) - w'_{(n)}(\xi, t) = 0$ must hold for a certain $0 < x = \xi < l$. The above inequality holds for any $x = \xi$, which means that

$$[w'(0, t) - w'_{(n)}(0, t)]^2 \leqslant l\varepsilon, \qquad |w'(0, t) - w'_{(n)}(0, t)| \leqslant (l\varepsilon)^{1/2}.$$

This means that

$$|w'(x, t) - w'_{(n)}(x, t)| \leqslant 2(l\varepsilon)^{1/2},$$

and this again means that the absolute value of w' is convergent. The smallness of $w'_{(n)}$, which is assumed here, therefore implies that w' is small. We are therefore justified in the application of linear elasticity theory.

For *multidimensional* elastic bodies, such as plates, it is *not* always possible to make such estimates. For them, the convergence of the Galerkin method with respect to the norm of H often makes it only possible to conclude that w is bounded with respect to the quadratic mean. Therefore, for multidimensional bodies, the kinetic criterion, together with the algebraic method, is only conditionally valid, as was the energy criterion. In this case it may be necessary to introduce a new stability definition, as we did previously. This new condition would require that the displacements and their derivatives are bounded only in the mean. Under certain conditions, this may raise questions regarding the validity of the linear elasticity theory.

2.3.5 Examples for the Algebraic Approach to Elastostatic Stability Problems

Let us first consider *conservative* or *monogenic* problems. We already know that the *static stability criterion* holds for them. It can be applied in many ways, which will now be discussed.

One way is to assume that a *non-trivial equilibrium position* exists, which is justified for conservative problems, and then to apply the intersection method to the deflected position of the elastic body. The equilibrium conditions are then written down for the parts that have been cut out. This leads to the differential equation for the problem. The corresponding boundary conditions lead to the formulation of a boundary value problem. Since the equations contain the load parameter, which is decisive for the existence of the solution and of a non-trivial equilibrium position, the problem becomes an *eigenvalue problem*. The buckling load, which is the critical value of the load parameter, is the eigenvalue of the problem. This is the classical stability calculation used in elastostatics, which has been shown to have restricted validity as it can only be applied for conservative problems and some certain non-conservative problems.

Example 1: Let us consider the *buckling column* shown in Fig. 77. It is loaded by a unidirectional, vertical, concentrated load P. A column element having length dx is cut out. Let us formulate the equilibrium conditions for it, taking into account that $\cos \phi \approx 1$, $\sin \phi \approx \phi$.

Fig. 77. Euler buckling column

The equilibrium condition for all forces in the direction of the column axis results in $N = P$, $dN = 0$. Equilibrium of all forces perpendicular to the column axis results in $Q + dQ = P(\phi + d\phi)$, $Q = P\phi$ or $dQ = P\,d\phi$. The equilibrium of all moments results in $dM = Q\,dx$.

In addition, $\phi = dw/dx$, $d\phi = (d^2w/dx^2)\,dx$.

Since $M = -\alpha w''$, the relationship $dM/dx = Q$, $d^2M/dx^2 = dQ/dx$ results in

$$d^2M/dx^2 = dQ/dx = -\alpha\,d^4w/dx^4.$$

On the other hand,

$$dQ/dx = P\,d\phi/dx = P\,d^2w/dx^2,$$

so that the differential equation of the problem is

$$\alpha\,d^4w/dx^4 + P\,d^2w/dx^2 = 0. \qquad (2.3.70)$$

It has been assumed that the bending stiffness α of the column is constant.

The boundary value problem is completely formulated, once the corresponding conditions have been specified. They are

$$w(0) = w''(0) = w(l) = w''(l) = 0$$

for the column hinged at both ends shown in Fig. 77. For the other possible support possibilities, the boundary conditions are

$w(0) = w(l) = w'(0) = w'(l) = 0$ for the column clamped at both ends

$w(0) = w(l) = w'(0) = w''(l) = 0$ for the column clamped at the bottom and hinged at the top

$w(0) = w'(0) = w''(l) = 0, w'''(l) = -(P/\alpha)w'(l)$ for the column clamped at the bottom free at the top.

The general solution of Eq. (2.3.70) is

$$w = C_1 f_1(p, x) + C_2 f_2(p, x) + C_3 x + C_4, \qquad (2.3.71)$$

where $p = P/\alpha$ is the load parameter and is an *eigenvalue* of the problem. The functions

$$f_1(p, x) = \cos p^{1/2} x, \qquad f_2(p, x) = \sin p^{1/2} x$$

have been introduced.

When (2.3.71) is substituted into a set of boundary conditions, we obtain

$$a_{ik} C_k = 0, \qquad i, k = 1, 2, 3, 4,$$

which is an algebraic system for the determination of the integration constants C_k. Since it is homogeneous, it will only have non-trivial solutions for the C_k and consequently there will only be a non-trivial equilibrium position for the column if det $a_{ik} = 0$. The determinant det a_{ik} is called *the buckling determinant*. The condition

$$\det a_{ik} = 0 \qquad (2.3.72)$$

is the equation for determining the eigenvalue p_{crit}. It therefore determines the buckling load (Euler load) $P_E = \alpha p_{\text{crit}}$.

For example, let us consider the column clamped at the bottom and hinged at the top. The corresponding boundary conditions result in

$$w(0) = C_1 + C_4 = 0,$$

$$w(l) = C_1 \cos p^{1/2} l + C_2 \sin p^{1/2} l + C_3 l + C_4 = 0,$$

$$w'(0) = C_2 p^{1/2} + C_3 = 0,$$

$$w''(l) = -C_1 p \cos p^{1/2} l - C_2 p \sin p^{1/2} l = 0,$$

and Eq. (2.3.72) is given by

$$\begin{vmatrix} 1 & 0 & 0 & 1 \\ \cos p^{1/2} l & \sin p^{1/2} l & l & 1 \\ 0 & p^{1/2} & 1 & 0 \\ -p \cos p^{1/2} l & -p \sin p^{1/2} l & 0 & 0 \end{vmatrix} = 0,$$

which results in the following equation for determining the eigenvalue p:

$$\tan(p^{1/2} l) - p^{1/2} l = 0.$$

The smallest value that satisfies this transcendental equation is $p^{1/2} l = 4.493$. The smallest buckling load is

$$P_E = \alpha p_{\text{crit}} = 4.493^2 \alpha / l^2.$$

The method of applying the equilibrium method used here assumes that the differential equation of the problem can be solved exactly. If this is not possible, which is often the case, it is necessary to use the 'algebraic' method in the stability calculation, i.e., the Galerkin method. Since we have assumed a conservative problem, the Galerkin method could also be used in its special form, the *Ritz method*.

Example 2: Let us consider the buckling column that carries uniformly distributed *longitudinal forces* which have a fixed direction q (for example, column weight). This case can be solved exactly by means of Bessel functions. The advantage of this example is that the accuracy of the algebraic method can be compared to the exact result.

Let us first determine the differential equation and the boundary conditions. This can either be done by consideration of the deflected column and the intersection principle, or we can employ the principle of virtual work. In this case, it has the simple form $\delta\Pi = 0$. We have

$$\Pi = \frac{1}{2} \int_0^l [\alpha w''^2 - q(l-x)w'^2] \, dx.$$

Therefore

$$\delta\Pi = \int_0^l [\alpha w'' \, \delta w'' - q(l-x)q' \, \delta q'] \, dx = 0. \qquad (2.3.73)$$

Partial integration results in

$$\int_0^l [\alpha w^{IV} + q(l-x)w'' - qw'] \, \delta w \, dx + [\alpha w'' \, \delta w'$$

$$- \alpha w''' \, \delta w - q(l-x)w' \, \delta w]_0^l = 0.$$

The condition that the boundary value expression

$$R(w) = [\alpha w'' \, \delta w' - \alpha w''' \, \delta w - q(l-x)w' \, \delta w]_0^l$$

be zero and the fact that the geometric boundary conditions

$$w(0) = w(l) = 0$$

hold for the column pinned at both ends, leads to the dynamic boundary conditions

$$w''(0) = w''(l) = 0,$$

and the differential equation

$$L(w) = \alpha w^{IV} + q(l-x)w'' - qw' = 0.$$

This follows from the relationship (2.3.73) obtained by partial integration.

Let us set $w = \rho_1\phi_1(x) + \rho_2\phi_2(x)$ and $\phi_1 = \sin(\pi/l)x$, $\phi_2 = \sin 2(\pi/l)x$. These are coordinate functions that satisfy all the boundary conditions. According to (2.3.50), the elements of the matrix **A** are

$$a_{ik} = \frac{1}{\mu} \int_0^l L(\phi_k)\phi_i \, dx.$$

In this case

$$a_{11} = \frac{1}{\mu} \int_0^l \left[\alpha \frac{\pi^4}{l^4} \sin \frac{\pi}{l} x - q(l-x) \frac{\pi^2}{l^2} \sin \frac{\pi}{l} x - \frac{\pi}{l} q \cos \frac{\pi}{l} x \right] \sin \frac{\pi}{l} x \, dx$$

$$= \frac{1}{\mu} \left(\frac{\alpha \pi^4}{2l^3} - q \frac{\pi^2}{4} \right),$$

$$a_{12} = \frac{1}{\mu} \int_0^l \left[\alpha \frac{\pi^4}{l^4} \sin \frac{\pi}{l} x - q(l-x) \frac{\pi^2}{l^2} \sin \frac{\pi}{l} x - \frac{\pi}{l} q \cos \frac{\pi}{l} x \right] \sin \frac{2\pi}{l} x \, dx$$

$$= \frac{1}{\mu} \left(-\frac{8}{9} q - \frac{4}{3} q \right) = -\frac{1}{\mu} \frac{20}{9} q.$$

$$a_{21} = \frac{1}{\mu} \int_0^l \left[\alpha \frac{16\pi^4}{l^4} \sin \frac{2\pi}{l} x - q(l-x) \frac{4\pi^2}{l^2} \sin \frac{2\pi}{l} x \right.$$

$$\left. - \frac{2\pi}{l} q \cos \frac{2\pi}{l} x \right] \sin \frac{\pi}{l} x \, dx = \frac{1}{\mu} \left(-\frac{32}{9} q + \frac{4}{3} q \right) = -\frac{1}{\mu} \frac{20}{9} q,$$

$$a_{22} = \frac{1}{\mu} \int_0^l \left[\alpha \frac{16\pi^4}{\mu} \sin \frac{2\pi}{l} x - q(l-x) \frac{4\pi^2}{l^2} \sin \frac{2\pi}{l} x \right.$$

$$\left. - \frac{2\pi}{l} q \cos \frac{2\pi}{l} x \right] \sin \frac{2\pi}{l} x \, dx = \frac{1}{\mu} \left(\frac{8\alpha \pi^4}{l^3} - q\pi^2 \right).$$

The static criterion requires that

$$\det(a_{ik}) = \frac{1}{\mu} \begin{vmatrix} \dfrac{\alpha \pi^4}{l^3} - \dfrac{\pi^2}{4} q & -\dfrac{20}{9} q \\[2mm] -\dfrac{20}{9} q & \dfrac{8\alpha \pi^4}{l^3} - q\pi^2 \end{vmatrix} = 0.$$

This leads to the quadratic equation $p^2 - 123.8\, p + 1955 = 0$ in which $p = ql^3/\alpha$ is the load parameter. The smallest critical value is the root of this equation, which is $p_{\text{crit}} = 18.58$. Therefore, the smallest buckling load is

$$q_E = 18.58 \, \alpha/l^3.$$

This value is very accurate compared to the exact value calculated by F. Willers, which is $q_E = 18.57 \, l^3/\alpha$.

If the boundary conditions are known some other way, so that the co-ordinate functions can be immediately selected, it is possible to directly use (2.3.73). The elements of the matrix **A** are then obtained from (2.3.48) as follows:

$$a_{jk} = \int_0^l \sum_i c_i D_i(\phi_k) D_i(\phi_j) \, dx,$$

where

$$i = 1, 2, \qquad c_1 = \alpha, \qquad c_2 = -q(l - x), \qquad D_1 = d^2/dx^2, \qquad D_2 = d/dx.$$

Therefore, for

$$\phi_1 = \sin(\pi/l)x, \qquad \phi_2 = \sin(2\pi/l)x,$$

we have

$$a_{11} = \int_0^l \left[\alpha \frac{\pi^4}{l^4} \sin^2 \frac{\pi}{l} x - q(l - x) \frac{\pi^2}{l^2} \cos^2 \frac{\pi}{l} x \right] dx$$

$$= -q \frac{\pi^2}{4} + \alpha \frac{\pi^4}{l^4} \frac{l}{2},$$

$$a_{12} = a_{21} = \int_0^l \left[\alpha \frac{4\pi^4}{l^4} \sin \frac{\pi}{l} x \sin \frac{2\pi}{l} x \right.$$

$$\left. - q(l - x) \frac{2\pi^2}{l^2} \cos \frac{\pi}{l} x \cos \frac{2\pi}{l} x \right] dx = -\frac{20}{9} q,$$

$$a_{22} = \int_0^l \left[\alpha \frac{16\pi^4}{l^4} \sin^2 \frac{2\pi}{l} x - q(l - x)4 \frac{\pi^2}{l^2} \cos^2 \frac{2\pi}{l} x \right] dx$$

$$= -q\pi^2 + \alpha \frac{\pi^4}{l^4} 8l,$$

so that the static criterion again is given by

$$\det(a_{ik}) = \begin{vmatrix} -q \dfrac{\pi^2}{4} + \alpha \dfrac{\pi^4}{2l^3} & -\dfrac{20}{9} q \\[2mm] -\dfrac{20}{9} q & -q\pi^2 + \alpha \dfrac{8\pi^4}{l^3} \end{vmatrix} = 0.$$

For $p = ql^3/\alpha$, the same quadratic equation

$$p^2 - 123.8p + 1955 = 0$$

is again obtained so that the smallest root is $p_{\text{crit}} = 18.58$ and the same buckling load $q_E = 18.58\alpha/l^3$ is obtained.

The third method is to use the classical *energy criterion*. As was shown in Sections 2.3.1 and 2.3.3, caution must be used when this criterion is applied to multidimensional problems. In spite of this, we will calculate with the energy method in the following example, in order to give an impression of the method normally used to calculate the stability boundary. A more rigorous theoretical basis for multidimensional problems will be the subject of the subsequent section 2.3.7. In any case, the condition $\delta(\delta^2 \Pi_g) > 0$ of the classical energy criterion, which is equivalent to saying that $\delta^2 \Pi_g$ is positive definite for stability, or has a minimum, is at least necessary. Instability occurs if it is violated.

Fig. 77a. Plate under compression loading

Example 3: Let us consider the *plate* shown in Fig. 77a. It is loaded by the edge forces N in the x direction. It is assumed that the plate is thin and that the Kirchhoff plate theory holds. The potential of the internal and external forces for the deflected plate, which is in a perturbed state with respect to the trivial equilibrium position, is given by

$$\Pi = \tfrac{1}{2} \delta^2 \Pi_g$$

$$= (D/2) \int_F [w_{xx}^2 + w_{yy}^2 + 2v w_{xx} w_{yy} + 2(1 - \nu) w_{xy}^2 - (N/D) w_x^2] \, dF. \qquad (2.3.74)$$

This also implies that the potential Π of the deflected plate is equal to one-half of the second variation of the potential Π_g of the plate in the original state.

The stability is given by $\delta\Pi \equiv \delta(\tfrac{1}{2} \delta^2 \Pi_g) = 0$, because it is exactly when this occurs that $\delta^2 \Pi_g$ is no longer definite and the stability is lost. It is, however, not impossible for Lyapunov stability to be lost before this, because the energy criterion is only necessary and not sufficient for the Lyapunov stability theory.

The condition $\delta\Pi = 0$ is identical to the condition that (2.3.74) has an extremum. The *Ritz method* will be used to find this extreme value. This is permissible in conservative problems like that with which we are dealing.

The trial solution

$$w(x, y) = \sum_m \sum_n \rho_{mn} \sin \frac{m\pi x}{a} \sin \frac{n\pi y}{b} \qquad (2.3.75)$$

can be used for the freely supported plate. It satisfies all the geometric boundary conditions, because w is zero for $x = 0$, $x = a$, $y = 0$, $y = b$, at the edges of the plate.

The expression (2.3.74) can be rewritten as

$$\tfrac{1}{2} \delta^2 \Pi_g = D/2 \int_F [(w_{xx} + w_{yy})^2 - 2(1 - \nu)(w_{xx} w_{yy} - w_{xy}^2) - (N/D) w_x^2] \, dF. \qquad (2.3.76)$$

When (2.3.75) is substituted into (2.3.76), it can be seen that the term multiplied by $2(1 - \nu)$ is zero. The remainder of this expression is

$$\frac{1}{2} \delta^2 \Pi_g = \frac{D}{2} \int \left[\sum_m \sum_n \rho_{mn} \left(\frac{m^2 \pi^2}{a^2} + \frac{n^2 \pi^2}{b^2} \right) \sin \frac{m\pi x}{a} \sin \frac{n\pi y}{b} \right]^2 dx \, dy$$

$$- \frac{N}{2} \int \left[\sum_m \sum_n \rho_{mn}^2 \left(\frac{m^2 \pi^2}{a^2} \right) \cos^2 \frac{m\pi x}{a} \sin^2 \frac{n\pi y}{b} \right] dx \, dy.$$

256

After the integrations have been carried out, we obtain

$$\frac{1}{2}\delta^2\Pi_g = \frac{ab}{8}D\sum_m\sum_n\rho_{mn}^2\left(\frac{m^2\pi^2}{a^2}+\frac{n^2\pi^2}{b^2}\right)^2 - \frac{ab}{8}N\sum_m\sum_n\rho_{mn}^2\frac{m^2\pi^2}{a^2}. \qquad (2.3.77)$$

The conditions that require that $\delta^2\Pi_g$ takes on an extreme value are

$$(\partial/\partial\rho_{mn})(\delta^2\Pi_g) = 0.$$

Applied to (2.3.77), this results in

$$\frac{ab}{4}D\rho_{mn}\left(\frac{m^2\pi^2}{a^2}+\frac{n^2\pi^2}{b^2}\right)^2 - \frac{ab}{4}N\rho_{mn}\frac{m^2\pi^2}{a^2} = 0.$$

Since $\rho_{mn} \neq 0$, this is satisfied for

$$N_{\text{crit}} = D\frac{\pi^2a^2}{m^2}\left(\frac{m^2}{a^2}+\frac{n^2}{b^2}\right)^2 = \frac{\pi^2D}{b^2}\left(\frac{mb}{a}+\frac{n^2a}{mb}\right)^2. \qquad (2.3.78)$$

The smallest critical value is obtained for $n = 1$ and the desired buckling load is

$$N_{\text{crit,min}} = \frac{k\pi^2D}{b^2}, \qquad k = \left(\frac{mb}{a}+\frac{a}{mb}\right)^2 = (\xi+\xi^{-1})^2, \qquad \xi = \frac{mb}{a}.$$

From this it can be seen that the plate can have several half-wave dents in the load direction, and their number depends on m. There is only *one* half-wave perpendicular to the load direction (because of $n = 1$).

The smallest buckling load is obtained for the smallest possible value of k. It is obtained from

$$dk/d\xi = 2(\xi + \xi^{-1})(1 - \xi^{-2}) = 0,$$

where $\xi = 1$. It is $(N_{\text{crit,min}})_{\text{min}} = 4\pi^2D/b^2$.

For fixed values of m, k is a function of the side ratio a/b of the plate. Fig. 78 shows that dependence of k for different values of m. The diagram is used

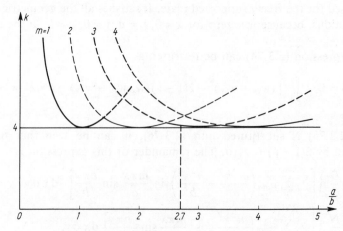

Fig. 78. Denting diagram of a rectangular plate

in the following way. For example, let $a/b = 2.7$. The ordinate along the lower boundary curve (solid line) is found which corresponds to this value. The values $m = 3$ and $k = 4.04$ are read off. This means that the plate will deform in load direction in the form of three half-waves, and the buckling load is $N_{crit,min} = 4.04\pi^2 D/b^2$.

The result (2.3.78) is also obtained if another formulation of the energy criterion is used, stating that $\delta^2\Pi_g$ is positive definite. This can only be the case if each term of the double series (2.3.77) is itself positive. Instability will certainly occur if each term of the series is zero. This occurs for

$$N_{crit} = D\,\frac{\pi^2 a^2}{m^2}\left(\frac{m^2}{a^2} + \frac{n^2}{b^2}\right)^2.$$

This is again the relationship (2.3.78), and everything else proceeds as before.

Let us now consider non-conservative or polygenic problems. Let us first consider the case where there is a *strongly asymmetric* matrix, which means that the *kinetic* criterion applies.

Example 4: Let us consider the column clamped at one end with a uniformly distributed, *tangential load* (Fig. 79a). Equation (2.3.27) holds. where

$$Q = q(l - x),$$

$$R(w) = [\alpha w_{xx}\,\delta w_x - a w_{xxx}\,\delta w - q(l - x)w_x\,\delta w]_0^l$$

so that

$$L(w) = \alpha w_{xxxx} + q(l - x)w_{xx}.$$

The condition $R(w) = 0$ results in the geometrical boundary conditions $w(0, t) = w_x(0, t) = 0$ and the dynamic condition

$$w_{xx}(l, t) = w_{xxx}(l, t) = 0.$$

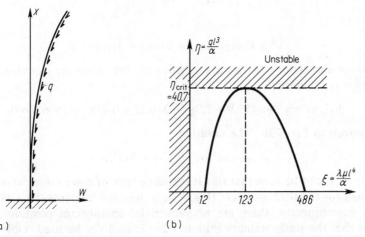

Fig. 79. (a) Buckling column with non-conservative load. (b) Eigenvalue curve

The *'abbreviated'* problem is given by

$$\mu\omega^2\phi(x) = \alpha\phi^{IV}(x) = 0,$$

$$\phi(0) = \phi'(0) = \phi''(l) = \phi'''(l) = 0,$$

which is obtained from $\mu\ddot{w} + L(w) = 0$ and the boundary conditions by substituting $w = e^{\omega^t}\phi(x)$ and crossing out the term $q(l - x)w_{xx}$ because it disturbs the self-adjoint property. The orthonormal eigenfunctions are

$$\phi_k(x) = \varkappa_k\left[c_{1k}\left(\cos\frac{m_k x}{l} - \cosh\frac{m_k x}{l}\right) - c_{2k}\left(\sin\frac{m_k x}{l} - \sinh\frac{m_k x}{l}\right)\right],$$

$k = 1, 2,$ $m_1 = 1.875,$ $m_2 = 4.694,$ $c_{11} = 4.148,$ $c_{21} = 3.037,$
$c_{12} = 53.640,$ $c_{22} = 54.631,$ $\varkappa_1 = (17.259l)^{-\frac{1}{2}}$ $\varkappa_2 = (2868.283l)^{-\frac{1}{2}}$

The elements a_{ik} of the matrix $\mathbf{A}$ can be obtained from the equation

$$a_{ik} = (1/\mu)\int_0^l L(\phi_k)\phi_i\,dx$$

which are

$$a_{11} = \frac{1}{17.259\,\mu l}\left(213.356\frac{\alpha}{l^3} + 7.5q\right),$$

$$a_{12} = \frac{264.435}{(17.259\cdot2868.283)^{\frac{1}{2}}}\frac{q}{\mu l}, \qquad a_{21} = -\frac{972.229}{(17.259\cdot2868.283)^{\frac{1}{2}}}\frac{q}{\mu l},$$

$$a_{22} = \frac{1}{2868.283\,\mu l}\left(1,392,542.792\frac{\alpha}{l^3} - 19,077.477q\right).$$

The characteristic equation $\det(\mathbf{A} - \lambda\mathbf{E}) = 0$ or

$$\begin{vmatrix} a_{11} - \lambda & a_{12} \\ a_{21} & a_{22} - \lambda \end{vmatrix} = 0$$

results in

$$\lambda^2 - \lambda(a_{11} + a_{22}) + a_{11}a_{22} - a_{12}a_{21} = 0.$$

When the expressions for a_{ik} are substituted, the following *eigenvalue curve* equation is obtained:

$$F(\xi, \eta) = \eta^2 + \eta(55.9 + 2.7\xi) - 216.2\xi + 0.43\xi^2 + 2606.0 = 0.$$

It is shown in Fig. 79b. The notation

$$\eta = ql^3/\alpha, \qquad \xi = \lambda\mu l^4/\alpha$$

was used. It can be seen that this is the same type of eigenvalue curve as was obtained for the Beck problem (Fig. 73). It does not intersect the η axis (load axis). Consequently, there are no non-trivial equilibrium positions, which means that the static stability criterion can indeed not be used. Only kinetic instability of the column is possible. This occurs for $q > 40.7\alpha/l^3$, as can be

259

seen from Fig. 79b. The critical load value

$$q_{\text{crit,min}} = 40.7\alpha/l^3$$

can either be found by determining the maximum of the eigenvalue curve or by applying the algebraic kinetic stability criterion.

In the first case, the condition

$$F(\xi, \eta) = 0, \qquad \partial F/\partial \xi = 0$$

must be used. The condition $\partial F/\partial \xi = 0$ results in $\eta^* = 80.1 - 0.32\xi$, so that $F(\xi, \eta^*) = 0$ becomes $\xi^2 + 210\xi - 40{,}908 = 0$. The positive solution $\xi^* = 123$ is obtained from which $\eta_{\text{crit}} = 80.1 - 0.32\xi^*$ or $\eta_{\text{crit}} = 40.7$ is calculated. The buckling load obtained is the one mentioned above $q_{\text{crit}} = 40.7\alpha/l^3$ and follows from the definition of η.

In the second case, the discriminant D of the characteristic equation must be used, where

$$D = (a_{11} - a_{22})^2 + 4a_{12}a_{21} = 0.$$

In round numbers, the relationships for the a_{ik} result in

$$\left(7\frac{q}{\mu l} - 473\frac{\alpha}{\mu l^4}\right)^2 - 21\frac{q^2}{\mu^2 l^2} = 0,$$

so that we are again led to the following quadratic equation with $\eta = ql^3/\alpha$

$$\eta^2 - 237\eta + 7990 = 0.$$

The smallest solution is $\mu = 40.7$, which is the already known critical value.

Let us now consider non-conservative problems for which the matrix is slightly asymmetric

Example 5: Let us consider the column clamped at both ends with a continuously distributed *tangential load* (Fig. 80a). W. Hauger provided this example. Just as in the preceding case, we have $L(w) = \alpha w_{xxxx} + q(l - x)w_{xx}$.

Fig. 80. (a) Buckling column with non-conservative load. (b) Eigenvalue curve

The boundary conditions are now

$$w(0, t) = w_x(0, t) = w(l, t) = w_x(l, t) = 0,$$

so that the following orthonormal eigenfunctions of the 'abbreviated' problem

$$\mu\omega^2\phi(x) + \alpha\phi^{IV}(x) = 0,$$

$$\phi(0) = \phi'(0) = \phi(l) = \phi'(l) = 0$$

must be used. The two functions

$$\phi_k(x) = \varkappa_k\left[c_{1k}\left(\cos\frac{m_k x}{l} - \cosh\frac{m_k x}{l}\right) - c_{2k}\left(\sin\frac{m_k x}{l} - \sinh\frac{m_k x}{l}\right)\right],$$

$$k = 1, 2, \qquad m_1 = 4.730, \qquad m_2 = 7.853, \qquad c_{11} = 57.646, \qquad c_{21} = 56.637,$$

$$c_{12} = 1285.985, \qquad c_{22} = 1286.984, \qquad \varkappa_1 = (3.323 \times 10^3 l)^{-\frac{1}{2}},$$

$$\varkappa_2 = (1.654 \times 10^6 l)^{-\frac{1}{2}}$$

are used in the calculation.

The elements

$$a_{ik} = (1/\mu)\int_0^l L(\phi_k)\phi_i \, dx$$

of the matrix **A** are obtained from them,

$$a_{11} = \frac{1}{3.323 \cdot 10^3\mu l}\left(1.663 \cdot 10^6\frac{\alpha}{l^3} - 2.044 \cdot 10^4 q\right),$$

$$a_{12} = -\frac{4.488 \cdot 10}{(3.323 \cdot 1.654 \cdot 10^9)^{\frac{1}{2}}}\frac{q}{\mu l}, \qquad a_{21} = -\frac{4.955 \cdot 10^5}{(3.323 \cdot 1.654 \cdot 10^9)^{\frac{1}{2}}}\frac{q}{\mu l},$$

$$a_{22} = \frac{1}{1.654 \cdot 10^6\mu l}\left(6.290 \cdot 10^9\frac{\alpha}{l_3} - 3.808 \cdot 10^7 q\right).$$

From det $(\mathbf{A} - \lambda\mathbf{E}) = 0$ and the notations $\eta = ql^3/\alpha$, $\xi = \lambda\mu l^4/\alpha$, the following *eigenvalue curve* equation is obtained

$$F(\xi, \eta) = \xi^2 + 2 \cdot 15.110\xi\eta + 141.631\eta^2$$

$$- 2 \cdot 2152.05\xi - 2 \cdot 17{,}461\eta + 1{,}903{,}920 = 0.$$

The two branches of this curve are shown in Fig. 80b. Owing to the symmetry in the loads and the boundary conditions, the situation is essentially the same as is found for conservative problems (Fig. 67). The eigenvalue curve intersects the load axis (η axis) and the lowest intersection point results in the critical load value

$$q_{\text{crit,min}} = 81.37\alpha/l^3,$$

which is the buckling load.

Since there are intersection points with the load axis and because the eigenvalue curve shows that in this case there cannot be kinetic instability, the critical load value be determined by the static stability criterion det $\mathbf{A} = 0$. It would have been possible to determine it directly from

$$\det \mathbf{A} = a_{11}a_{22} - a_{12}a_{21} = 0$$

without determining the eigenvalue curve. When the values of the a_{ik} given above are substituted into this equation, and if $\eta = ql^3/\alpha$, the following equation is obtained, in round numbers:

$$\eta^2 - 248\eta + 13{,}570 = 0.$$

The smallest root is

$$\eta_{\text{crit}} = 81.5$$

which compares quite well with the more exact value $\eta_{\text{crit}} = 81.37$ mentioned before.

Finally, Table I summarizes the non-conservative load cases of the elastic column treated up to the present.

Let us now consider another non-conservative system where a *slightly asymmetric* matrix occurs.

Example 6: This example was given by G. Hermann and R. W. Bungay. It is a *pinned column* (Fig. 81) with a *subtangential or supertangential buckling load P* (subtangential for $\alpha < 1$, supertangential for $\alpha > 1$).

The kinetic energy T of the system is

$$T = \tfrac{1}{2}\, ml^2 (3\dot{\phi}_1^2 + 2\dot{\phi}_1\dot{\phi}_2 + \dot{\phi}_2^2).$$

The potential energy of the restoring moments is

$$U = \tfrac{1}{2}\, c(2\phi_1^2 - 2\phi_1\phi_2 + \phi_2^2),$$

and the load results in the following generalized forces:

$$Q_1 = Pl(\phi_1 - \alpha\phi_2), \qquad Q_2 = Pl(1 - \alpha)\phi_2.$$

The Lagrange equations of the second kind are

$$\frac{\mathrm{d}}{\mathrm{d}t}\left(\frac{\partial T}{\partial \dot{\phi}_k}\right) + \frac{\partial U}{\partial \phi_k} = Q_k$$

and result in the equations of motion. They are

$$3ml^2\ddot{\phi}_1 + ml^2\ddot{\phi}_2 + (2c - Pl)\phi_1 + (\alpha Pl - c)\phi_2 = 0,$$

$$ml^2\ddot{\phi}_1 + ml^2\ddot{\phi}_2 - c\phi_1 + [c - (1 - \alpha)Pl]\phi_2 = 0.$$

The trial solution $\phi_1 = A_1 e^{\omega t}$, $\phi_2 = A_2 e^{\omega t}$ results in the following algebraic system of equations:

$$(2c - Pl + 3ml^2\omega^2)A_1 + (\alpha Pl - c + ml^2\omega^2)A_2 = 0,$$

$$(-c + ml^2\omega^2)A_1 + [c - (1 - \alpha)Pl + ml^2\omega^2]A_2 = 0,$$

Table I[1]

Support conditions	Load	Tangential concentrated force	Tangential disbuted load; $q = \text{const}$	Tangential disbuted load; $q = q_0(l - x)$
	Buckling load	$P_k = 20.05\alpha/l^2$	$q_k = 40.7\alpha/l^3$	$q_{0k} = 158.2\alpha/l^3$
	Eigenvalue curve			
	Stability criterion:	Kinetic	Kinetic	Kinetic
	Author:	M. Beck	H. Leipholz	W. Hauger
	Load:	Tangential concentrated load	$q = \text{const}$	$q = q_0(l - x)$
	Buckling load:	$P_k = P_E = 9.87\alpha/l^2$	$q_k = 18.96\alpha/l^2$	$q_{0k} = 62.28\alpha/l^4$
	Eigenvalue curve			
	Stability criterion:	Static	Static	Static
	Authors:	—	E. Abody and A. Petur, A. Pflüger, H. Leipholz	W. Hauger
	Load:	Tangential concentrated load	$q = \text{const}$	$q = q_0(l - x)$
	Buckling load:	$P_k = P_E = 20.2\alpha/l^2$	$q_k = 57.95\alpha/l^3$	$q_{0k} = 402.3\alpha/l^4$
	Eigenvalue curve:			
	Stability criterion:	Static	Static	Kinetic
	Author:	—	W. Hauger	W. Hauger

263

Table I (*Continued*)

Support conditions	Load	Tangential concentrated force	Tangential disbuted load; $q = \text{const}$	Tangential disbuted load; $q = q_0(l - x)$
	Load:	Tangential concentrated force	$q = \text{const}$	$q = q_0(l - x)$
	Buckling load:	$P_k = P_E = 39.5\alpha/l^2$	$q\int = 81.36\alpha/l^3$	$q_{0k} = 328.0\alpha/l^4$
	Eigenvalue curve:			
	Stability criterion:	Static	Static	Static
	Author:	—	W. Hauger	W. Hauger

[1] l is the length, and α the bending stiffness of the column.

which can be written in the form $(\mathbf{A} + \Omega^2\mathbf{C})\mathbf{X} = 0$, where

$$p = \frac{Pl}{c}, \qquad \Omega^2 = \frac{ml^2\omega^2}{c},$$

$$\mathbf{A} = \begin{pmatrix} 2-p & \alpha p - 1 \\ -1 & 1-(1-\alpha)p \end{pmatrix}, \quad \mathbf{C} = \begin{pmatrix} 3 & 1 \\ 1 & 1 \end{pmatrix}, \quad \mathbf{X} = \begin{pmatrix} A_1 \\ A_2 \end{pmatrix}.$$

In order to determine the stability of the non-trivial equilibrium position of the jointed system, the eigenvalue problem

$$\det(\mathbf{A} - \lambda\mathbf{C}) = 0, \qquad \lambda = -\Omega^2 \tag{2.3.79}$$

Fig. 81. Jointed rod with subtangential or supertangential buckling load

must be discussed. The problem is slightly asymmetric, because all the assumptions regarding this case described in Section 2.3.4 are satisfied., as we will now show. However, it must be realized that instead of $\mathbf{E}$ in (2.3.79) there is the more general, positive definite matrix $\mathbf{C}$. It is now necessary to formulate all results for the eigenvalue $\lambda_{A,C}$ of the pair of matrices $\mathbf{A}$, $\mathbf{C}$.

It can be seen that $\mathbf{A}$ is symmetric for $\alpha = 0$ and that it is even positive definite for $p = 0$. In addition, (2.3.79) continues to have only real eigenvalues $\lambda_{A,C}$ for sufficiently small α. It may be shown that this is at least the case for $0 \le \alpha \le \alpha_{g,min} = 0.345$, so that the static stability criterion must remain in force at least in this α interval. This means that the conditions for weak asymmetry are satisfied.

The proof is carried out as follows. Equation (2.3.79) is multiplied on the left by

$$\mathbf{C}^{-1} = \begin{pmatrix} \frac{1}{2} & -\frac{1}{2} \\ -\frac{1}{2} & \frac{3}{2} \end{pmatrix}$$

so that

$$(\mathbf{C}^{-1}\mathbf{A} - \lambda\mathbf{E})\mathbf{X} = \begin{pmatrix} \frac{1}{2}(3 - p) & \frac{1}{2}(p - 2) \\ \frac{1}{2}(p - 5) & \frac{1}{2}[4 + (2\alpha - 3)p] \end{pmatrix} \begin{pmatrix} A_1 \\ A_2 \end{pmatrix} = 0 \quad (2.3.80)$$

is obtained. The eigenvalue of interest $\lambda_{A,C}$ is obtained as the root of the characteristic equation $\det(\mathbf{C}^{-1}A - \lambda\mathbf{E}) = 0$. If $\mathbf{B} = \mathbf{C}^{-1}A = (b_{ik})$ and

$$b_{11} = \tfrac{1}{2}(3 - p), \qquad b_{12} = \tfrac{1}{2}(p - 2), \qquad b_{21} = \tfrac{1}{2}(p - 5), \qquad b_{22} = \tfrac{1}{2}[4 + (2\alpha - 3)p]$$

$$(2.3.81)$$

the characteristic equation is

$$\lambda_{A,C}^2 - \lambda_{A,C}(b_{11} + b_{22}) + b_{11}b_{22} - b_{12}b_{21} = 0. \quad (2.3.82)$$

Complex eigenvalues $\lambda_{A,C}$ are only possible if the discriminant of (2.3.82) is smaller than zero. This would be the case for

$$D = (b_{11} - b_{22})^2 + 4b_{12}b_{21} < 0.$$

The limiting case is given by $D = 0$, which is also the *kinetic stability criterion*. When $D = 0$ is calculated from (2.3.81), the result is

$$p = \frac{8 - \alpha \pm \{(8 - \alpha)^2 - 41[1 + (1 - \alpha)^2]\}^{1/2}}{2[1 + (1 - \alpha)^2]}. \quad (2.3.83)$$

The relationship (2.3.83) is the connection between the α, p values at the kinetic instability boundary. Figure 82 shows the corresponding boundary curve on the stability diagram based on this law. In the following discussion, we will only consider the positive quadrant of the α, p plane. The diagram shows that there are α values for which *no* real p values can be found that would satisfy (2.3.83). This is the case for

$$(8 - \alpha)^2 - 41[1 + (1 - \alpha)^2] < 0. \quad (2.3.84)$$

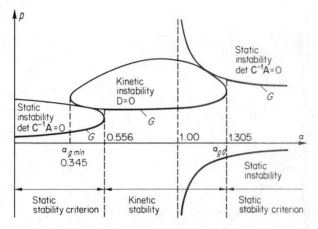

Fig. 82. Stability diagram

It is easily seen that (2.3.84) is satisfied for $\alpha < \alpha_{g,\min} = 0.345$ and for $\alpha < \alpha_{g,0} = 1.305$. It immediately follows that there are no complex $\lambda_{A,C}$ for $0 \leqslant \alpha \leqslant 0.345$ according to our assumption.

The *static stability criterion* is given by $\det \mathbf{C}^{-1}\mathbf{A} \equiv 0$ and therefore by $b_{11}b_{22} - b_{12}b_{21} = 0$. Making use of (2.3.81), the following equation is obtained

$$p^2(1 - \alpha) - 3p(1 - \alpha) + 1 = 0.$$

This is the boundary curve for static instability shown in Fig. 82.

The curve G is drawn as a thick curve and delimits the important lower region of stability. The stability diagram shows that for $\alpha = 0.556$ and $\alpha = 1.305$ there is a *jump* in the stability conditions. It can also be seen that the regions of static and kinetic stability alternate as the parameter α is varied.

It should be noted that in the intervals $0.345 < \alpha < 0.556$ and $1.00 < \alpha < 1.305$, it is possible for both criteria to hold simultaneously. The *decisive* one will always be the one that results in a portion of the heavy boundary curve G. Therefore, the static criterion holds for $0.345 < \alpha < 0.556$ and the kinetic criterion holds for $1.00 < \alpha < 1.305$. This becomes apparent, when, for example, the eigenvalue curve

$$p^2 - (3 + 18\lambda_{A,C})p + 70\lambda_{A,C} - 20\lambda_{A,C}^2 - 10 = 0,$$

which follows from (2.3.82) is plotted for $\alpha = 1.1$ (Fig. 83). The course of the $\lambda_{A,C}$ curve in the complex plane is marked with the corresponding p values (Fig. 84). It it is assumed that negative p values are not possible, the curve begins at $p_0 = 0$ on the real axis and stays along this axis as p increases. At $p_1 = 2.19$, it turns into the complex plane. The kinetic instability begins exactly at this point. At $p_2 = 4.64$, the image point of $\lambda_{A,C}$ again reaches the real axis and follows it until it goes through the origin for $p_3 = 5$. At this point static instability occurs. Since kinetic instability occurs first in this case, because the

Fig. 83. Eigenvalue curve

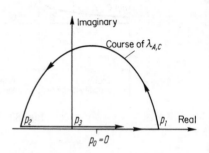

Fig. 84. Path of the eigenvalue $\lambda_{A,C}$ in the complex plane

image point turns into the complex plane before passing through the origin, the kinetic stability criterion holds for $\alpha = 1.1$ as well as for all α values of the interval $1 < \alpha < 1.305$.

Let us show the *effect of mass distribution* following an investigation by W. Hauger.

Example 7: Let us consider the *column*, shown in Fig. 85, which is clamped at the bottom. There is a *concentrated mass m* at the point $x = l_1$ whose weight is $G = mg$. There is a *tangential concentrated load P_t* at this point. At the free end, $x = l$, there is a *vertical concentrated load P_v*. The total mass of the column is $M = \mu l$. Its constant bending stiffness is α.

Fig. 85. Influence of mass distribution

The principle of virtual work results in the equation

$$\mu\left[\int_0^{l_1} \ddot{w}_1\, \delta w_1\, dx + \int_{l_1}^{l} \ddot{w}_2\, \delta w_2\, dx\right]$$
$$+ \delta(V_i + V_a) + [m\ddot{w}(l_1, t) + P_t w_x(l_1, t)\, \delta w(l_1, t)] = 0, \quad (2.3.85)$$

where

$$V_i = \frac{1}{2}\left[\int_0^{l_1} \alpha w_{1xx}^2\, dx + \int_{l_1}^{l} \alpha w_{2xx}^2\, dx\right],$$

$$V_a = -\frac{1}{2}\left[(P_v + P_t + G)\int_0^{l_1} w_{1x}^2\, dx + P_v\int_{l_1}^{l} w_{2x}^2\, dx\right],$$

and the boundary conditions are

$$w_1(0, t) = w_{1x}(0, t) = w_{2xx}(l, t) = 0, \quad (2.3.86)$$

$$w_{2xxx}(l, t) = -(P_v/\alpha)w_{2x}(l, t). \quad (2.3.87)$$

The transition conditions are

$$w_1(l_1, t) = w_2(l_1, t) = w(l_1, t),\ \ddot{w}_1(l_1, t) = \ddot{w}_2(l_1, t) = \ddot{w}(l_1, t),$$

$$w_{1x}(l_1, t) = w_{2x}(l_1, t) = w_x(l_1, t),\ w_{1xx}(l_1, t) = w_{2xx}(l_1, t) = w_{xx}(l_1, t). \quad (2.3.88)$$

The deflection of the column axis is given by $w_1(x, t)$ in the interval $0 \le x \le l_1$ and by $w_2(x, t)$ in the interval $l_1 \le x \le l$.

Taking the boundary conditions (2.3.86) into account, the equation (2.3.85) becomes

$$\int_0^{l_1} [\mu\ddot{w}_1 + \alpha w_{1xxxx} + (P_v + P_t + G)w_{1xx}]\, \delta w_1(x, t)\, dx + \int_{l_1}^{l} [\mu\ddot{w}_2$$

$$+ \alpha w_{2xxxx} + P_v w_{2xx}]\, \delta w_2(x, t)\, dx + [m\ddot{w}(l_1, t) + P_t w_x(l_1, t)$$

$$- (P_t + G)w_x(l_1, t) - \alpha w_{1xxx}(l, t) + \alpha w_{2xxx}(l_1, t)]\, \delta w_1(l_1, t)$$

$$- [\alpha w_{2xxx}(l, t) + P_v w_2(l, t)]\, \delta w_2(l, t) = 0 \quad (2.3.89)$$

after the variation is carried out.

In order to apply the *Galerkin method*, the coordinate functions must be selected so that (2.3.86) is indeed satisfied. This is easy to do because the eigenfunctions for free rod oscillations satisfy them. The functions are also orthogonal, which is useful in the calculations. It can be shown that the two-term trial solution

$$w(x, t) \equiv w_1(x, t) \equiv w_2(x, t) = e^{i\omega t}[\rho\phi_1(x) + \rho_2\phi_2(x)] \quad (2.3.90)$$

can be used, where ϕ_1, ϕ_2 are the mentioned eigenfunctions. By means of

(2.3.90), Eq. (2.3.89) can first be transformed into

$$
\int_0^{l_1} [\alpha(\rho_1\phi_1^{IV} + \rho_2\phi_2^{IV}) - \mu\omega^2(\rho_1\phi_1 + \rho_2\phi_2) + (P_v + P_t + G)(\rho_1\phi_1'' + \rho_2\phi_2'')]
$$

$$
\times (\delta\rho_1\phi_1 + \delta\rho_2\phi_2)\,dx + \int_{l_1}^{l} [\alpha(\rho_1\phi_1^{IV} + \rho_2\phi_2^{IV}) - \mu\omega^2(\rho_1\phi_1 + \rho_2\phi_2)
$$

$$
+ P_v(\rho_1\phi_1'' + \rho_2\phi_2'')] (\delta\rho_1\phi_1 + \delta\rho_2\phi_2)\,dx
$$

$$
- \{G[\rho_1\phi_1'(l_1) + \rho_2\phi_2'(l_1)] + m\omega^2[\rho_1\phi_1(l_1) + \rho_2\phi_2(l_1)]\}
$$

$$
\times [\delta\rho_1\phi_1(l_1) + \delta\rho_2\phi_2(l_1)] - \{\alpha[\rho_1\phi_1'''(l) + \rho_2\phi_2'''(l)]
$$

$$
+ P_v[\rho_1\phi_1'(l) + \rho_2\phi_2'(l)]\}[\delta\rho_1\phi_1(l) + \delta\rho_2\phi_2(l)] = 0.
$$

After multiplication and noting that the variations $\delta\rho_1, \delta\rho_2$ are independent, the following algebraic system of equations $\alpha_{ik}\rho_k = 0$ is obtained, whose coefficients are

$$
\alpha_{ik} = \int_0^{l} [\alpha\phi_k^{IV}\phi_i + P_v\phi_k''\phi_i - \mu\omega^2\phi_k\phi_i]\,dx + \int_0^{l_1} (P_t + G)\phi_k''\phi_i\,dx
$$

$$
- [G\phi_k'(l_1)\phi_i(l_1) + m\omega^2\phi_k(l_1)\phi_i(l_1) + \alpha\phi_k'''(l)\phi_i(l) + P_v\phi_k'(l)\phi_i(l)]. \quad (2.3.91)
$$

The eigenvalue equation is obtained from the condition $\det(a_{ik}) = 0$. It is given by

$$
a\xi^2 + 2b\xi\eta + c\eta^2 + 2d\xi + 2e\eta + f = 0, \quad (2.3.92)
$$

where

$$
\xi = \mu l^4\omega^2/\alpha, \qquad \eta = P_t l^2/\alpha.
$$

The coefficients of (2.3.92) have the following meaning:

$$
a = \det(c_{ij}), \qquad b = \tfrac{1}{2}(b_{11}c_{22} + b_{22}c_{11} - b_{12}c_{21} - b_{21}c_{12}),
$$

$$
c = \det(b_{ij}), \qquad d = \tfrac{1}{2}(c_{11}e_{22} + c_{22}e_{11} - c_{12}e_{21} - c_{21}e_{12}),
$$

$$
e = \tfrac{1}{2}(b_{11}e_{22} + b_{22}e_{11} - b_{12}e_{21} - b_{21}e_{12}),
$$

$$
f = \det(e_{ij}),
$$

$$
a_{ij} = k_i^4 A_{ij} + t(B_{ij} - F_{ij}), \qquad b_{ij} = C_{ij},
$$

$$
c_{ij} = -(A_{ij} + RD_{ij}), \qquad d_{ij} = R(C_{ij} - E_{ij}), \qquad e_{ij} = a_{ij} + d_{ij}\zeta,
$$

$$
A_{ij} = (1/l)\int_0^{l} \phi_i\phi_j\,dx, \qquad B_{ij} = l\int_0^{l} \phi_i''\phi_j\,dx, \qquad C_{ij} = l\int_0^{l} \phi_i''\phi_j\,dx,
$$

$$
D_{ij} = \phi_i(l_1)\phi_j(l_1), \qquad E_{ij} = l\phi_i'(l_1)\phi_j(l_1), \qquad F_{ij} = l\phi_i'(l)\phi_j(l).
$$

In addition, the following data are required:

$$
t = P_v l^2/\alpha, \qquad R = m/\mu l \equiv G/\mu lg, \qquad \zeta = \mu l^3 g/\alpha, \qquad k_1 = 1.875, \qquad k_2 = 4.694.
$$

The critical load parameters may be found from the kinetic stability criterion.

Fig. 86. Calculation of the optimum mass ratio R

It is the value for which (2.3.92) has a horizontal tangent (see, for example, Fig. 79b):

$$\eta_{\text{crit}} = -\frac{1}{ac - b^2}\left[(ae - bd) \pm \{(ae - bd)^2 - (ac - b^2)(af - d^2)\}^{1/2}\right]. \quad (2.3.93)$$

From (2.3.93) it follows that the buckling load is $P_{t,\text{crit}} = \eta_{\text{crit}}\alpha/l^2$.

W. Hauger evaluated this for $l_1 = l$, $t = 1$, $\zeta = 0.01$. This ζ value corresponds approximately to a steel column 2.2 m long having a circular cross section with radius $r = 4$ cm. When (2.3.93) was evaluated, he found that there is an optimum value of R and therefore an optimum ratio of concentrated mass m to total mass $M = \mu l$ of the column, for which the buckling load P_t has a maximum value. Figure 86 shows this relationship.

A large number of non-conservative stability problems occur in *aeroelasticity*. There are a number of structures which are loaded by aerodynamic forces ('follower' forces). Let us consider one example and also refer to the book by Bisplinghoff and Ashley (13) which contains this example as well as a number of others.

Example 8: Let us consider the 'flutter' of a plane, thin plate, which is infinitely long in the y direction and has length l in the x direction. We will show that this flutter is a kinetic instability due to the effect of aerodynamic forces at supersonic speeds. The original state is the plane, unstressed plate.

If it is assumed that the perturbed state is essentially given by deflections $W(x, t)$ perpendicular to the plane of the plate, the variational equations are given by

$$\frac{\partial^4 W}{\partial \xi^4} + p\frac{\partial W}{\partial \xi} + \varkappa\frac{\partial^2 W}{\partial t^2} = 0. \quad (2.3.94)$$

$\xi = x/l$ is a dimensionless variable, $\varkappa = \mu l^4/D$, $p = 2ql^3/D(M^2 - 1)^{1/2}$, μ is the mass density, D is the stiffness of the plate, M the Mach number, $q = \rho_\infty U^2/2$ the stagnation pressure, ρ_∞ the density of the undisturbed stream, U the stream velocity. The derivation of (2.3.94) is given by Bisplinghoff and Ashley

(13). The influence of damping by the aerodynamic forces has been ignored for purposes of simplicity.

If we consider a perturbation of the original state in only one direction, perpendicular to the plane of the plate, we are only dealing with one case of conditional stability. However, this case certainly includes the important features of the process.

The trial solution $W(\xi, t) = w(\xi)\, e^{\omega t}$ is made for (2.3.94), so that stability will prevail as long as ω is imaginary. Equation (2.3.94) becomes

$$w^{IV} + pw' + \varkappa\, \omega^2 w = 0. \qquad (2.3.95)$$

The trial solution $w = \rho_i w_i$ is used for (2.3.95), and it is assumed that the coordinate functions satisfy *all* boundary conditions. Then the Galerkin method in the restricted sense can be carried out and we have the following according to (1.6.67):

$$\sum_k \rho_k \int_0^l L^*(w_k) w_j\, d\xi = 0, \qquad L^*(y) = \frac{d^4 y}{d\xi^4} + p\frac{dy}{d\xi} + \varkappa\omega^2 y. \qquad (2.3.96)$$

We will assume that the plate is freely supported at the edges. The boundary conditions are then $w(0) = w''(0) = w(l) = w''(l) = 0$. We also restrict ourselves to a two-term trial solution w. We have

$$w_1(\xi) = \sqrt{2}\, \sin \pi\xi \qquad w_2(\xi) = \sqrt{2}\, \sin 2\pi\xi. \qquad (2.3.97)$$

As usual, $(\mathbf{A} - \lambda\mathbf{E})\mathbf{X} = 0$ is written for (2.3.96). The relationships $\mathbf{X} = (\rho_k)$ and $\mathbf{A} = (a_{ik})$ hold, where

$$a_{ik} = \int_0^l L(w_k) w_i\, d\xi, \qquad L(y) = \frac{d^4 y}{d\xi^4} + p\frac{dy}{d\xi}. \qquad (2.3.98)$$

Furthermore, $\lambda = -\varkappa\omega^2$, so that λ must be positive and real for stability.

The following is calculated from (2.3.98):

$$a_{11} = \pi^4, \qquad a_{12} = -(8/3)p, \qquad a_{21} = +(8/3)p, \qquad a_{22} = 16\pi^4.$$

The static stability criterion $\det \mathbf{A} = 0$ results in

$$16\pi^8 + (64/9)p^2 = 0, \qquad p = \pm i(3/2)\pi^4,$$

so that there is apparently no real critical load parameter. Therefore, the criterion fails, and the kinetic calculation must be carried out.

The characteristic equation $\det(\mathbf{A} - \lambda\mathbf{E}) = 0$ is

$$\lambda^2 - 17\pi^4\lambda + 16\pi^8 + (64/9)p^2 = 0. \qquad (2.3.99)$$

The eigenvalue curve

$$p = \pm\left(-\frac{9}{64}\lambda^2 + \frac{153}{64}\pi^4\lambda - \frac{9}{4}\pi^8\right)^{1/2} \qquad (2.3.100)$$

is derived and is shown in Fig. 87. The course of this curve shows that the static

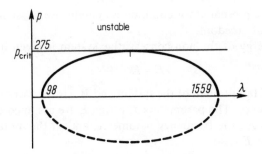

Fig. 87. Eigenvalue curve

criterion cannot be satisfied, because there is no intersection point with the load axis (p axis).

The kinetic stability criterion requires that the discriminant of (2.3.99) is zero. This is identical to the condition $(a_{11} - a_{22})^2 + 4a_{12}a_{21} = 0$, so that, because of the meaning of a_{ik}, we have

$$15^2\pi^8 - (4 \cdot 64/9)p^2 = 0$$

and therefore

$$p_{crit} = (3 \cdot 15\pi^4)/(2 \cdot 8) = 275.$$

The critical flow velocity U_{crit} is obtained form p_{crit}, which follows from the relationships between p and q and q and U. Flutter of the plate and kinetic instability must begin at U_{crit}.

2.3.6 Nonconservative Stability Problems and the Application of an Extended Version of the Energy Method

The treatment of classical problems related with conservative mechanical systems by means of the energy method is based on the application of the Hamiltonian which is a first integral of such systems. Yet, for non-conservative follower force systems such a procedure is not possible any more, because the Hamiltonian ceases to be a first integral of those systems.

In a series of papers, H. Leipholz[1,2,3] has shown that an energy approach to non-conservative systems remains feasible if one adds to the Hamiltonian the work done by the non-conservative forces. These papers were written with a finite number of degrees of freedom in mind. Here, such a generalized energy

---

[1]Leipholz H., On the Application of the Energy Method to the Stability Problem of Nonconservative Mechanical Systems, *Acta Mech.* **28**, pp. 113–138, 1977.

[2]Leipholz, H., An energy Approach to Stability Problems of Nonconservative Systems, *Developm. in Theor. and Appl. Mech., Proc. of Ninth Sectam, Nashville, 1978*, pp. 427–437, 1978.

[3]Leipholz, H., Stability Analysis of Nonconservative Systems Via Energy Considerations, *Mechanics Today*, **5**, pp. 193–214, 1980.

approach will be presented for continuous systems, i.e., systems with infinitely many degrees of freedom.

The total energy of the non-conservative system is given by the expression

$$E = H - W_N, \qquad (2.3.101)$$

where H is the Hamiltonian of the system and W_N is the work done by the non-conservative forces. The parameters affecting E are assumed to be g, the load parameter; $\lambda = \omega^2$, the frequency parameter; and y, the mode shape. Hence, the variation of E reads

$$\delta E = \frac{\delta H}{\delta y} \delta y + \frac{\partial H}{\partial g} \delta g + \frac{\partial H}{\partial \lambda} \delta \lambda - \frac{\partial W_N}{\partial g} \delta g - \delta W_N. \qquad (2.3.102)$$

In (2.3.102), it has been taken into account that W_N depends explicitly only on g. Moreover, $\delta H/\delta y$ is a functional derivative so that integration by parts (integral theorem of Gauss) yields easily

$$\frac{\delta H}{\delta y} \delta y = \int_\Omega [D(y) + N] \, \delta y \, d\Omega \qquad (2.3.103)$$

where

$$\frac{\delta H}{\delta y} \delta y \equiv \int_\Omega \frac{\delta \mathcal{H}}{\delta y} \delta y \, d\Omega \equiv \int_\Omega \delta \mathcal{H} \, d\Omega \qquad (2.3.104)$$

and $\int_\Omega \ldots d\Omega$ being either a line, surface, or volume integral depending on the dimension of the continuous system under investigation. In (2.3.103), $D(y)$ is a differential operator related to the differential equation modelling the system, and N represents the non-conservative forces. The meaning of (2.3.103) and (2.3.104) can be explained as follows:

The basic expression to be considered is

$$\frac{\delta H}{\delta y} \delta y \equiv \int_\Omega \frac{\delta \mathcal{H}}{\delta y} \delta y \, d\Omega \equiv \int_\Omega \delta \mathcal{H} \, d\Omega, \qquad (2.3.104)$$

where $\mathcal{H}$ is the density of the Hamiltonian.

For a rod:

$$\int_\Omega \ldots d\Omega \equiv \int_0^l \ldots dx \quad \text{and} \quad \mathcal{H} = -\tfrac{1}{2}\mu\omega^2 y^2 + \tfrac{1}{2}[\alpha(y'')^2 - g(l-x)(y')^2].$$

Therefore,

$$\delta \mathcal{H} = -\mu\omega^2 y \, \delta y + \alpha y'' \, \delta y'' - g(l-x)y' \, \delta y'$$

and

$$\frac{\delta H}{\delta y} \delta y = \int^l [-\mu\omega^2 y \, \delta y + \alpha y'' \, \delta y'' - g(l-x)y' \, \delta y'] \, dx. \qquad (*)$$

For all possible boundary conditions, (*) yields through integration by parts

$$\frac{\delta H}{\delta y} \delta y = \int_0^l [-\mu\omega^2 y + \alpha y''' + g(l-x)y'' - gy'] \, \delta y \, dx.$$

However,

$$D(y) = -\mu\omega^2 y + \alpha y'''' + g(l - x)y'' \qquad \text{and} \qquad N = -gy'.$$

Hence, indeed,

$$\frac{\delta H}{\delta y}\,\delta y = \int_0^l [D(y) + N]\,\delta y\,\mathrm{d}x.$$

as stated in (2.3.103).

For a rectangular, fully supported plate:

$$\int_\Omega \ldots \mathrm{d}\Omega = \int_F \ldots \mathrm{d}x_1\,\mathrm{d}x_2,$$

where F is the surface of the plate. Moreover, for a plate with compressive follower forces parallel to the x_1-direction,

$$\mathcal{H} = -\tfrac{1}{2}\mu\omega^2 y^2$$

$$+ \frac{R}{2}\left[(y_{x_1x_1} + y_{x_2x_2})^2 + 2(1 - \nu)(y_{x_1x_2}^2 - y_{x_1x_1}y_{x_2x_2}) - \frac{g}{R}(a - x_1)y_{x_1}^2\right]$$

where y is the deflection perpendicular to the middle surface of the plate, R the flexural rigidity, q the load parameter, a the side-length of the plate in the x_1-direction, μ the density per unit surface, and ν Poisson's parameter.

Obviously,

$$\int_F \frac{\delta\mathcal{H}}{\delta y}\,\delta y\,\mathrm{d}x_1\,\mathrm{d}x_2 = \int_F \delta\mathcal{H}\,\mathrm{d}x_1\,\mathrm{d}x_1 = \int_F [(-\mu\omega^2 y + R\,\Delta^2 y)\,\delta y$$

$$+ g(a - x_1)y_{x_1}\,\delta y_{x_1}]\,\mathrm{d}x_1\,\mathrm{d}x_2. \qquad (\dagger)$$

But

$$\int_F g\,(a - x_1)y_{x_1}\,\delta y_{x_1}\,\mathrm{d}x_1\,\mathrm{d}x_2 + \int_F [g(a - x_1)y_{x_1x_1} - gy_{x_1}]\,\delta y\,\mathrm{d}x_1\,\mathrm{d}x_2$$

$$+ \int_C [g(a - x_1)y_{x_1}\,\delta y]\,\mathrm{d}C, \qquad (\ddagger)$$

where C indicates that the respective expression is to be taken along the contour of the plate. Due to the boundary conditions of the plate along C,

$$\int_C [g(a - x_1)y_{x_1}\,\delta y]\,\mathrm{d}C = 0. \qquad (\S)$$

Hence, with ($\ddagger$) and ($\S$) in ($+$), and because of (2.3.104) one obtains

$$\frac{\delta H}{\delta y}\,\delta y = \int_F (-\mu\omega^2 y + R\,\Delta^2 y + g(a - x_2)y_{x_1x_1} - gy_{x_1})\,\delta y\,\mathrm{d}x_1\,\mathrm{d}x_2.$$

274

However,

$$D(y) = -\mu\omega^2 y + R\,\Delta^2 y + g(a - x_1)y_{x_1 x_1} \qquad \text{and} \qquad N = -gy_{x_1}.$$

Hence, again,

$$\frac{\delta H}{\delta y}\,\delta y = \int_F [D(y) + N]\,\delta y\,dx_1\,dx_2,$$

confirming (2.3.103).

Referring back to (2.3.102), it has to be pointed out that for a rod problem,

$$\frac{\delta H}{\delta y}\,\delta y = \frac{\partial H}{\partial y''}\,\delta y'' + \frac{\partial H}{\partial y'}\,\delta y' + \frac{\delta H}{\delta y}\,\delta y, \qquad \delta W_N = \frac{\partial W_N}{\partial y}\,\delta y + \frac{\partial W_N}{\partial y'}\,\delta y'.$$

$$(2.3.105)$$

Also, it should be noted that for a rod loaded with *follower forces*,

$$D(y) = 0 \qquad (2.3.106)$$

by definition.

With (2.3.103), (2.3.105), and (2.3.106) in (2.3.102), one arrives at

$$\delta E = \int_\Omega N\,\delta y\,d\Omega + \frac{\partial H}{\partial g}\,\delta g + \frac{\partial H}{\partial \lambda}\,\delta\lambda - \frac{\partial W_N}{\partial g}\,\delta g - \frac{\partial W_N}{\partial y}\,\delta y - \frac{\partial W_N}{\partial y'}\,\delta y'.$$

$$(2.3.107)$$

Since the variation of y is due to a change in g and λ, one has

$$\delta y = \frac{\partial y}{\partial g}\,\delta g + \frac{\partial y}{\partial \lambda}\,\delta\lambda, \qquad \delta y' = \frac{\partial y'}{\partial g}\,\delta g + \frac{\partial y'}{\partial \lambda}\,\delta\lambda. \qquad (2.3.108)$$

Using (2.3.108) in (2.3.107) yields

$$\delta E = \left[\int_\Omega N\frac{\partial y}{\partial g}\,d\Omega + \frac{\partial H}{\partial g} - \frac{\partial W_N}{\partial g} - \frac{\partial W_N}{\partial y}\frac{\partial y}{\partial g} - \frac{\partial W_N}{\partial y'}\frac{\partial y'}{\partial g}\right]\delta g$$

$$+ \left[\int_\Omega N\frac{\partial y}{\partial \lambda}\,d\Omega + \frac{\partial H}{\partial \lambda} - \frac{\partial W_N}{\partial y}\frac{\partial y}{\partial \lambda} - \frac{\partial W_N}{\partial y'}\frac{\partial y'}{\partial \lambda}\right]\delta\lambda. \qquad (2.3.109)$$

Quantity E is the *total* energy of the system. Therefore,

$$\delta E \equiv 0. \qquad (2.3.110)$$

Hence, it follows from (2.3.109) by virtue of (2.3.110) that

$$\frac{\delta g}{\delta\lambda} = -\frac{\int_\Omega N\frac{\partial y}{\partial \lambda}\,d\Omega + \frac{\partial H}{\partial \lambda} - \frac{\partial W_N}{\partial y}\frac{\partial y}{\partial \lambda} - \frac{\partial W_N}{\partial y'}\frac{\partial y'}{\partial \lambda}}{\int_\Omega N\frac{\partial y}{\partial g}\,d\Omega + \frac{\partial H}{\partial g} - \frac{\partial W_N}{\partial g} - \frac{\partial W_N}{\partial y}\frac{\partial y}{\partial g} - \frac{\partial W_N}{\partial y'}\frac{\partial y'}{\partial g}}. \qquad (2.3.111)$$

A critical situation is indicated by $\delta g/\delta\lambda = 0$. Consequently, one has as the *flutter criterion* the condition

$$\int_\Omega N\frac{\partial y}{\partial\lambda}\,d\Omega + \frac{\partial H}{\partial\lambda} - \frac{\partial W_N}{\partial y}\frac{\partial y}{\partial\lambda} - \frac{\partial W_N}{\partial y'}\frac{\partial y'}{\partial\lambda} = 0. \qquad (2.3.112)$$

Even for follower load systems, flutter is not the only mode of instability, depending on the boundary conditions. For example, so-called completely supported rods, see Fig. 63, exhibit instability also by *divergence*, even if they were loaded with follower forces. Let it therefore be shown how also respective divergence criteria can be derived by means of the previously presented theory.

Let firstly the case of completely supported systems be investigated. To fix the idea of complete support, the reader may have those above mentioned rods in mind that are supported at both ends. Due to such support,

$$W_N \equiv 0. \qquad (2.3.113)$$

Moreover, *divergence occurs* for $\lambda \equiv 0$, so that

$$\delta\lambda \equiv 0. \qquad (2.3.114)$$

Again due to the boundary conditions, also

$$\int_\Omega N\,d\Omega \equiv 0 \qquad (2.3.115)$$

holds.

By virtue of (2.3.113) and (2.3.114), equation (2.3.102) becomes

$$\delta E = \frac{\delta H}{\delta y}\delta y + \frac{\partial H}{\partial g}\delta g. \qquad (2.3.116)$$

Using (2.3.104) in (2.3.116) yields

$$\delta E = \int_\Omega [D_0(y) + N]\,\delta y\,d\Omega + \frac{\partial H}{\partial g}\delta g, \qquad (2.3.117)$$

where D_0 is the operator into which the operator D changes for $\lambda \equiv 0$.

Let now the mean value theorem of integration be applied to the integral on the right-hand side of (2.3.117). Then, one has instead of (2.3.117)

$$\delta E = \int_\Omega [D_0 y + N]\,d\Omega\,\delta\bar{y} + \frac{\partial H}{\partial g}\delta g, \qquad (2.3.118)$$

where $\delta\bar{y}$ is an appropriate mean value of δy. But, one must keep in mind that (2.3.118) is only valid as long as the function $D_0(y) + N$ is sign definite. However, this is fortunately the case for $g \in [0, g_{div}]$ as has been shown for some rod problems explicitly.[1,2]

---

[1]Leipholz H., On the Spillover Effect in the Control of Continuous Elastic Systems. *Mech. Res. Commun.* **11**, 217–226, 1984.
[2]Leipholz, H., On the Control of Structural Systems with Distributed Parameters, *Proc. 3rd Int. Conf. on Systems Engineering, Wright State Univ., Dayton, Ohio*, pp. 160–166, 1984.

Using (2.3.101) as well as (2.3.115) in (2.3.118) and rearranging yields

$$\frac{\delta g}{\delta \bar{y}} = \frac{\displaystyle\int_\Omega D_0(y)\,\mathrm{d}\Omega}{-\dfrac{\partial H}{\partial g}}. \tag{2.3.119}$$

A critical situation, i.e. divergence, occurs for $\delta g/\delta \bar{y} = 0$. Hence, (2.3.119) yields as the divergence criterion

$$\int_\Omega D_0(y)\,\mathrm{d}\Omega = 0, \qquad D_0(y) = 0. \tag{2.3.120}$$

This criterion can be brought into the form

$$\int_\Omega D_0(y)y\,\mathrm{d}\Omega = 0. \tag{2.3.121}$$

Setting

$$D_0(y) = D_E(y) - gD_L(y), \tag{2.3.122}$$

if follows from (2.3.121) that

$$g_{\mathrm{div}} = \frac{\displaystyle\int_\Omega D_E(y)y\,\mathrm{d}\Omega}{\displaystyle\int_\Omega D_L(y)y\,\mathrm{d}\Omega}, \tag{2.3.123}$$

where g_{div} is the divergence load, $D_E(y)$ is the operator modelling the elastic forces, and where $-gD_L(y)$ models the external forces. Hence, (2.3.123) assumes the rôle of Rayleigh's quotient which is common in the case of conservative systems.

It may briefly be mentioned that by virtue of (2.3.113), the flutter criterion (2.3.112) can be simplified for completely supported systems to read

$$\int_\Omega N\frac{\partial y}{\partial \lambda}\,\mathrm{d}\Omega + \frac{\partial H}{\partial \lambda} = 0. \tag{2.3.124}$$

Let now the case of conservative systems be discussed. For these, relations (2.3.113) and (2.3.114) remain certainly valid. But, condition (2.3.115) can be replaced by the stronger condition

$$D_0(y) + N = \tilde{D}_0 = 0. \tag{2.3.125}$$

Following the same argumentation as for completely supported rods and keeping in addition (2.3.125) in mind, one has

$$\delta E = \int_\Omega \tilde{D}_0(y)\,\delta y\,\mathrm{d}\Omega + \frac{\partial H}{\partial g}\delta g \tag{2.3.126}$$

instead of (2.3.117) and

$$\delta E = \int_\Omega \tilde{D}_0(y)\,\mathrm{d}\Omega\,\delta \bar{y} + \frac{\partial H}{\partial g}\delta g \tag{2.3.127}$$

instead of (2.3.118), where (2.3.127) holds as long as the functional $\tilde{D}_0(y)$ is sign definite.

Using (2.3.110) in (2.3.127) and rearranging yields

$$\frac{\delta g}{\delta \bar{y}} = \frac{\displaystyle\int_\Omega \tilde{D}_0(y)\,d\Omega}{-\dfrac{\partial H}{\partial g}}. \qquad (2.3.119^*)$$

As before, divergence occurs for $\delta g/\delta \bar{y} = 0$ yielding

$$\int_\Omega \tilde{D}_0(y)\,d\Omega = 0, \qquad \tilde{D}_0(y) = 0. \qquad (2.3.120^*)$$

Hence, one can substitute (2.3.120*) by

$$\int_\Omega \tilde{D}_0(y)y\,d\Omega = 0. \qquad (2.3.121^*)$$

Now, one has

$$\tilde{D}_0(y) = D_E(y) - g\tilde{D}_L(y) \qquad (2.3.128)$$

instead of (2.3.122), where $\tilde{D}_L$ is a *self-adjoint* operator. Hence,

$$g_{\mathrm{div}} = \frac{\displaystyle\int_\Omega D_E(y)y\,d\Omega}{\displaystyle\int_\Omega \tilde{D}_L(y)y\,d\Omega}. \qquad (2.3.129)$$

By virtue of (2.3.125), the flutter criterion for a conservative system would require according to (2.3.124) that $\partial H/\partial \lambda = 0$. However,

$$\frac{\partial H}{\partial \lambda} = \int_\Omega \mu y^2\,d\Omega \neq 0, \qquad (2.3.130)$$

where μ is the always positive unit mass density. Therefore, conservative systems do not exhibit flutter.

Interpretation of the flutter and divergence criteria for rod problems

For elastic rods of length l, flexural stiffness α, linear mass density μ and distributed follower forces q, the differential equation for the lateral deflections $w(x, t)$ reads

$$\mu\ddot{w} + \alpha w'''' + q(l - x)w'' = 0. \qquad (2.3.131)$$

However,

$$w(x, t) = e^{i\omega t}y(x), \qquad (2.3.132)$$

where $y(x)$ represents vibration modes. Therefore, with (2.3.132), equation

278

(2.3.131) changes into

$$-\mu\omega^2 y + \alpha y'''' + g(l-x)y'' = D(y) = 0. \qquad (2.3.133)$$

The boundary conditions are

$$y(0) = y''(0) = y(l) = y''(l) = 0 \text{ for the pinned–pinned rod}, \qquad (2.3.134)$$

$$y(0) = y'(0) = y(l) = y'(l) = 0 \text{ for the clamped–clamped rod}, \qquad (2.3.135)$$

$$y(0) = y'(0) = y(l) = y''(l) = 0 \text{ for the clamped–pinned rod}, \qquad (2.3.136)$$

and

$$y(0) = y'(0) = y''(l) = y'''(l) = 0 \text{ for the clamped–free rod}. \qquad (2.3.137)$$

Let first the *clamped–free rod* be discussed. For this system,

$$E = \tfrac{1}{2} \int_0^l [-\mu\omega^2 y + \alpha y'''' + g(l-x)y'']\, y\, dx$$

and integration by part yields

$$E = \tfrac{1}{2} \int_0^l [-\mu\omega^2 y^2 + \alpha(y'')^2 - g(l-x)y'^2]\, dx + \tfrac{1}{2} \int_0^l gy'\, y\, dx. \qquad (2.3.138)$$

In (2.3.138),

$$\tfrac{1}{2} \int_0^l [-\mu\omega^2 y^2 + \alpha(y'')^2 - g(l-x)y'^2]\, dx = H, \qquad (2.3.139)$$

$$\tfrac{1}{2} \int_0^l gy'\, y\, dx = -\tfrac{1}{2} \int_0^l (-gy')y\, dx = -\tfrac{1}{2} \int_0^l Ny\, dx = -W_N, \qquad (2.3.140)$$

and

$$N = -gy' \qquad (2.3.141)$$

represents the non-conservative follower forces. Hence, with (2.3.139) and (2.3.140), the relationship (2.3.138) yields, as predicted, Eq. (2.3.101), i.e.,

$$E = H - W_N$$

Furthermore, according to (2.3.103) and (2.3.139),

$$\frac{\delta H}{\delta y}\, \delta y = \int_0^l (\alpha y''\, \delta y'' - g(l-x)y'\, \delta y' - \mu\omega^2 y\, \delta y)\, dx, \qquad (2.3.142)$$

an expression which by integration by parts can be transformed into

$$\frac{\delta H}{\delta y}\, \delta y = \int_0^l [-\mu\omega^2 y + \alpha y'''' + g(l-x)y'' - gy']\, \delta y\, dx. \qquad (2.3.143)$$

By virtue of (2.3.133) and (2.3.141), Eq. (2.3.143) can be written as

$$\frac{\delta H}{\delta y} \delta y = \int_0^l [D(y) + N] \, \delta y \, dx,$$

which confirms the correctness of (2.3.104).

By means of (2.3.140) one has for expression (2.3.105) explicitly

$$-\delta W_N = \tfrac{1}{2} \int_0^l gy' \, \delta y \, dx + \tfrac{1}{2} \int_0^l gy \, \delta y' \, dx. \qquad (2.3.144)$$

Moreover,

$$\frac{\partial W_N}{\partial g} = -\tfrac{1}{2} \int_0^l y' y \, dx, \qquad (2.3.145)$$

and, because of (2.3.139), and keeping in mind that $\lambda \equiv \omega^2$, also

$$\frac{\partial H}{\partial g} = -\frac{1}{2} \int_0^l (l-x)(y')^2 \, dx, \qquad \frac{\partial H}{\partial \lambda} = -\frac{1}{2} \int_0^l \mu y^2 \, dx. \qquad (2.3.146)$$

With (2.3.141), (2.3.144), (2.3.145), (2.3.146), Eq. (2.3.107) reads

$$\delta E = \tfrac{1}{2} \int_0^l \left[-gy' \, \delta y + gy \, \delta y' - (l-x)(y')^2 \, \delta g - \mu y^2 \, \delta \lambda + y' y \, \delta g \right] \, dx.$$
$$(2.3.147)$$

Using (2.3.108) in (2.3.147) yields

$$\delta E = \frac{1}{2} \int_0^l \left[-gy' \frac{\partial y}{\partial g} \delta g - gy' \frac{\partial y}{\partial \lambda} \delta \lambda + gy \frac{\partial y'}{\partial g} \delta g + gy \frac{\partial y'}{\partial \lambda} \delta \lambda \right.$$
$$\left. - (l-x)(y')^2 \, \delta g - \mu y^2 \, \delta \lambda + y' y \, \delta g \right] \, dx. \qquad (2.3.148)$$

By virtue of (2.3.110), Eq. (2.3.148) yields after rearranging

$$\frac{\delta g}{\delta \lambda} = \frac{\int_0^l \left[\mu y^2 - gy \frac{\partial y'}{\partial \lambda} + gy' \frac{\partial y}{\partial \lambda} \right] \, dx}{\int \left[gy \frac{\partial y'}{\partial g} - gy' \frac{\partial y}{\partial g} + (l-x)y'' y \right] \, dx}, \qquad (2.3.149)$$

which can easily be identified as representing (2.3.111). It may be mentioned though, that in order to obtain (2.3.149) from (2.1.147), the identity

$$\int_0^l \left[-(l-x)(y')^2 + y' y \right] \, dx \equiv \int_0^l (l-x)y'' y \, dx \qquad (2.3.150)$$

has also been applied.

From (2.3.149) follows the flutter criterion

$$\int_0^l \mu y^2 \, dx = \int_0^l \left[gy \frac{\partial y'}{\partial \lambda} - gy' \frac{\partial y}{\partial \lambda} \right] \, dx. \qquad (2.3.151)$$

Let it be shown that (2.3.151) corresponds to (2.3.112). For this purpose write

280

(2.3.112) using (2.3.139), (2.3.140) and (2.3.141). One then obtains

$$\int_0^l \left[(-gy') \frac{\partial y}{\partial \lambda} - \frac{1}{2}\mu y^2 + \frac{1}{2} gy' \frac{\partial y}{\partial \lambda} + \frac{1}{2} gy \frac{\partial y'}{\partial \lambda} \right] dx = 0.$$

This expression can be rewritten as

$$-\frac{1}{2}\int_0^l \left[\mu y^2 + gy' \frac{\partial y}{\partial \lambda} - gy \frac{\partial y'}{\partial \lambda} \right] dx = 0,$$

from which (2.3.151) follows immediately.

Condition (2.3.151) shall be used later on to derive the smallest flutter load of the clamped–free rod numerically.

Let now the completely supported rods be considered for which the boundary conditions (2.3.134), (2.3.135), and (2.3.136) hold. By virtue of these boundary conditions, indeed,

$$W_N = \frac{1}{2} g \int_0^l y' y \, dx = -\frac{1}{4} g[y^2(l) - y^2(0)] = 0,$$

which corresponds to (2.3.113), and

$$\int_0^l N \, dx = -g \int_0^l y' y \, dx = -\frac{1}{2} g[y^2(l) - y^2(0)] = 0,$$

which corresponds to (2.3.115).

Let now the condition for divergence be checked, which occurs for $\lambda = \omega^2 \equiv 0$, $\delta\lambda \equiv 0$. Going back to (2.3.102), one has first because of (2.3.113) and (2.3.114),

$$\delta E = \frac{\delta H}{\delta y}\delta y + \frac{\partial H}{\partial g}\delta g. \qquad (2.3.152)$$

With (2.3.139) and for $\lambda = \omega^2 \equiv 0$, one finds that

$$\frac{\delta H}{\delta y}\delta y = \int_0^l [\alpha y'' \, \delta y'' - g(l-x)y' \, \delta y'] \, dx. \qquad (2.3.153)$$

Integration by parts performed on (2.3.153) yields

$$\frac{\delta H}{\delta y}\delta y = \int_0^l [\alpha y'''' + g(l-x)y'' - gy'] \, \delta y \, dx. \qquad (2.3.154)$$

The operator D_0 is supposed to follow from operator D for $\lambda = \omega^2 \equiv 0$. Equation (2.3.133) shows, that therefore

$$D_0(y) = \alpha y'''' + g(l-x)y''. \qquad (2.3.155)$$

By means of (2.3.155) and (2.3.141) one realizes that (2.3.154) yields

$$\frac{\delta H}{\delta y}\delta y = \int_0^l [D_0(y) + N] \, \delta y \, dx,$$

which, substituted in (2.3.152), leads to

$$\delta E = \int_0^l [D_0(y) + N] \, \delta y \, dx + \frac{\partial H}{\partial g} \, \delta g \qquad (2.3.156)$$

confirming (2.3.117).

For $\lambda = \omega^2 \equiv 0$, (2.3.139) yields

$$\frac{\partial H}{\partial g} = -\frac{1}{2} \int_0^l (l - x)(y')^2 \, dx. \qquad (2.3.157)$$

With (2.3.154) and (2.3.157), relationship (2.3.152) becomes equal to

$$\delta E = \int_0^l [\alpha y'''' + g(l - x)y'' - gy'] \, \delta y \, dx - \tfrac{1}{2} \int_0^l (l - x)(y')^2 \, dx \, \delta g = 0. \qquad (2.3.158)$$

Applying (2.3.110) and using the mean value theorem of integration allows one to change (2.3.158) into

$$\frac{\delta g}{\delta \bar{y}} = 2 \, \frac{\displaystyle\int_0^l [\alpha y'''' + g(l - x)y'' - gy'] \, dx}{\displaystyle\int_0^l (l - x)(y')^2 \, dx}. \qquad (2.3.158^*)$$

Yet,

$$\int_0^l gy' \, dx = g[y(l) - y(0)] = 0 \qquad (2.3.159)$$

due to the boundary conditions. Therefore, relationship $(2.3.158^*)$ can be simplified to read

$$\frac{\delta g}{\delta \bar{y}} = \frac{\displaystyle\int_0^l [\alpha y'''' + g(l - x)y''] \, dx}{\displaystyle\int_0^l (l - x)(y')^2} \qquad (2.3.160)$$

which corresponds by virtue of (2.3.155) and (2.3.157) to (2.3.119).

Divergence occurs for $\delta g/\delta \bar{y} = 0$. Hence,

$$\int_0^l [\alpha y'''' + g(l - x)y''] \, dx = 0, \qquad \alpha y'''' + g(l - x)y'' = 0 \qquad (2.3.161)$$

must hold simultaneously. Conditions (2.3.161) confirm (2.3.120).

Conditions (2.3.161) are equivalent to

$$\int_0^l [\alpha y'''' + g(l - x)y''] y \, dx = 0, \qquad (2.3.162)$$

which corresponds to (2.3.121). One realizes easily, that

$$g_{\text{div}} = \frac{\displaystyle\int_0^l \alpha y'''' y \, dx}{-\displaystyle\int_0^l (l - x)y'' y \, dx} \equiv \frac{\displaystyle\int_0^l \alpha (y'')^2 \, dx}{\displaystyle\int_0^l (l - x)(y')^2 \, dx}. \qquad (2.3.163)$$

Since $D_E(y) = \alpha y''''$ and $D_L = -(l-x)y''$, relationship (2.3.163) confirms (2.3.123).

As far as the flutter criterion (2.3.124) for completely supported systems is concerned, it reads for rods with (2.3.139) and (2.3.141)

$$\int_0^l \mu y^2 \, dx + 2 \int_0^l gy' \frac{\partial y}{\partial \lambda} \, dx = 0. \qquad (2.3.164)$$

Whether flutter will occur in addition to divergence for completely supported rods depends on the behavior of the functional

$$F = \int_0^l \mu y^2 \, dx + 2 \int_0^l gy' \frac{\partial y}{\partial \lambda} \, dx \qquad (2.3.165)$$

on the left-hand side of (2.3.164). If this functional is not sign definite and can assume the value zero, then flutter will take place. However, if the functional is sign definite, condition (2.3.164) cannot be satisfied, and flutter is ruled out. This happens for example for the pinned–pinned rod which is a sheer divergence system.

Finally, the case of conservative systems shall be investigated. For these systems, (2.3.113), (2.3.114) and (2.3.125) hold, i.e.

$$W_N = -\tfrac{1}{2} \int_0^l gy' y \, dx \equiv 0, \qquad \delta\lambda = \delta\omega^2 \equiv 0$$

and

$$D_0(y) + N = \alpha y'''' + g(l-x)y'' - gy' = \alpha y'''' + g[(l-x)y']' \qquad (2.3.125^*)$$
$$= \tilde{D}_0(y) = 0.$$

Concluding from (2.3.125*) that

$$D_E(y) = \alpha y'''', \qquad D_L = -[(l-x)y']',$$

one has for the relationship (2.3.129) explicitly

$$g_{\mathrm{div}} = \frac{\displaystyle\int_0^l \alpha y'''' y \, dx}{\displaystyle\int_0^l -[(l-x)y']' y \, dx} = \frac{\displaystyle\int_0^l \alpha (y'')^2 \, dx}{\displaystyle\int_0^l (l-x)(y')^2 \, dx}, \qquad (2.3.166)$$

which is indeed Rayleigh's quotient for the buckling load g_{div}.

Numerical Examples

Let first the flutter criterion (2.3.151) be applied. For this purpose consider the clamped–free rod. The coordinate functions to be used in a Galerkin approach are given in H. Leipholz, *Direct Variational Methods and Eigenvalue Problems of Engineering*, Noordhoff International Publ., Leyden, 1977, p.

96. By means of ϕ_1 and ϕ_2 one can approximate

$$y = a_1\phi_1 + \phi_2. \qquad (2.3.167)$$

With this assumption for y, one can set up the Galerkin equations

$$\left.\begin{array}{l} a_1(\Delta_1 + g\Phi_{11}) + g\Phi_{12} = 0, \\ (\Delta_2 + g\Phi_{22}) + a_1 g\Phi_{21} = 0, \end{array}\right\} \qquad (2.3.168)$$

where

$$\left.\begin{array}{lll} \Delta_1 = -\mu\omega^2 + \alpha(12.4l^{-4}), & \Delta_2 = -\mu\omega^2 + \alpha(485.5l^{-4}), & \\ & & \\ \Phi_{11} = 0.43l^{-1}, & \Phi_{22} = -6.65l^{-1}, & \Phi_{ii} = \displaystyle\int_0^l (l-x)\phi_i''\phi_i \,\mathrm{d}x, \\ & & \\ \phi_{12} = -4.34l^{-1}, & \Phi_{21} = 1.18l^{-1}, & \Phi_{ki} = \displaystyle\int_0^l (l-x)\phi_i''\phi_k \,\mathrm{d}x, \\ & & \\ \psi_{ki} \displaystyle\int_0^l \phi_i'\phi_k \,\mathrm{d}x, & \psi_{21} - \psi_{12} = \phi_{21} - \phi_{12}. & \end{array}\right\} \qquad (2.3.169)$$

The numerical values in (2.3.169) have been determined using the information in the above cited book by H. Leipholz on Direct Methods, where it can be found on pages 96 and 97.

Setting

$$\Delta_i + g\Phi_{ii} = D_i, \qquad (2.3.170)$$

the first equation in (2.3.168) yields

$$a_1 = -g\frac{\Phi_{12}}{D_1}. \qquad (2.3.171)$$

Moreover, one has

$$\frac{\partial a_1}{\partial \lambda} = -\mu g\frac{\Phi_{12}}{D_1^2}. \qquad (2.3.172)$$

By virtue of (2.3.167) and (2.3.171), and the orthonormality of the ϕ_i,

$$\int_0^l \mu y^2 \,\mathrm{d}x = \mu \int_0^l (a_1\phi_1 + \phi_2)^2 \,\mathrm{d}x = \mu(a_1^2 + 1). \qquad (2.3.173)$$

Using in (2.3.173) the relationship (2.3.171) yields

$$\int_0^l \mu y^2 \,\mathrm{d}x = \mu\frac{g^2\Phi_{12}^2 + D_1^2}{D_1^2}. \qquad (2.3.174)$$

Again, by virtue of (2.3.167), the orthonormality of the ϕ_i, and in addition due to the definition of the ψ_{ki} in (2.3.169), one obtains

$$\int_0^l \left[gy\frac{\partial y'}{\partial \lambda} - gy'\frac{\partial y}{\partial \lambda}\right] \mathrm{d}x = g\frac{\partial a_1}{\partial \lambda}(\Psi_{21} - \Psi_{12}) = \mu g^2\frac{\Phi_{12}(\Psi_{12} - \Phi_{21})}{D_1^2}.$$

$$(2.3.175)$$

284

Hence, (2.3.151) reads

$$\frac{\mu}{D_1^2}(g^2\Phi_{12}^2 + D_1^2) = \frac{\mu}{D_1^2}g^2\phi_{12}(\Psi_{12} - \Psi_{21}). \qquad (2.3.176)$$

After rearranging and using the last identity in (2.3.169), one arrives at

$$g^2\Phi_{12}\Phi_{21} + D_1^2 = 0$$

as the flutter criterion. It is obvious that this relationship can be generalized to read

$$g^2\Phi_{12}\Phi_{21} + D_i^2 = 0, \qquad i = 1, 2. \qquad (2.3.177)$$

This is the final form of the flutter criterion.

Introducing the new notation

$$\gamma_i = \Phi_{ii}^2 + \Phi_{12}\Phi_{21}, \qquad (2.3.178)$$

as well as (2.3.170), criterion (2.3.177) can be transformed into

$$g = \frac{\Delta_i}{\gamma_i}\left[-\Phi_{ii} \pm (-\Phi_{12}\Phi_{21})^{1/2}\right]. \qquad (2.3.179)$$

Using in (2.3.179) the values given in (2.3.169) for Δ_i, γ_i, Φ_{ii}, Φ_{12} and Φ_{21}, one can derive the following sets of equations

$$g = \begin{cases} 0.372\mu\omega^2 l - 4.606\alpha l^{-3}, & \text{Line e} \\ -0.547\mu\omega^2 l + 6.768 l^{-3}, & \text{Line f} \end{cases} \qquad (2.3.180)$$

and

$$g = \begin{cases} -0.114\mu\omega^2 l + 54.469\alpha l^{-3}, & \text{Line j} \\ -0.232\mu\omega^2 l + 110.664\alpha l^{-3}, & \text{Line k.} \end{cases} \qquad (2.3.181)$$

Equating the right-hand sides of (2.3.180,1) and (2.3.181,1) yields

$$\omega^2 = 121.55\alpha\mu^{-1}l^{-4} \qquad (2.3.182)$$

as a flutter frequency. Equating the right-hand sides of (2.3.180, 2) and (2.3.181, 2) yields correspondingly the flutter frequency

$$\omega^2 = -329.829\alpha\mu^{-1}l^{-4}. \qquad (2.3.183)$$

Using (2.3.182) either in (2.3.180, 1) or in (2.3.181, 1) yields the flutter load

$$g_{fl} = 40.61\alpha l^{-3}.$$

Using correspondingly (2.3.183) in either (2.3.180, 2) or (2.3.181, 2) yields the other flutter load

$$g_{fl} = 187.18\alpha l^{-3}. \qquad (2.3.185)$$

These results are in good agreement with those given in the cited book on pages

99 and 100 which have been derived in a different way. Such agreement of results confirms the applicability of the flutter criterion (2.3.151).

Equations in (2.3.180) and (2.3.181) represent straight lines in the eigenvalue plane, i.e. in the g, ω^2 plane. Therefore, a simple geometrical interpretation of the calculation of flutter loads is possible, providing the designing engineer with a convenient tool: the points in the eigenvalue plane having flutter loads and flutter frequencies as coordinates are obtained as the points of intersection of the above mentioned lines. This is shown in Fig. 88.

Secondly, the flutter criterion (2.3.164) for completely supported rods shall be discussed. Considering the specific boundary conditions for rods, one realizes easily that

$$\int_0^l gy' \frac{\partial y}{\partial \lambda} \, dx = - \int_0^l gy \frac{\partial y'}{\partial \lambda} \, dx.$$

For this reason, criterion (2.3.164) can be written as

$$\int_0^l \mu y^2 \, dx = \int_0^l gy \frac{\partial y'}{\partial \lambda} \, dx - \int_0^l gy' \frac{\partial y}{\partial \lambda} \, dx,$$

which is identical with (2.3.151). Therefore, all the calculations carried out before remain valid, and the result will be again (2.3.171), i.e.

$$g = \frac{\Delta_i}{\gamma_i} \left[-\Phi_{ii} + (-\Phi_{12}\Phi_{21})^{\frac{1}{2}} \right].$$

However, the last relationship in (2.3.169), i.e. $\Psi_{21} - \Psi_{12} = \Phi_{21} - \Phi_{12}$, has now to be replaced by $\Phi_{21} - \Phi_{12} = -2\Psi_{12}$. This fact is again true due to the boundary conditions of completely supported rods.

From H. Leipholz, Direct Methods, pages 96 and 97, one obtains for the clamped–pinned rod

$$\Delta_1 = -\mu\omega^2 + \alpha(237.72l^{-4}), \qquad \Delta_2 = -\mu\omega^2 + \alpha(2496.49l^{-4}),$$

$$\Phi_{11} = -4.29l^{-1}, \qquad \Phi_{22} = -18.41l^{-1}, \qquad \Phi_{12} = 0.92l^{-1}, \qquad \Phi_{21} = -5.08l^{-1},$$

and therefore

$$\gamma_1 = \Phi_{12}\Phi_{21} + \Phi_{11}^2 = 13.73l^{-2}, \qquad \gamma_2 = \Phi_{12}\Phi_{21} + \Phi_{22}^2 = 334.26l^{-2}.$$

$$(2.3.186)$$

Using these values in (2.3.171) yields

$$g = \begin{cases} -0.471\mu\omega^2 l + 111.675\alpha l^{-3}, \\ -0.155\mu\omega^2 l + 36.87\alpha l^{-3}, \end{cases} \qquad (2.3.187)$$

and

$$g = \begin{cases} -0.049\mu\omega^2 l + 121.37\alpha l^{-3}, \\ -0.062\mu\omega^2 l + 153.64\alpha l^{-3}. \end{cases} \qquad (2.3.188)$$

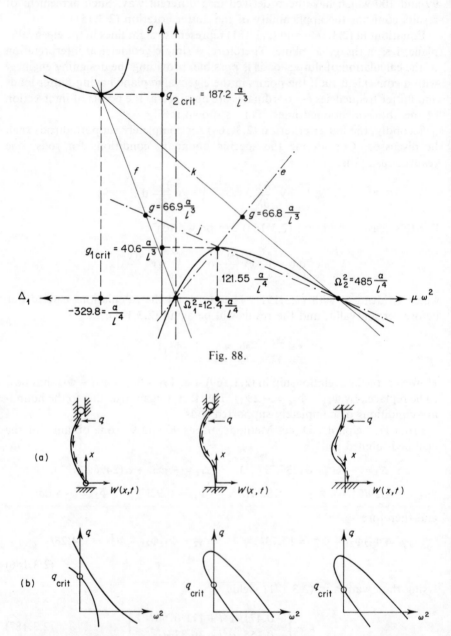

Fig. 88.

Fig. 88. (a) Sufficiently supported and appropriately directed rods subjected to follower forces (b) Eigenvalue curves of the rods in Fig. 88a

Fig. 89. Illustration of the lower bound theorem

Equation (2.3.187, 1) and (2.3.188, 1) yields

$$\omega^2 = -22.974\alpha\mu^{-1}l^{-4}, \tag{2.3.189}$$

and equating (2.3.187, 2) with (2.3.188, 2) yields

$$\omega^2 = -1255.59\alpha\mu^{-1}l^{-4}. \tag{2.3.190}$$

Using these flutter frequencies either in (2.3.187, 1), (2.3.188, 1), respectively, or in (2.3.187, 2), (2.3.188, 2) respectively, yields the flutter loads

$$g_{fl} = 122.495\alpha l^{-3} \qquad \text{and} \qquad g_{fl} = 231.49\alpha l^{-3}.$$

Fig. 90. Sufficiently supported and appropriately directed plate

The first one has been reported in the cited book on page 100 in Fig. 20. It corresponds well to the above one calculated here.

If one tried to repeat such calculations for the pinned–pinned rod, one would find with the numerical values reported in the book, i.e.

$$\Phi_{12} = -1.78 l^{-1}, \qquad \Phi_{21} = -7.12 l^{-1},$$

that (2.3.171) would yield complex values for g. The conclusion must be that *no flutter loads exist* for the pinned–pinned rod. This result has in fact been established in different ways in the cited book. The pinned–pinned rod is indeed an example for so-called pseudo-divergence systems (see the book, pp. 65–74) and conservative systems of the second kind (see the book, pp. 70–74). These are systems which fail by divergence only, although non-conservative follower forces are present.

Finally, let the divergence criterion (2.3.163) for completely supported rods be checked. In terms of the previously introduced notations it reads

$$g_{\text{div}} = \frac{\alpha \lambda_i^4 l^{-4}}{-\Phi_{ii}}.$$

Setting $i = 1$ and using the numerical values given in the cited book on pages 96 and 97, one has

$$g_{\text{div}} = \frac{\alpha 237.72 l^{-4}}{4.29 l^{-1}} = 55.4 \frac{\alpha}{l^3}.$$

In spite of the need to work with an extended version of the energy method in the case of non-conservative systems with follower forces, there are situations in which the Hamiltonian alone is adequate to yield lower bounds for the instability range of the follower load parameter q. This will be shown in the following:

Let the elastic systems to be investigated be characterized by a differential equation of the kind

$$\mu \ddot{w} \frac{\delta P}{\delta w} + M = 0, \qquad M = q w_1, \tag{2.3.191}$$

and a set of boundary conditions

$$[U(w)]_B = 0. \tag{2.3.192}$$

In (2.3.191), μ is the mass per unit surface, w is the displacement, q is the follower force acting in the x_1-direction, the expression

$$P = \int_S \mathscr{P} \, dS,$$

where $\mathscr{P}$ = potential density, S = surface, is the potential energy, and for derivatives, the following notations are being used:

$$\ddot{w} = \frac{\partial^2 w}{\partial t^2}, \qquad w_i \equiv \frac{\partial w}{\partial x_i}, \qquad w_{ij} = \frac{\partial^2 w}{\partial x_i \partial x_i}, \qquad i, j = 1, 2. \tag{2.3.193}$$

Consider the functional

$$\int_{t_0}^{t} \int_{S} \left(\mu \ddot{w} + \frac{\delta P}{\delta w} + M \right) \dot{w} \, dS \, dt = 0. \tag{2.3.194}$$

Integrating the first two terms in (2.3.194) with respect to t yields

$$\int_{S} \left[\tfrac{1}{2} \mu (\dot{w})^2 + \mathcal{P} \right] dS + \int_{t_0}^{t} \int_{S} M\dot{w} \, dS \, dt = 0. \tag{2.3.195}$$

The first term in (2.3.195) is obviously the Hamiltonian

$$H = \int_{S} \mathcal{H} \, dS = \int_{S} \left[\frac{1}{2} \mu (\dot{w})^2 + \mathcal{P} \right] dS. \tag{2.3.196}$$

Hence, differentiating (2.3.195) with respect to time and rearranging yields

$$\frac{dH}{dt} = - \int_{S} M\dot{w} \, dS = - \int_{S} q w_1 \dot{w} \, dS. \tag{2.3.197}$$

However, due to the boundary conditions, standing waves will develop in the elastic system. Therefore, in general, not excluding the possibility of flutter beforehand,

$$w(x, t) = u + iv, \tag{2.3.198}$$

where

$$u = f(t)\phi(x), \qquad v = g(t)\phi(x). \tag{2.3.199}$$

The correctness of these assumptions becomes apparent as follows: In the case of flutter, the frequency of vibration of the elastic system is complex. Therefore, the eigenfunction must also be complex: hence, (2.3.198). Using (2.3.198) in (2.3.191) and (2.3.192) yields the two sets of equations

$$\mu \ddot{u} + \frac{\delta P}{\delta u} + q u_1 = 0, \qquad [U(u)]_B = 0 \tag{2.3.200}$$

and

$$\mu \ddot{v} + \frac{\delta P}{\delta v} + q v_1 = 0, \qquad [U(v)]_B = 0. \tag{2.3.201}$$

From (2.3.200) and (2.3.201) follows immediately the fact that u and v can only differ by a factor in t, while they must have the same factor in x.
Thus, (2.3.199).

In the case of *flutter*, u and v are both different from zero, and *the time factors $f(t)$ and $g(t)$ are both unboundedly increasing functions.*

Setting

$$q u_1 = M(u), \tag{2.3.202}$$

(2.3.200) becomes

$$\mu \ddot{u} + \frac{\delta P(u)}{\delta u} + M(u) = 0, \qquad [U(u)]_B = 0, \tag{2.3.203}$$

and in analogy to (2.3.191), (2.3.194), (2.3.195), and (2.3.196) one now has the sequence of equations in terms of u:

$$\int_{t_0}^{t} \int_{S} \left[\mu\ddot{u} + \frac{\delta P(u)}{\delta u} + M(u) \right] \dot{u} \, \mathrm{d}S \, \mathrm{d}t = 0, \qquad (2.3.204)$$

$$\int_{S} \left[\tfrac{1}{2}\mu(\dot{u})^2 + \mathscr{P}(u) \right] \mathrm{d}S + \int_{t_0}^{t} \int_{S} M(u)\dot{u} \, \mathrm{d}S \, \mathrm{d}t = 0, \qquad (2.3.205)$$

and

$$H_u = \int_{S} \left[\tfrac{1}{2}\mu(\dot{u})^2 + \mathscr{P}(u) \right] \mathrm{d}S. \qquad (2.3.206)$$

It can be shown that for the problems under investigation,

$$\mathscr{P}(u) = D(u) - qF(x)u_1^2 \qquad (2.3.207)$$

where D is an appropriate differential operator and $F(x)$ is an appropriate function. These quantities will become obvious later on in a section dealing with examples. Using (2.3.207) in (2.3.206) yields

$$H_u = \int_{S} \left| \tfrac{1}{2}\mu(\dot{u})^2 + D(u) - qF(x)u_1^2 \right| \mathrm{d}S. \qquad (2.3.208)$$

In the case of unidirectional loading of the same magnitude as the follower forces q, the Euler buckling load q_c of the elastic systems follows from Rayleigh's quotient

$$q_c \leqslant \frac{\displaystyle\int_{S} D(u) \, \mathrm{d}S}{\displaystyle\int_{S} F(x)u_1^2 \, \mathrm{d}S}, \qquad (2.3.209)$$

in which the *real* function u has been used as a comparison function. Hence,

$$\int_{S} D(u) \, \mathrm{d}S \geqslant q_c \int_{S} F(x)u_1^2 \, \mathrm{d}S. \qquad (2.3.210)$$

Using (2.3.210) in (2.3.208) yields

$$H_u \geqslant \int \left[\tfrac{1}{2}\mu(\dot{u})^2 + (q_c - q)F(x)u_1^2 \right] \mathrm{d}S. \qquad (2.3.211)$$

Obviously,

$$H_u \geqslant 0, \text{ positive definite, if } q \leqslant q_c. \qquad (2.3.212)$$

From (2.3.205) and (2.3.206) follows

$$H_u + \int_{t_0}^{t} \int_{S} M(u)\dot{u} \, \mathrm{d}S \, \mathrm{d}t = 0. \qquad (2.3.213)$$

Differentiating in (2.3.213) and using (2.3.202) yields

$$\frac{\mathrm{d}H_u}{\mathrm{d}t} = -\int_{S} qu_1\dot{u} \, \mathrm{d}S. \qquad (2.3.214)$$

Substituting in (2.3.214) the expression for u given in (2.3.199) results in

$$\frac{dH_u}{dt} = -qf\dot{f}\int_S \phi\phi_1 \, dS. \tag{2.3.215}$$

Yet,

$$\int_S \phi\phi_1 \, dS = \tfrac{1}{2}\int_B \phi^2 \cos(n, x_1) \, dB, \tag{2.3.216}$$

where the integral on the right-hand side is a contour integral over the boundary B of the elastic system. Therefore, finally,

$$\frac{dH_u}{dt} = -\tfrac{1}{2}qf\frac{df}{dt}\int_B \phi^2 \cos(n, x_1) \, dB. \tag{2.3.217}$$

Subsequently, two cases will be investigated:
(i) The elastic system is 'sufficiently supported and appropriately directed.' In this case, the contour B is supposed to be supported in such a way and/or directed in such a way that

$$\phi^2 \cos(n, x_1) = 0, \tag{2.3.218}$$

because

$$\phi^2|_{B_1} = 0, \qquad \cos(n, x_1)|_{B_2} = 0, \qquad B_1 + B_2 = B. \tag{2.3.219}$$

Consequently,

$$\frac{dH_u}{dt} = 0, \qquad H_u = \text{const.}, \tag{2.3.220}$$

for a sufficiently supported and appropriately directed system.
(ii) The elastic system is 'not completely supported but appropriately directed.' In this case, the contour B is supposed not to be supported along a part $B^* \subset B$ on which however

$$\cos(n, x_1)|_{B^*} = +1 \tag{2.3.221}$$

holds true. Then,

$$\phi^2|_{B-B^*} = 0, \qquad \int_{B^*} \phi^2 \cos(n, x_1) \, dB = p, \tag{2.3.222}$$

where $p > 0$. Consequently,

$$\frac{dH_u}{dt} = -\frac{1}{2}pqf\frac{df}{dt} \tag{2.3.223}$$

and

$$H_u = -\tfrac{1}{2}pqf^2 + K, \tag{2.3.224}$$

where $K > 0$.

Let first *case (i)* be pursued: It is possible to keep the initial perturbation of the elastic system so small that

$$H_u \leqslant \varepsilon^2 \qquad \text{for } t = t_0. \qquad (2.3.225)$$

However, according to (2.3.220), then, also

$$H_u \leqslant \varepsilon^2 \qquad \text{for } t \geqslant t_0. \qquad (2.3.226)$$

From (2.3.211) follows that

$$0 \leqslant \int_S (q_c - q)F(x)u_1^2 \, dS < H_u \leqslant \varepsilon^2 \qquad \text{for } t \geqslant t_0 \text{ and } q \leqslant q_c. \qquad (2.3.227)$$

Hence,

$$\int_s u_1^2 \, dS = \| u_1 \|^2 < \frac{\varepsilon^2}{(q_c - q)M}, \qquad M = F(\xi), \qquad 0 < \xi < x_{1,\max}, \qquad (2.3.228)$$

for $t \geqslant t_0$ and $q \leqslant q_c$.
But then, according to (2.3.199),

$$f^2 \int_S \left(\frac{\partial \phi}{\partial x_1} \right)^2 \, dS = f^2 C^2 < \frac{\varepsilon^2}{(q_c - q)M}, \qquad \text{for } \begin{cases} t \geqslant t_0, \\ q \leqslant q_c, \end{cases} \qquad (2.3.229)$$

where C is a positive constant. Finally,

$$f < \frac{\varepsilon}{C\sqrt{(q_c - q)M}} = \varepsilon_f, \text{ for } \begin{cases} t \geqslant t_0, \\ q \leqslant q_c, \end{cases} \qquad (2.3.230)$$

where ε_f is an arbitrary small, positive constant. The result is that within the load range $0 < q \leqslant q_c$, the time factor f remains bounded, even arbitrarily small.

Word by word, the same reasoning as it has been applied to u can be applied to v. As a result, one arrives at

$$g < \varepsilon_g \qquad \text{for } t \geqslant t_0, \qquad q \leqslant q_c. \qquad (2.3.231)$$

As a consequence of (2.3.230) and (2.3.231) it follows that functions u and v, and therefore function w, are *bounded* functions. Hence, there is stability in the load range $0 < q \leqslant q_c$, and flutter is ruled out.

Let now *case (ii)* be investigated: At this time, the fact that H is positive definite for $0 < q \leqslant q_c$ (see (2.3.212)) has to be brought into concordance with (2.3.224). Both relationships for H_u are only in agreement with each other, if

$$f^2 < \frac{2K}{pq} \qquad \text{for } t \geqslant t_0 \text{ and } q \leqslant q_c. \qquad (2.3.232)$$

Hence,

$$f^2 < \sqrt{\frac{2K}{pq}} = C_f \qquad \text{for } t \geqslant t_0 \text{ and } q \leqslant q_c \qquad (2.3.233)$$

must hold true for f at any time in the load range $0 < q \leqslant q_c$.

By the same reasoning, one arrives at

$$g < C_g \qquad \text{for } t \geq t_0 \text{ and } q \leq q_c. \tag{2.3.234}$$

Based on (2.3.233) and (2.3.234) one can conclude that functions u, v, and therefore w, are *bounded* functions. Thus, there is again stability in the load range $0 < q \leq q_c$ but no flutter.

The final result may be given in the form of the following *theorem*:

For elastic systems of type (2.3.191), (2.3.192) subjected to follower forces q, which are either 'sufficiently supported and appropriately directed' or 'not completely supported but appropriately directed', there can be no flutter in the load range $0 < q \leq q_c$. Here q_c is the Euler buckling load of the elastic system for the case that it is loaded with unidirectional forces of the same magnitude as the follower forces q.

The preceding deliberations were rather abstract. Therefore, it may be desirable to see that a number of well-known, practical problems are covered by the results obtained in the preceding section of this paper.

Rod Problems:

For elastic rods,

$$P = \int_0^l \mathscr{P}\,dx, \qquad \mathscr{P} = \tfrac{1}{2}\alpha(w'')^2 - \tfrac{1}{2}q(l-x)(w')^2. \tag{2.3.235}$$

Comparing the expression for $\mathscr{P}$ with (2.3.207) shows that

$$D \equiv \frac{1}{2}\alpha\left(\frac{\partial^2}{\partial x^2}\right)^2, \qquad F(x) \equiv \tfrac{1}{2}(l-x). \tag{2.3.236}$$

Moreover,

$$\frac{\delta P}{\delta w} = \frac{\partial \mathscr{P}}{\partial w} - \frac{\partial}{\partial x}\left(\frac{\partial \mathscr{P}}{\partial w'}\right) + \frac{\partial^2}{\partial x^2}\left(\frac{\partial \mathscr{P}}{\partial w''}\right) = q(l-x)w'' - qw' + \alpha w''''. \tag{2.3.237}$$

Using (2.3.237) in (2.3.191) yields therefore, with

$$M = qw', \tag{2.3.238}$$

the differential equation

$$\mu\ddot{w} + \alpha w'''' + q(l-x)w'' = 0. \tag{2.3.239}$$

According to (2.3.194), (2.3.237), and (2.238), one has to evaluate

$$\int_{t_0}^t \int_0^l (\mu\ddot{w} + \alpha w'''' + q(l-x)w'')\dot{w}\,dx\,dt = 0, \tag{2.3.240}$$

which can easily be found to be

$$\int_0^l [\tfrac{1}{2}\mu(\dot{w})^2 + \tfrac{1}{2}\alpha(w'')^2 - \tfrac{1}{2}q(l-x)(w')^2]\,dx + \int_{t_0}^t \int_0^l qw'\dot{w}\,dx\,dt = 0. \tag{2.3.241}$$

The first term on the left-hand side in (2.3.241) is the Hamiltonian H of the

294

rod. hence

$$H = -\int_{t_0}^{t} \int_{0}^{l} q w' \dot{w} \, dx \, dt, \qquad (2.3.242)$$

and, in agreement with (2.3.197) as well as with statements in[1],

$$\frac{dH}{dt} = -\int_{0}^{l} q w' \dot{w} \, dx. \qquad (2.3.243)$$

(i) *Sufficiently supported and appropriately directed rod systems* are obviously the rods shown in Fig. 88a. Moreover, for these rods, the lower bound theorem given in[2] and illustrated in Fig. 89, can be invoked. According to this theorem,

$$q_{\text{crit}} > q_c, \qquad (2.3.244)$$

where q_{crit} is the lowest critical load of the rods. Hence, in the domain $0 < q < q_c$, the eigenfunction $w \equiv u$ is real and according to this fact and to (2.3.220),

$$\frac{dH}{dt} = 0, \qquad H = \text{const.} \qquad (2.3.245)$$

Therefore, the following stability considerations are possible: let

$$H < \bar{\bar{\varepsilon}}^2 \qquad \text{at } t = t_0. \qquad (2.3.246)$$

But, then also

$$H < \bar{\bar{\varepsilon}}^2 \qquad \text{for } t \geqslant t_0 \qquad (2.3.247)$$

because of $\dot{H} = 0$.

From (2.3.236) and (2.3.211) follows

$$\tfrac{1}{2}(q_c - q) \int_{0}^{l} (l - x)(w')^2 \, dx < H < \bar{\bar{\varepsilon}}^2 \quad \text{for } t \geqslant t_0. \qquad (2.3.248)$$

Hence,

$$\int_{0}^{l} (w')^2 \, dx \leqslant \frac{\bar{\bar{\varepsilon}}^2}{(q_c - q)M} = \bar{\bar{\bar{\varepsilon}}}^2, \qquad \text{where } M = l - \xi, \qquad 0 \leqslant \xi \leqslant l. \qquad (2.3.249)$$

Moreover, according to Schwarz' inequality,

$$[w(x)]^2 = \left(\int_{0}^{x} w' \, dx \right)^2 \leqslant l \int_{0}^{l} (w')^2 \, dx. \qquad (2.3.250)$$

[1]Leipholz, H., *Stability of Elastic Systems*, Sitjhoff and Noordhoff, Alphen aan den Rijn, The Netherlands, pp. 45, 265, 1980.
[2]Leipholz H. and Polzin, T., On a Lower Bound Theorem for the Buckling Load of Elastic Beams, *Acta Mechanica* 25, 257–277, 1977.

Combining (2.3.250) with (2.3.249) yields

$$[w(x)]^2 < l\bar{\varepsilon}^2 \qquad (2.3.250)$$

and finally,

$$w(x) < \bar{\varepsilon}\sqrt{l} = \varepsilon \qquad \text{for } t \geqslant t_0 \text{ and } q \leqslant q_c. \qquad (2.3.251)$$

Condition (2.3.251) expresses for sufficiently supported rod systems shown in Fig. 88a the fact that there is stability in the following sense:

Let $q \leqslant q_c$, and let the initial perturbation be of such a limited intensity that inequality (2.3.246) is being satisfied. Thus, the ensuing motion of the rods is stable in so far as at any time $t \geqslant t_0$, the lateral deflections $w(x)$ of the rods remain below the bound ε in (2.3.251). This bound is actually prescribed by the initial condition (2.3.246). Hence, there is stability in the sense of Lyapunov.

(ii) *Not completely supported but appropriately directed rod systems*. There is *one* example for such systems, the clamped–free rod shown in Fig. 79. For this rod, the trace of the eigenvalue curve indicates that instability occurs at the critical load q_{fl} and is by flutter.

For the sake of generality assume that $q_{\mathrm{fl}} < q_c$ where possible, because for the clamped–free rod, the previously mentioned lower bound theorem does not hold. Then, in the load range $0 < q \leqslant q_c$, the expressions

$$w(x, t) = u + iv, \qquad u = f(t)\phi(x), \qquad v = g(t)\phi(x)$$

may at least be partly valid, corresponding to (2.3.198) and (2.3.199). Due to this correspondence, and pursuing pertinent arguments as before, one arrives at

$$H_u \geqslant \int_0^l [\tfrac{1}{2}\mu(\dot{u})^2 + \tfrac{1}{2}(q_c - q)(l - x)(u')^2]\,\mathrm{d}x \qquad (2.3.253)$$

in correspondence to (2.3.211). Hence,

$$H_u \geqslant 0, \text{ positive definite for } 0 < q \leqslant q_c, \qquad (2.3.254)$$

corresponding to (2.3.212). Moreover,

$$\frac{\mathrm{d}H_u}{\mathrm{d}t} = -\tfrac{1}{2}qf\frac{\mathrm{d}f}{\mathrm{d}t}[\phi^2(x)]_0^l, \qquad (2.3.255)$$

which corresponds to (2.3.217). But $\phi(0) = 0$ due to the boundary conditions of the rod. Hence,

$$H_u = -\tfrac{1}{4}qf^2\phi^2(l) + K, \qquad (2.3.256)$$

where K is a positive constant of integration. Condition (2.3.256) has to be reconciled with condition (2.3.254). This is only possible if function $f(t)$ were not to grow unboundedly so that the negative term on the right-hand side of (2.3.256) could not exceed K. But, then, $f(t)$ must be bounded and *flutter would not occur*.

By the same token it could be shown that function $g(t)$ had to be bounded too, thus, again excluding flutter.

Therefore, due to the absence of flutter in the load range $0 < q \leqslant q_c$, expressions (2.3.252) can be abandoned and substituted by

$$w(x, t) = f(t)\phi(x), \tag{2.3.257}$$

where all functions are *real* ones, and $f(t)$ is bounded. Consequently, one has instead of (2.3.254) and (2.3.255),

$$H \geqslant 0 \qquad \text{for } 0 < q \leqslant q_c \tag{2.3.258}$$

and

$$\frac{\mathrm{d}H}{\mathrm{d}t} = -\tfrac{1}{2}q\phi^2(l)f\frac{\mathrm{d}f}{\mathrm{d}t}, \tag{2.3.259}$$

where

$$H = \int_0^l \left[\tfrac{1}{2}\mu(\dot{w})^2 + \tfrac{1}{2}\alpha(w'')^2 - \tfrac{1}{2}q(l-x)(w')^2\right]\mathrm{d}x \tag{2.3.260}$$

is the true Hamiltonian of the rod, and where

$$\int_0^l \alpha(w'')^2 \,\mathrm{d}x \geqslant q_c \int_0^l (l-x)(w')^2 \,\mathrm{d}x. \tag{2.3.261}$$

These relationships are indeed meaningful, as w has just before been established as a *real* function.

Integrating in (2.3.259) yields

$$H = -\tfrac{1}{2}q\phi^2(l)f^2 + K. \tag{2.3.262}$$

Because of (2.3.258),

$$\tfrac{1}{2}q\phi^2(l)f^2 \leqslant K \tag{2.3.263}$$

must hold true. But then, (2.3.262) and (2.3.263) yield

$$0 \leqslant H < K \qquad \text{for } t \geqslant t_0. \tag{2.3.264}$$

That means, *the Hamiltonian H of the clamped–free rod remains bounded at any time in the load range $0 < q \leqslant q_c$.*

The boundedness of H implies stability in the following way: From (2.3.260) and (2.3.261) follows

$$\tfrac{1}{2}\int_0^l (q_c - q)(l-x)(w')^2 \,\mathrm{d}x < H. \tag{2.3.265}$$

If

$$H < K \qquad \text{for } t \geqslant t_0, \tag{2.3.266}$$

then also

$$\tfrac{1}{2}\int_0^l (q_c - q)(l-x)(w')^2 \,\mathrm{d}x < K \qquad \text{for } t \geqslant t_0, \tag{2.3.267}$$

and

$$\int_0^l (w')^2 \, dx < \frac{2K}{(q_c - q)M}, \qquad M = l - \xi, \qquad 0 \leqslant \xi \leqslant l, \qquad \text{for } t \geqslant t_0.$$
(2.3.268)

Therefore, by virtue of (2.3.250),

$$[w(x)]^2 < \frac{2lK}{(q_c - q)M} \qquad \text{for } t \geqslant t_0.$$
(2.3.269)

Finally,

$$w(x) < \sqrt{\frac{2lK}{(q_c - q)M}} \qquad \text{for } t \geqslant t_0 \text{ and } q \leqslant q_c.$$
(2.3.270)

That is indeed stability in the load range $0 < q \leqslant q_c$, yet not exactly in the sense of Lyapunov.

Plate Problems:

First, let us show how to operate in the case of plates with the general notations introduced earlier. For this purpose, the plate shown in Fig. 90 will be used as a model. For this plate,

$$P = \int_S \mathscr{P} \, dS,$$
(2.3.271)

where S is the surface of the plate and

$$\mathscr{P} = \frac{B}{2} (\Delta w)^2 + B(1 - \sigma)(w_{12}^2 - w_{11} w_{22}) - \tfrac{1}{2} q(E_3 - x_1) w_1^2.$$
(2.3.272)

In (2.3.272), B is the bending stiffness, σ is Poisson's ratio, and $E_3 = E_3(x_2)$ is the supported curved edge at the right-hand side of the plate. By comparison with (2.3.207), one finds that

$$D \equiv \frac{B}{2} (\Delta)^2 + B(1 - \sigma) \left[\left(\frac{\partial^2}{\partial x_1 \partial x_2} \right)^2 - \frac{\partial^2}{\partial x_1^2} \frac{\partial^2}{\partial x_2^2} \right],$$
(2.3.273)

$$F(x) \equiv \tfrac{1}{2}(E_3 - x_1).$$
(2.3.274)

Moreover,

$$\frac{\delta P}{\delta w} = \frac{\partial \mathscr{P}}{\partial w} - \frac{\partial}{\partial x_1} \left(\frac{\partial \mathscr{P}}{\partial w_1} \right) - \frac{\partial}{\partial x_2} \left(\frac{\partial \mathscr{P}}{\partial w_2} \right) + \frac{\partial^2}{\partial x_1^2} \left(\frac{\partial \mathscr{P}}{\partial w_{11}} \right)$$
$$+ \frac{\partial^2}{\partial x_2^2} \left(\frac{\partial \mathscr{P}}{\partial w_{22}} \right) + \frac{\partial^2}{\partial x_1 \partial x_2} \left(\frac{\partial \mathscr{P}}{\partial w_{12}} \right).$$
(2.3.275)

Applying (2.3.275) to (2.3.272) yields

$$\frac{\delta P}{\delta w} = B\Delta^2 w + q(E_3 - x_1) w_{11} - q w_1.$$
(2.3.276)

298

Furthermore, one has

$$M = qw_1. \tag{2.3.277}$$

Substituting (2.3.276) and (2.3.277) in (2.3.191), i.e. in this case in

$$\mu\ddot{w} + \frac{\delta\mathscr{P}}{\delta w} + M = 0, \tag{2.3.278}$$

yields

$$\mu\ddot{w} + B\Delta^2 w + q(E_3 - x_1)w_{11} = 0, \tag{2.3.279}$$

which is indeed the differential equation of the plate in Fig. 90.

From a previous paper[1], the following information on the boundary conditions of the plate in Fig. 90 is obtained:

along edge E_1: $w = 0$, $\partial w/\partial n = 0$.   (2.3.280)

along edge E_2: $w = 0$, $\Delta w = 0$,   (2.3.281)

along edge E_3: $w = 0$, $\Delta w = (\sigma - 1)\rho^{-1}(\partial w/\partial n)$,   (2.3.282)
ρ = radius of curvature of E_3

along edge E_4: $\Delta w = (1 - \sigma)w_{11}$, $\partial\Delta w/\partial n = (1 - \sigma)(\partial w_{12}/\partial s)$.   (2.3.283)

According to (2.3.194) and (2.3.195), the relationship

$$\int_S \frac{d}{dt}(\mathscr{P})\, dS = \int_S \frac{\delta P}{\delta w}\,\dot{w}\, dS \tag{2.3.284}$$

must hold true. Using the previously introduced boundary conditions, the correctness of (2.3.284) will now be shown. This is important for convincing oneself that the theory introduced earlier is in fact applicable to plates.

First, let us state that the expression

$$P = \int_S \left[\frac{B}{2}(\Delta w)^2 + B(1 - \sigma)(w_{12}^2 - w_{11}w_{22}) - \tfrac{1}{2}q(E_3 - x_1)w_1^2 \right] dS, \tag{2.3.285}$$

which is obtained from (2.3.271) and (2.3.272), can be substituted by

$$P = \int_S \left[\frac{B}{2}(\Delta w)^2 - \tfrac{1}{2}q(E_3 - x_1)w_1^2 \right] dS + \frac{B}{2}(1 - \sigma)\int_E F\, dE, \tag{2.3.286}$$

where

$$\left.\begin{aligned}
F &= 0 &&\text{along } E_1 \text{ and } E_2, \\
F &= \rho^{-1}\left(\frac{\partial w}{\partial n}\right)^2 &&\text{along } E_3, \\
F &= -w_2 w_{22} + w_1 w_{12} &&\text{along } E_4.
\end{aligned}\right\} \tag{2.3.287}$$

[1]Leipholz, H., Lyapunov's Second Method and Its Application to Continuous Systems, *SM Archives* 1, 367–444, 1976.

This statement follows again from information contained in[1].

For (2.3.284) to be correct, one has to show that

$$\int_S [B\Delta^2 w + q(E_3 - x_1)w_{11} - qw_1]\,\dot{w}\,dS =$$

$$\int_S \frac{d}{dt}\left[\frac{B}{2}(\Delta w)^2 - \frac{1}{2}q(E_3 - x_1)w_1^2\right]dS + \frac{B}{2}(1 - \sigma)\int_E \frac{dF}{dt}\,dE =$$

$$\int_S [B(\Delta w)(\Delta \dot{w}) - q(E_3 - x_1)w_1\dot{w}_1]\,dS + \frac{B}{2}(1 - \sigma)\int_E \dot{F}\,dE. \qquad (2.3.288)$$

This is done in various steps:

(a) $$\int_S B(\Delta w)(\Delta \dot{w})\,dS = \int_S B\Delta^2 w\dot{w}\,dS - B\int_E \frac{\partial \Delta w}{\partial n}\,\dot{w}\,dE + B\int_E \Delta w\frac{\partial \dot{w}}{\partial n}\,dE.$$
$$(2.3.289)$$

By means of the boundary conditions (2.3.280) – (2.3.283), equation (2.3.289) can be changed into

$$\int_S B(\Delta w)(\Delta \dot{w})\,dS = \int_S B\Delta^2 w\dot{w}\,dS - B\int_{E_3} (1 - \sigma)\rho^{-1}\frac{\partial w}{\partial n}\frac{\partial \dot{w}}{\partial n}\,dE$$

$$- B\int_{E_4} (1 - \sigma)\frac{\partial w_{12}}{\partial s}\,\dot{w}\,dE + B\int_{E_4} (1 - \sigma)w_{11}\frac{\partial \dot{w}}{\partial n}\,dE. \qquad (2.3.290)$$

However, along E_4

$$\frac{\partial w_{12}}{\partial s} = -w_{121}, \qquad \frac{\partial \dot{w}}{\partial n} = \dot{w}_2. \qquad (2.3.291)$$

Hence,

$$\int_S B(\Delta w)(\Delta \dot{w})\,dS = \int_S B\Delta^2 w\dot{w}\,dS - B\int_{E_3} (1 - \sigma)\rho^{-1}\frac{\partial w}{\partial n}\frac{\partial \dot{w}}{\partial n}\,dE$$

$$+ B\int_{E_4} (1 - \sigma)w_{121}\dot{w}\,dE + B\int_{E_4} (1 - \sigma)w_{11}\dot{w}_2\,dE. \qquad (2.3.292)$$

(b) $$\int_S q(E_3 - x_1)w_1\dot{w}_1\,dS = \int_E q(E_3 - x_1)w_1\dot{w}\cos(n, x_1)\,dE$$

$$- \int_S [q(E_3 - x_1)w_{11} - qw_1]\,\dot{w}\,dS.$$

But, along E_1, E_2, and E_3, $\dot{w} \equiv 0$, and along E_4 $\cos(n, x_1) = 0$. Hence,

$$\int_E q(E_3 - x_1)w_1\dot{w}_1\cos(n, x_1)\,dE = 0.$$

---

[1]Refer to Footnote 1 on p. 298.

300

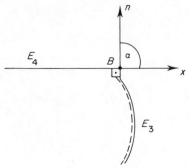

$\cos(n,x_1) = \cos \alpha = \cos 90° = 0$ Fig. 91. Corner detail at B for the plate in Fig. 90

Consequently,

$$\int_S q(E_3 - x_1)w_1\dot{w}_1 \, dS = -\int_S [q(E_3 - x_1)w_{11} - qw_1]\dot{w} \, dS. \qquad (2.3.293)$$

$$(c) \quad \int_E \dot{F} \, dE = \begin{cases} 0 & \text{along } E_1 \text{ and } E_2, \\[2mm] 2\int_E \rho^{-1}\dfrac{\partial w}{\partial n}\dfrac{\partial \dot{w}}{\partial n} \, dE & \text{along } E_3, \\[4mm] \dfrac{d}{dt}\int_E F \, dE = \dfrac{d}{dt}\int_E (-w_2 w_{11} + w_1 w_{12}) \, dE & \\[2mm] & \text{along } E_4. \end{cases} \qquad (2.3.294)$$

But,

$$\int_{E_4} w_1 w_{12} \, dE = [w_1 w_2]_A^B - \int_{E_4} w_{11} w_2 \, dE. \qquad (2.3.295)$$

At point A (see Fig. 90), as seen from E_1, $w_1 = 0$, $w_2 = 0$.   (2.3.296)

At point B (see Fig. 91), as seen from E_3, $\partial w/\partial s = 0$.

Hence,

$$\frac{\partial w}{\partial s} = -w_1 \sin \alpha + w_2 \cos \alpha = 0,$$

and

$$w_1 = w_2 \cotan \alpha = w_2 \frac{\cos \alpha}{\sin \alpha} = w_2 \frac{\cos 90°}{\sin 90°} = 0. \qquad (2.3.297)$$

Using (2.3.296) and (2.3.297) in (2.3.295) yields

$$\int_{E_4} w_1 w_{12} \, dE = -\int_{E_4} w_{11} w_2 \, dE. \qquad (2.3.298)$$

Therefore,

$$\frac{\mathrm{d}}{\mathrm{d}t}\int_{E_4} F\,\mathrm{d}E = \frac{\mathrm{d}}{\mathrm{d}t}\int_{E_4}(-2w_2w_{11})\,\mathrm{d}E \qquad \text{along } E_4, \qquad (2.3.299)$$

which is also

$$\frac{\mathrm{d}}{\mathrm{d}t}\int_{E_4} F\,\mathrm{d}E = \int_{E_4}\dot{F}\,\mathrm{d}E = -2\int_{E_4}\dot{w}_2 w_{11}\,\mathrm{d}E - 2\int_{E_4} w_2\dot{w}_{11}\,\mathrm{d}E. \qquad (2.3.300)$$

Yet,

$$\int_{E_4} w_2\dot{w}_{11}\,\mathrm{d}E = [w_2\dot{w}_1]_A^B - [w_{21}\dot{w}]_A^B + \int_{E_4} w_{211}\dot{w}\,\mathrm{d}E,$$

and, as before, $[w_2\dot{w}_1]_A^B = 0$, as well as $[w_{21}\dot{w}]_A^B = 0$ because of $\dot{w}(A) = \dot{w}(B) = 0$. Hence,

$$\int_{E_4} w_2\dot{w}_{11}\,\mathrm{d}E = \int_{E_4} w_{211}\dot{w}\,\mathrm{d}E. \qquad (2.3.301)$$

By means of (2.3.301), expression (2.3.300) can be changed into

$$\int_{E_4}\dot{F}\,\mathrm{d}E = -2\int_{E_4}\dot{w}_2 w_{11}\,\mathrm{d}E - 2\int_{E_4} w_{211}\dot{w}\,\mathrm{d}E. \qquad (2.3.302)$$

Substituting (2.3.302) in (2.3.294) yields finally

$$\int_E \dot{F}\,\mathrm{d}E = 2\int_{E_3}\rho^{-1}\frac{\partial w}{\partial n}\frac{\partial \dot{w}}{\partial n}\,\mathrm{d}E - 2\int_{E_4}\dot{w}_2 w_{11}\,\mathrm{d}E - 2\int_{E_4} w_{211}\dot{w}\,\mathrm{d}E. \quad (2.3.303)$$

Now, it is possible to show that (2.3.288) is indeed an identity. For this purpose insert the expressions (2.3.292), (2.3.293), and (2.3.303) in the right hand side of (2.3.288). This step makes the two sides in (2.3.288) equal to each other.

It is easy to realize that if (2.3.284) is satisfied, the general theory presented earlier holds true for the plate problem and can be applied, since in addition operator D and function F have been identified through Eqs. (2.3.273) and (2.3.274).

The plate in Fig. 90 can be classified as being a 'sufficiently supported and appropriately directed' one. Therefore, flutter will not occur in the load range $0 < q < q_c$. Consequently, one may set

$$w \equiv u = f(t)\phi(x_1, x_2), \qquad (2.3.304)$$

where both functions f as well as ϕ are real and where $f(t)$ is bounded. Hence, correspondong to (2.3.220) and (2.3.272),

$$H = \int_S [\tfrac{1}{2}\mu(\dot{w})^2 + \mathscr{P}]\,\mathrm{d}S = \int_S [\tfrac{1}{2}\mu(\dot{w})^2 + \tfrac{1}{2}B(\Delta w)^2$$
$$+ B(1-\sigma)(w_{12}^2 - w_{11}w_{22}) - \tfrac{1}{2}q(E_3 - x_1)w_1^2]\,\mathrm{d}S = \text{const.} \qquad (2.3.305)$$

302

As shown in[1]

$$(\Delta w)^2 + 2(1 - \sigma)(w_{12}^2 - w_{11}w_{22}) = \alpha(1 - \sigma)(w_{11}^2 + 2w_{12}^2 + w_{22}^2), \qquad (2.3.306)$$

where

$$0 < \sigma < 1 \text{ and } \alpha > 1. \qquad (2.3.307)$$

Hence, also

$$H = \int_S \left[\tfrac{1}{2}\mu(\dot{w})^2 + \frac{\alpha B(1 - \sigma)}{2}\,(w_{11}^2 + 2w_{12}^2 + w_{22}^2) - \tfrac{1}{2}q(E - x_1)w_1^2 \right] \mathrm{d}S = \text{const.} \qquad (2.3.308)$$

Moreover,

$$H < \int_S \left[\tfrac{1}{2}\mu\dot{w}^2 + \frac{\alpha B(1 - \sigma)}{2}\,(w_{11}^2 + 2w_{12}^2 + w_{22}^2) \right] \mathrm{d}S. \qquad (2.3.309)$$

By means of

$$\varkappa_{(2)} = \max\left(\frac{\mu S}{2},\; \frac{\alpha B(1 - \sigma)S}{2} \right) \qquad (2.3.310)$$

and

$$\| w \|_{(2)}^2 = \sup(\dot{w})^2 + \sup(w_{11})^2 + \sup 2(w_{12})^2 + \sup(w_{22})^2, \qquad (2.3.311)$$

inequality (2.3.309) yields

$$H < \varkappa_{(2)} \| w \|_{(2)}^2. \qquad (2.3.312)$$

Again, from [1] the inequalities

$$\int_S \psi^2\,\mathrm{d}S \leqslant l^2 \int_S \psi_1^2\,\mathrm{d}S < l^2 \int_S (\psi_1^2 + \psi_2^2)\,\mathrm{d}S \qquad (2.3.313)$$

are taken, which hold true for sufficiently smooth functions ψ that vanish along E_1. In (2.3.313), l is the maximum length of the plate in the x_1-direction.

Applying (2.3.313) successively to $w_1 \equiv \psi$ and $w_2 \equiv \psi$ yields after some manipulation

$$\int_S (w_{11}^2 + 2w_{12}^2 + w_{22}^2)\,\mathrm{d}S > l^{-2} \int_S (w_1^2 + w_2^2)\,\mathrm{d}S. \qquad (2.3.314)$$

From (2.3.308) follows

$$H > \int_S \left[\frac{B(1 - \sigma)}{2}\,(w_{11}^2 + 2w_{12}^2 + w_{22}^2) - \tfrac{1}{2}ql(w_1^2 + w_2^2) \right] \mathrm{d}S. \qquad (2.3.215)$$

By means of (2.3.314), this inequality can be changed into

$$H > \frac{B(1 - \sigma) - ql^3}{2l^2} \int_S (w_1^2 + w_2^2)\,\mathrm{d}S. \qquad (2.3.316)$$

---

[1] Refer to Footnote 1 on p. 298.

Using in (2.3.313) $w \equiv \psi$ yields

$$\int_S w^2 \, dS < l^2 \int (w_1^2 + w_2^2) \, dS. \qquad (2.3.317)$$

Using (2.3.317) in (2.3.316) leads to

$$H > \frac{B(1-\sigma) - ql^3}{2l^2} \int_S w^2 \, dS, \qquad (2.3.318)$$

where

$$q < \frac{B(1-\sigma)}{l^3} \qquad (2.3.319)$$

must be required.

Set

$$\varkappa_{(1)} = \frac{B(1-\sigma) - ql^3}{2l^2} \qquad (2.3.320)$$

and

$$\|w\|_{(1)}^2 = \int_S w^2 \, dS. \qquad (2.3.321)$$

Then, from (2.3.318), (2.3.319), and (2.3.321) follows the inequality

$$H > \varkappa_{(1)} \|w\|_{(1)}^2. \qquad (2.3.322)$$

Combining (2.3.312) and (2.3.322) yields

$$\varkappa_{(1)} \|w\|_{(1)}^2 < \varkappa_{(2)} \|w\|_{(2)}^2 \qquad \text{for } t \geqslant t_0. \qquad (2.3.323)$$

The validity of inequality (2.3.323) is due to the fact that H *is constant*.

As a result, the following stability statement can be made: If for the plate in Fig. 90 the follower load q is in the range

$$0 < q < \tilde{q}, \qquad \tilde{q} = \min \left[q_c, \frac{B(1-\sigma)}{l^3} \right], \qquad (2.3.324)$$

and if at $t = t_0$ the initial perturbation was such that $\|w\|_{(2)}^2 < \varepsilon^2$, then,

$$\|w\|_{(1)} < \varepsilon \sqrt{\frac{\varkappa_{(2)}}{\varkappa_{(1)}}} \qquad \text{for } t \geqslant t_0. \qquad (2.3.325)$$

That means, *the deflections w of the plate remain small at any time in the mean square.*

Consider subsequently a plate which belongs to the class of the 'not completely supported but appropriately directed' ones. Such a plate is shown in Fig. 92.

For a plate of this class, one knows from previous considerations that H *is positive definite,* and that it is at the same time *bounded* in the load range

Fig. 92.

$0 < q < q_c$, see (2.3.212) and (2.3.224). Therefore,

$$K > H > \int_S \left[\frac{B(1-\sigma)}{2} (w_{11}^2 + 2w_{12}^2 + w_{22}^2) - \tfrac{1}{2} ql(w_1^2 + w_2^2) \right] dS, \qquad (2.3.326)$$

where K is the bound of H.

Since the plate in Fig. 92 is clamped along edge E_1, the previously mentioned inequalities (2.3.314) and (2.3.317) apply. Thus, one is successively led to

$$K > H > \frac{B(1-\sigma) - ql^3}{2l^2} \int_S (w_1^2 + w_2^2)\, dS \qquad (2.3.327)$$

and to

$$K > H > \frac{B(1-\sigma) - ql^3}{2l^4} \int_S w^2\, dS. \qquad (2.3.328)$$

From (2.3.328), the following conclusion concerning stability can be drawn:

$$\int_S w^2\, dS < \frac{2l^4 K}{B(1-\sigma) - ql^3} \qquad \text{for } t \geqslant t_0.$$

That means, *the deflections w of the plate remain bounded at any time in the mean square if q is in the load range*

$$0 < q < \tilde{q}, \ \tilde{q} = \text{Min} \left[q_c, \frac{B(1-\sigma)}{l^3} \right].$$

The preceding deliberations are specifically important from the point of view of design. Since for the class of elastic systems defined by (2.3.191) and

(2.3.192), no flutter occurs in the load range $0 < q < q_c$, and since their stability can be evaluated by means of the Hamiltonian H, those design criteria remain valid, which have been developed for the same systems subjected to unidirectional forces of equal magnitude as the follower forces. The limit of validity of these design criteria is q_c, which is the respective Euler buckling load.

2.3.7 Lyapunov Stability of Continuous Elastic Systems

Consider the stability of a finite-dimensional mechanical system as in Section 1.5. After having determined its generalized coordinates, and after having cast its equation of motion into a first-order system in terms of these coordinates and the corresponding generalized momenta, one will be concerned with finding an appropriate function of the coordinates and momenta as well as the time derivative of that function. For the calculation of the function's derivative, the equations of motion of the mechanical system may be used. 'Appropriate' means here: the function ought to be sign-definite in the state space, and its derivative with respect to time should vanish or should also be definite, however, with a sign opposite to that of the funciton itself. A function of this character is called a *Lyapunov function*.

The conditions imposed in order to establish the definiteness and the required time behaviour of the function yield in a respective parameter space the stability domain of the system. Let the system for example be an elastic one, and let one of the parameters be a compressive load. Then, one of the above mentioned conditions will give the buckling load. The conclusion to be drawn is that for all values of the load below that of the buckling load, the system will be stable in the sense that coordinates and momenta remain bounded uniformly with respect to time after an initial perturbation of the system. The boundedness of coordinates and momenta follows from an inspection of the Lyapunov function taking its time behaviour into account. Since this boundedness is an unalterable consequence of the nature of the Lyapunov function, one will in practise conclude the stability investigation already with establishing the definiteness conditions for the Lyapunov function and its detivative assuming tacitly the boundedness of coordinates and momenta in the case of satisfied stability conditions without checking further. One is indeed entitled to proceed in that way as the boundedness of the state space variables is to be measured in the sense of the state space norm. But, since this space is supposed to have a finite number of dimensions, it has essentially one single norm, for example the *uniform norm* $\| f \|_\infty$, i.e.,

$$\left. \begin{array}{l} \| f \|_\infty = \text{lub } \{ \, | f(p) | : p \in A \}, \\ \text{lub} = \text{lowest upper bound}, \\ A = \text{domain on which } f \text{ is defined}, \end{array} \right\}$$

while all other norms are equivalent to it and can therefore be substituted by it. Hence, after having determined when stability will prevail, by having determined the conditions for an appropriate Lyapunov functional, one may immediately conclude that in the case of stability the image points of the system's states remain in a bounded domain around the origin of the state space. This is so, because, regardless which other, equivalent norm one may use to evaluate the system's momentary states, it will all end up with a result yielded by the $\| \ldots \|_{\infty}$ norm. Thus, why bother with investigating the topological nature of stability at all?

In this spirit, attempts have been made to apply Lyapunov's technique right away to stability problems of continuous systems just replacing the Lyapunov *function* by a Lyapunov *functional* and determining a respective buckling load from the conditions rendering the said functional and its derivative appropriately definite. Also, the conclusion had then been drawn that the continuous system was stable for load intensities below the level set by the buckling load — meaning that deflection, rates of displacements etc. would remain controllably small. In this way, for example, one had proceeded implicitly in classical books on plate buckling (1.6, 8.5). Taking for a conservative plate problem the Hamiltonian as a Lyuponv functional is a most obvious choice, as the Hamiltonian is time-invariant for conservative problems. Therefore, one has only to search for the conditions making the Hamiltonian positive definite. Definiteness will already prevail if the total potential energy (strain energy plus potential energy of the compressive forces) is positive. Hence, one can restrict oneself to consider the total potential energy only. Positive definiteness is lost and the stability limit is reached if the potential energy becomes zero. This is therefore the condition yielding the *buckling load*. However, all these classical investigations stop short of considering the consequences of the positive definiteness of the Hamiltonian. Let it therefore be briefly indicated here that such considerations are by no means trivial ones and should not be neglected.

For an elastic plate, the Hamiltonian reads

$$H = \int_{\Omega} (\tfrac{1}{2}\mu \dot{w}^2 + \mathscr{E} - \tfrac{1}{2} N w_1^2) \, d\Omega, \qquad (2.3.329)$$

where Ω is the surface of the plate, μ its mass density per unit surface, w its deflection, $\dot{w} = \partial w / \partial t$, $w_i = \partial w / \partial x_i$, $i = 1$, (in general $i = 1, 2$), N the compressive edge load in x_1 direction, and

$$\mathscr{E} = \tfrac{1}{2} D [w_{11}^2 + w_{22}^2 + 2 w_{11} w_{22} + 2(1 - \sigma)(w_{12}^2 - w_{11} w_{22})], \qquad (2.3.330)$$

where D is the plate's flexural stiffness, σ is the Poisson ratio, and

$$w_{ij} = \partial^2 w / \partial x_i \partial x_j, \qquad i, j = 1, 2.$$

Rayleigh's principle yields

$$\int_{\Omega} \mathscr{E} \, d\Omega \geq \tfrac{1}{2} \int_{\Omega} N_c w_1^2 \, d\Omega, \qquad (2.3.331)$$

where N_c is the lowest of the so-called buckling loads.

From (2.3.329) follows

$$H > U = \int_\Omega [\mathcal{E} - \tfrac{1}{2} N w_1^2]\, d\Omega, \qquad (2.3.332)$$

and H is positive definite if the potential energy U of the plate is positive definite. Let $0 < \eta < 1$ be used. Then,

$$U = \int_\Omega [\eta \mathcal{E} + (1 - \eta)\mathcal{E} - \tfrac{1}{2} N w_1^2]\, d\Omega. \qquad (2.3.333)$$

From (2.3.331) follows

$$\int_\Omega (1 - \eta)\mathcal{E}\, d\Omega \geq \frac{1 - \eta}{2} \int_\Omega N_c w_1^2\, d\Omega. \qquad (2.3.334)$$

Using (2.3.334) in (2.3.333) yields under condition

$$N < (1 - \eta)N_c \qquad (2.3.335)$$

the inequalities

$$U > \int_\Omega \left[\eta \mathcal{E} + \left(\frac{1 - \eta}{2} N_c - \frac{1}{2} N \right) w_1^2 \right] d\Omega > 0. \qquad (2.3.336)$$

Thus, condition (2.3.335) guarantees by virtue of (2.3.332) and (2.3.336) the positive definiteness of H.

Again, due to (2.3.335), also

$$U > \int_\Omega \eta \mathcal{E}\, d\Omega$$

which by means of (2.3.330) yields

$$U > \int_\Omega \frac{\eta D}{2} [w_{11}^2 + w_{22}^2 + 2w_{11}w_{22} + 2(1 - \sigma)(w_{12}^2 - w_{11}w_{22})]\, d\Omega. \qquad (2.3.337)$$

Yet,

$$w_{11}^2 + w_{22}^2 + 2w_{11}w_{22} + 2(1 - \sigma)(w_{12}^2 - w_{11}w_{22}) > (1 - \sigma)(w_{11}^2 + 2w_{12}^2 + w_{22}^2).$$

$$(2.3.338)$$

Hence, combining (2.3.337) and (2.3.338), results in

$$U > \int_\Omega \frac{\eta(1 - \sigma)D}{2} (w_{11}^2 + 2w_{12}^2 + w_{22}^2)\, d\Omega. \qquad (2.3.339)$$

According to [1],

$$(\| w \|_2^2)^2 < k \left\{ \int_\Omega (w_{11}^2 + 2w_{12}^2 + w_{22}^2)\, d\Omega + \int_B w^2\, dB \right\} \qquad (2.3.340)$$

[1] Rektorys, K., *Variational methods in Mathematics, Science and Engineering* D. Reidel Publishing Co., Dordrecht–Boston, p. 341, 1977.

for all w in Ω. In (2.3.340), $\| w \|^2_{\frac{}{}}$ is a Sobolev norm, B is the contour of the plate and k is an appropriate constant.

Assume that the plate is completely supported (supported along the whole contour). Then,

$$\int_B w^2 \, dB = 0,$$

and (2.3.340) becomes

$$(\| w \|^2_{\frac{}{}})^2 < k \int_\Omega (w^2_{11} + 2w^2_{12} + w^2_{22}) \, d\Omega. \tag{2.3.341}$$

Moreover, according to[2],

$$\sup |w| \leqslant C \| w \|^2_{\frac{}{}}. \tag{2.3.342}$$

Using (2.3.339), (2.3.341), and (2.3.342) simultaneously yields

$$\sup |w|^2 \leqslant C^2 k \int_\Omega (w^2_{11} + 2w^2_{12} + w^2_{22}) \, d\Omega \leqslant \frac{2C^2 k}{\eta(1-\sigma)D} \, U. \tag{2.3.343}$$

Hence,

$$\sup |w| < C \left(\frac{2kU}{\eta(1-\sigma)D} \right)^{\frac{1}{2}}. \tag{2.3.344}$$

This result indicates that for a positive definite Hamiltonian, one can indeed assume smallness of the plate's deflections w. However, the result is much weaker than usually assumed: firstly, it holds so far only for a completely supported plate. What happens, if part of the plate's contour were free? Secondly, the result (2.3.334) holds only under condition (2.3.335), i.e., for a compressive load N *smaller than the so-called buckling load* N_c. Is therefore the classical buckling load N_c meaningful at all?

The preceding deliberations may have shown that a thorough discussion of the nature of the stability is necessary. This may be done by means of a proper evaluation of the real significance of the Lyapunov functional's definiteness. The result of such a discussion might, for example, explain the discrepancy between experimental evidence and expectations of stability based solely on the so-called buckling load q_c.

Since such deliberations involve to a large extent concepts related to the notion of a normed topological space, it may be understandable if it is claimed that, at least when it comes to the stability of *continuous* elastic systems, the topological aspect of stability should be taken into account.

In section 1.1.2 it was shown that, even in the case of an ordinary situation for which the classical form of Lyapunov's method holds, a topological formulation of the problem can be adopted. For finite-dimensional systems such

---

[2]Oden, J. T., *Applied Functional Analysis*, Prentice-Hall, Englewood Cliffs, N.Y., p. 289, 1977.

a formulation may not be necessary, yet it may still be useful as it will show in which way a generalisation of the method will be possible for an application of it to continuous systems.

Consider a $2n$-dimensional *phase space* M with coordinates q_i, p_i, $i = 1, 2, \ldots, n$. A $2n$-tuple $\{q_i, p_i\}$ is a *phase point* P_i. Space M is the set of all phase points, i.e., $M = \{P_i\}$.

Let the first-order system of equations of motion be given by

$$\dot{q}_i = f_i(q, p), \qquad \dot{p}_i = g_i(q, p). \qquad (2.3.345)$$

Then, the system of equations

$$\mathrm{d}q_i/\mathrm{d}p_i = f_i/g_i \qquad (2.3.346)$$

derived from (2.3.345) may be integrated yielding the solution

$$q_i = \Phi^{(j)}(p_i). \qquad (2.3.347)$$

In (2.3.347), the superscript j indicates that one has a manifold of such solutions following from the various initial conditions. Expression (2.3.347) can be rewritten to yield

$$P^{(j)}(q, p) = k_j = \text{const.} \qquad (2.3.348)$$

which is said to be a *phase curve*. Obviously,

$$P^{(j)} \subset \{P_i\} = M,$$

and

$$P^{(j)} = \left\{ \begin{array}{l} P_i : P_i \text{ are those phase points whose coordinates} \\ \quad \text{satisfy } q_i = \Phi^{(j)}(p_i). \end{array} \right\}$$

Consider a positive definite function $V : M \to \mathbf{R}$, where $\mathbf{R}$ is the set of real numbers. Then,

$$V_m(q_i, p_i) = c_m = \text{const.}$$

are closed surfaces in the M space for sufficiently small, positive values of the c_m. Hence, the subsets of M given by

$$V_k < c_k, \ c_k \in [0, c_m]$$

are a topology on the subspace $V_m \subset M$. In other words, the Lyapunov functions V_m induce a topology on an appropriately chosen subspace of M.

Let the inequality

$$K_l \rho(P_i) < V_k < K_u \rho(P_i), \qquad (2.3.349)$$

obtain, where K_l and K_u are appropriate constants and ρ is a norm. Since M is a finite-dimensional space, all norms are equivalent, and one may choose

$$\rho(P_i) = \max |q, p| \qquad (2.3.350)$$

as 'the' norm.

310

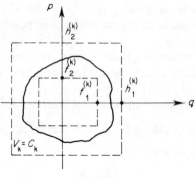

Fig. 93

Inequality (2.3.349) can be visualized in Fig. 93. From this figure it follows with

$$h_k = \max(h_1^{(k)}, h_2^{(k)}), \qquad f_k = \min(f_1^{(k)}, f_2^{(k)})$$

that for all $P_i \in V_k$

$$f_k < |q, p| < h_k. \qquad (2.3.351)$$

Hence, (2.3.349) holds true for

$$K_u > \frac{c_k}{f_k} \text{ and } K_l > \frac{c_k}{h_k}. \qquad (2.3.352)$$

The Lyapunov function V_k itself may not be a norm, as this function may not be a homogeneous one. Also, it may not satisfy the triangle inequality. Therefore, inequality (2.3.349) does not necessarily express *norm equivalence*, but it implies *topological equivalence* for ρ and V. Thanks to this topological equivalence, stability considerations can be carried out in the following way: Take a $P_i^{(j)}(t_0) \in \mathscr{P}^{(j)}$ inside the domain V_k. Then, according to (2.3.349)

$$V_k[P_i^{(j)}(t_0)] < K_u \rho[P_i^{(j)}(t_0)] \qquad (2.3.353)$$

Also, by assumption,

$$\dot{V}_k = \frac{\partial V_k}{\partial q_i} \dot{q}_i + \frac{\partial V_k}{\partial p_i} \dot{p}_i = \frac{\partial V_k}{\partial q_i} f_i + \frac{\partial V_k}{\partial p_i} g_i \leqslant 0. \qquad (2.3.354)$$

Hence, for any t,

$$V_k[P_i^j(t)] \leqslant V_k[P_i^j(t_0)]. \qquad (2.3.355)$$

Again, from (2.3.349) follows

$$K_l \rho[P_i^{(j)}(t)] < V_k[P_i^{(j)}(t)]. \qquad (2.3.356)$$

The three inequalities (2.3.353), (2.3.355), and (2.3.356) together yield

$$K_l \rho[P_i^{(j)}(t)] < V_k[P_i^{(j)}(t_0)] < K_u \rho[P_i^{(j)}(t_0)]. \qquad (2.3.357)$$

Thus,

$$\rho[P_i^{(j)}(t)] < \frac{K_u}{K_l}\rho[P_i^{(j)}(t_0)].\qquad(2.3.358)$$

Interpreting inequality (2.3.358), one arrives at the following statements:
  If

$$\rho[P_i^{(j)}(t_0)] < \eta,\qquad(2.3.359)$$

then, for any time $t > t_0$

$$\rho[P_i^{(j)}(t)] < \frac{K_u}{K_l}\eta.\qquad(2.3.360)$$

From (2.3.352) it follows that

$$\frac{K_u}{K_l} = \frac{uh_m}{lf_m},\qquad(2.3.361)$$

where u and l are constants satisfying the conditions

$$u > \frac{c_k h_m}{c_m f_k},\qquad l < \frac{c_k h_m}{c_m h_k}.\qquad(2.3.362)$$

Therefore, (2.3.360) can be rewritten by means of (2.3.361) to yield

$$\rho[P_i^{(j)}(t)] < \frac{u}{l}\frac{h_m}{f_m}\eta = \varepsilon.\qquad(2.3.363)$$

Obviously, η is a linear function of ε, i.e.,

$$\eta = \varkappa\varepsilon,\qquad \varkappa = lf_m/uh_m.\qquad(2.3.364)$$

On the basis of the preceding facts, it can be concluded that *stability in the sense of Lyapunov* prevails: Let ε be given. Then, an $\eta = \varkappa\varepsilon$ can be found such that when

$$\text{for } t = t_0,\ \rho[P_i^{(j)}(t_0)] < \eta\qquad(2.3.365)$$

holds, also,

$$\text{for any } t > t_0,\ \rho[P_i^{(j)}(t)] < \varepsilon(\eta).\qquad(2.3.366)$$

Strictly speaking, stability has been established in the sense of a norm. Yet, all norms are equivalent to norm (2.3.350). Hence, (2.3.365) and (2.3.366) yield under any circumstances the relations

$$\max|q, p| < \varepsilon \qquad \text{for all } t > t_0,$$

if

$$\max|q, p| < \eta \qquad \text{for } t = t_0.$$

Therefore, the norm-concept could have been avoided and that is actually being done in most of the cases.

Based on the previous deliberations, an even more abstract presentation shall now be given for the stability phenomenon in conjunction with certain sets. These sets may be seen as idealisations in the context of two-dimensional elastic objects, for example plates. Let $\mathbf{R}$ be the set of real numbers and $\mathbf{R}^+$ the set of the positive real numbers. Consider the product set

$$\mathbf{R}^2 \times \mathbf{R}^+ = \{x_1, x_2, t\}$$

where $x_1 \in \mathbf{R}$, $x_2 \in \mathbf{R}$, and $t \in \mathbf{R}^+$

Having a plate in mind, one can also say that the product set

$$\mathbf{R}^2 \times \mathbf{R}^+ = \{\mathscr{P}, \mathscr{T}\}$$

is the 'position-time' set, where the pairs $\mathscr{P} = (x_1, x_2)$ define points on the surface Ω of the plate and where $\mathscr{T} = (t)$ are points on the time axis.

Consider the set of functions

$$\mathscr{W} = \{W, \dot{W}, W_1, W_2, W_{11}, W_{12}, W_{22}\}$$

where $\dot{W} = \partial W/\partial t$, $W_i = \partial W/\partial x_i$, $W_{ij} = \partial^2 W/\partial x_i \partial x_j$, $i, j = 1, 2$. If W is one of the solutions of the differential equation governing the motion of the elastic object, the set $\mathscr{W}$ involves the *characteristics* which describe at any time t the *state* of the elastic object.

The set $\mathscr{W}$ maps $\{\mathscr{P}, \mathscr{T}\}$ on $\mathbf{R}$:

$$\mathscr{W} : \{\mathscr{P}, \mathscr{T}\} \to \mathbf{R}.$$

The domain of $\mathscr{W}$ is $\Omega \times T$, where $\Omega \in \mathbf{R}^2$ and $T \in \mathbf{R}^+$. The range of $\mathscr{W}$ is $\Delta \subset \mathbf{R}^7$ and,

$$d_{\min} < \Delta < d_{\max},$$

where d is a set involving displacements, velocities, slopes, and curvatures. Obviously, the mapping $\mathscr{W}$ is surjective, and

$$\Delta \equiv \mathscr{W}\{\mathscr{P}, \mathscr{T}\} \equiv \{w\},$$

whence $\{w\}$ is the set of images of $\mathscr{W}$. Hence, $w \in \Delta \subset \mathbf{R}^7$.

Choose a fixed value $\tau \in \mathbf{R}^+$. Then,

$$\mathbf{R}^2 \times \tau = \{\mathscr{P}, \tau\} \subset \{\mathscr{P}, \mathscr{T}\}$$

and

$$\mathscr{W}[\mathscr{P}, \tau] = \{w_\tau\} = \{w : t = \tau\}$$

is the state of the elastic object at $t = \tau$. Since

$$w_\tau = \{\bar{w}(\tau), \dot{\bar{w}}(\tau), \bar{w}_1(\tau), \bar{w}_2(\tau), \bar{w}_{11}(\tau), \bar{w}_{12}(\tau), \bar{w}_{22}(\tau)\},$$

where $\bar{w}$ is a deflection, $\dot{\bar{w}}$ a velocity, etc., w_τ is a set of seven numbers $\in \mathbf{R}$ describing the state of the elastic object at a certain position (x_1, x_2) for a certain time $t = \tau$.

Since $\bar{w}(\tau) \in \mathbf{R}$, $\dot{\bar{w}}(\tau) \in \mathbf{R}$, etc., all these quantities represent points on $r_{\bar{w}}$ —,

$r_w^{\leftrightarrow}$—, etc. — lines which themselves are images of **R**. Hence, using these lines as axes, one can span an abstract *state space* $\mathscr{S}$. In this space, every state of the elastic object has an image point S_τ, $\tau \in \{t : t \in \mathbf{R}^+\}$,

$$\mathscr{S} = \{S_\tau\}.$$

Let $W^{(j)}$ be a specific solution of the elastic object's differential equation. Then, to this solution corresponds a 'motion'. The 'motion' is an operator $M^{(j)}$ that maps the state at τ into the state at $\tau + \Delta\tau$:

$$M_{\Delta\tau}^{(j)}(S_\tau^{(j)}) = S_{\tau+\Delta\tau}^j.$$

Obviously,

$$M_{\Delta\tau}^{(j)} \circ M_{\Delta\tau}^{(j)} = M_{2\Delta\tau}^{(j)}$$

$$M_0^{(j)} = i \text{ (identity operator)}.$$

Moreover, the 'motion' is irreversible. Hence M is a semi-group.

While the 'motion' evolves, a set of states $S_\tau^{(j)}$, $\tau \in \mathbf{R}^+$ is produced which may be called an 'orbit' $O^{(j)}$:

$$O^{(j)} \equiv \{\mathscr{S}_\tau^{(j)}, \tau = \{\tau_1, \tau_2, \ldots, \tau_k, \ldots\}, \tau_k \in \mathbf{R}^+\}.$$

There exists a manifold of orbits, one orbit for every $j \in I$, $I =$ index set.

Let a 'distance' d be defined in the space $\mathscr{S}$. Then, $\mathscr{S}$ becomes a metric space. Since a metric induces a topology, $\mathscr{S}$ is specifically a *topological metric space*.

By means of the 'distance' d, certain stability concepts can be created. For example, the following

(i) If

$$(\forall\tau)[d(S_\tau^{(i)}, S_\tau^{(j)}) < k],$$

The two 'orbits' $O^{(i)}$ and $O^{(j)}$ are said to be stable relative to each other.

(ii) If

$$(\forall\tau)(\forall i, j)[d(S_\tau^{(i)}, S_\tau^{(j)}) < k],$$

the manifold of all possible 'orbits' and thus the 'elastic object' itself is said to be globally stable.

Let there exist an elastic object which can assume a state called 'trivial equilibrium' and which is defined as

$$(\forall\tau)[S_\tau^{(e)} \equiv \mathbf{O}],$$

where $\mathbf{O}$ is the origin of the state space $\mathscr{S}$. Then, the corresponding 'orbit' $O^{(e)}$ degenerates into a single point, i.e., into the point $\mathbf{O}$.

Under the assumption that $S_\tau^{(r)}$ exists, one has

$$d(S_\tau^{(e)}, S_\tau^{(j)}) = d(\mathbf{O}, S_\tau^{(j)}) = \| S_\tau^{(j)} \|,$$

where $\| \cdots \|$ is said to be the norm of $\mathscr{S}$. If

$$(\forall\tau)(\forall j)[\| S_\tau^{(j)} \| < k],$$

314

the elastic object is globaly stable. This is so, because the preceding statement implies

$$(\forall\tau)(\forall j)[d(\mathbf{O}, S_\tau^{(j)}) < k].$$

Hence, the 'trivial equilibrium' is stable relative to any 'orbit' $O^{(j)}$. But, then, the 'trivial equilibrium' is unrestrictedly stable, which is global stability of the elastic system.

It is worthwhile to note that according to the previous deliberations one may work on stability with a *normed* space $\mathscr{S}$ instead of a metric *one*, if (i) the set of all 'orbits', i.e., $\{O^{(j)}\}$, includes $O^{(e)}$ and if (ii) the stability of any 'orbit' $O^{(j)} \neq O^{(e)}$ is to be evaluated with respect to $O^{(e)}$ and $O^{(e)}$ only.

Let the elastic object to be treated Lyapunov-like be a thin plate. The current state of the plate is described in the state space S by the seven quantities

$$\left.\begin{array}{l} \bar{w} = \text{deflection}, \qquad \dot{\bar{w}} = \text{velocity}, \\[1ex] \bar{w}_1 \ \bar{w}_2 = \text{slopes}, \\[1ex] \bar{w}_{11}, \bar{w}_{12}, \bar{w}_{22} = \text{'curvatures'} \end{array}\right\} \qquad (2.3.367)$$

henceforth called the *characteristics*.

The *first step* is to introduce a Lyapunov functional L involving either all or only some of the characteristics. Stability conditions are obtained by making the functional positive definite and its derivative with respect to time negative definite or equal to zero. In most of the cases, the Lyapunov functional does not involve all the characteristics. Then, the positive definiteness of the functional holds true only for a certain subspace of $\mathscr{S}_L \subset \mathscr{S}$ which is obtained by projecting $\mathscr{S}$ along those axes which correspond to the very characteristics missing in L. In $\mathscr{S}_L$, the Lyapunov functional represents (at least for a certain domain around the origin) a topology if L is the integral of a quadratic form of those characteristics contained in L.

The *second step* is to find an appropriate norm in $\mathscr{S}_L$ yielding an upper and a lower bound for L and, thus, establishing topological equivalence between L and this norm. However, in most of the cases, such a topological equivalence can not be found. One must be content with finding two *distinct* norms which do hold for subspaces different from each other and different from the subspace $\mathscr{S}_L$ on which L is a topology. Then,

$$(\| w \|_1)^2 < L < (\| w \|_2)^2 \qquad (2.3.368)$$

may be said to express *topological imbedding*.

The need to work with two distinct norms has already been pointed out by A. A. Movchan[1]. There is no doubt that in order to establish the chain of inequalities (2.3.368), one must possess a stock of adequate integral inequalities. Therefore, an essential part of the theory of stability of elastic

---

[1]Movchan A. A., Stability of processes with respect to two metrics, *J. Appl. Math. Mech.*, **24**, 1506–1524, 1960.

continuous objects ought to be devoted to the furnishing of such inequalities. It may be mentioned that for example Sobolev's imbedding theorems provide useful inequalities of that kind. Hence, again, there is a hint that stability of continuous systems can not be properly investigated without falling back upon topology and functional analysis.

The *third step* consists in interpreting the meaning of the stability implied by the above mentioned norms. By virtue of the initial conditions,

$$L(t_0) < (\| w(t_0) \|_2)^2 < \varepsilon. \tag{2.3.369}$$

By virtue of the established time behaviour of L,

$$L(t) \leqslant L(t_0) \qquad \text{for any } t > t_0 \tag{2.3.370}$$

But for any time,

$$(\| w(t) \|_1)^2 < L(t). \tag{2.3.371}$$

Hence, inequalities (2.3.369), (2.3.370), and (2.3.371) yield

$$(\| w(t) \|_1)^2 < \varepsilon \qquad \text{for any } t > t_0. \tag{2.3.372}$$

From (2.3.372), one is supposed to infer on the nature of the assumed stability. Unfortunately, in many cases the conclusions drawn on the basis of $\| w \|_1$ are of little practical significance. Then, again, the stock of inequalities must be reviewed, and one must come up with an additional inequality.

$$(\| w(t) \|_D)^2 < (\| w(t) \|_1)^2, \tag{2.3.373}$$

yielding in view of (2.3.372)

$$(\| w(t) \|_D)^2 < \varepsilon \qquad \text{for any } t > t_0. \tag{2.3.374}$$

Norm $\| w \|_D$ may be called the *design norm*. It must involve those characteristics in (2.3.367) which are important for the design of the elastic plate with respect to technical requirements, preferably, the deflection w. Then, (2.3.373) allows one to conclude on the permanent smallness of the *design characteristics* following an initial perturbation of the plate described by (2.3.372) and assuming that the *stability conditions*, which were required for securing the chain of inequalities (2.3.368) are being satisfied.

It is important to realize that norm $\| w \|_D$ may be, and is indeed mostly, defined in yet another subspace of $\mathscr{S}$ than L and the norms $\| w \|_1$ and $\| w \|_2$ are. Thus, one must dispose of a sophisticated set of integral inequalities, and to enrich the basic stock of such inequalities is indeed a very important task of a researcher engaged in investigating the stability of continuous elastic objects.

Consider first a conservative plate problem as shown in Fig. 94 involving a conservative distributed load of intensity q. The differential equation describing the plate's small vibrations reads

$$\mu \ddot{\bar{w}} + D \nabla^2 \nabla^2 \bar{w} + q [a - x_1 \bar{w}_1]_1 = 0. \tag{2.3.375}$$

316

Fig. 94. Completely supported plate subjected to conservative loading

Since the problem is conservative, Rayleigh's principle can be applied yielding

$$q_c \leqslant \frac{\int_\Omega D\bar{w} \nabla^2 \nabla^2 \bar{w} \, d\Omega}{-\int_\Omega [(a - x_1)\bar{w}_1]_1 \bar{w} \, d\Omega} \qquad (2.3.376)$$

for the smallest eigenvalue q_c of the boundary value problem

$$\left.\begin{array}{l} D\nabla^2 \nabla^2 \bar{w} + q[(a - x_1)\bar{w}_1]_1 = 0, \\[4pt] \bar{w} = 0 \text{ along } B, \\[4pt] \bar{w}_{11} + \nu\bar{w}_{22} = 0 \qquad \text{along } x_1 = 0, \ x_1 = a, \\[4pt] \bar{w}_{22} + \nu\bar{w}_{11} = 0 \qquad \text{along } x_2 = 0, \ x_2 = b. \end{array}\right\} \qquad (2.3.377)$$

However, for the boundary conditions in (2.3.377),

$$\int_\Omega D\bar{w} \nabla^2 \nabla^2 \bar{w} \, d\Omega = \int_\Omega D(\nabla^2 \bar{w})^2 \, d\Omega, \qquad (2.3.378)$$

$$\int_\Omega [(a - x_1)w_1] \bar{w} \, d\Omega = -\int_\Omega (a - x_1)\bar{w}_1^2 \, d\Omega. \qquad (2.3.379)$$

Hence,

$$q_c \leqslant \frac{\int_\Omega D(\nabla^2 \bar{w})^2 \, d\Omega}{\int_\Omega (a - x_1)\bar{w}_1^2 \, d\Omega}. \qquad (2.3.380)$$

For conservative plate problems there is no difficulty finding a Lyapunov functional. Set

$$L \equiv H,$$

where H is the Hamiltonian. Then,

$$L = \int_\Omega \left\{ \tfrac{1}{2}\mu\dot{\bar{w}}^2 + \tfrac{1}{2}D\left[\bar{w}_{11}^2 + 2\bar{w}_{11}\bar{w}_{22} + \bar{w}_{22}^2 + 2(1-\sigma)(\bar{w}_{12}^2 - \bar{w}_{11}\bar{w}_{22}) \right.\right.$$

$$\left.\left. - \frac{q}{D}(a-x_1)\bar{w}_1^2 \right) \right\} d\Omega. \quad (2.3.381)$$

For the assumed boundary conditions of the plate, expression (2.3.381) can be simplified to read

$$L = \int_\Omega \left\{ \tfrac{1}{2}\mu\dot{\bar{w}}^2 + \tfrac{1}{2}D\left[(\nabla^2\bar{w})^2 - \frac{q}{D}(a-x_1)\bar{w}_1^2 \right] \right\} d\Omega. \quad (2.3.382)$$

Hence, also

$$L > \int_\Omega \tfrac{1}{2}D\left[(\nabla^2\bar{w})^2 - \frac{q}{D}(a-x_1)\bar{w}_1^2 \right] d\Omega, \quad (2.3.383)$$

and by virtue of (2.3.380)

$$L > \int_\Omega \tfrac{1}{2}[q_c - q](a-x_1)\bar{w}_1^2] \, d\Omega. \quad (2.3.384)$$

Thus,

$$L \geqslant 0 \quad (2.3.385)$$

for

$$q \leqslant q_c. \quad (2.3.386)$$

Furthermore, since $\dot{H} = 0$ by hypothesis,

$$\dot{L} \equiv \dot{H} = 0.$$

All this proves that the choice of H as a Lyapunov functional is reasonable.

Yet, let us look at the situation more closely: the functional L, defined by (2.3.381) is even under stability condition (2.3.386) not positive definite in space $\mathscr{S}$ spanned by the characteristics (2.3.367) as $\bar{w}$ and $\bar{w}_2$ are missing in (2.3.381). It becomes *positive definite* under condition (2.3.386) only in the subspace $\mathscr{S}_L$ spanned by $\dot{\bar{w}}$, $\bar{w}_1$, $\bar{w}_{11}$, $\bar{w}_{22}$, and $\bar{w}_{12}$. In $\mathscr{S}_L$, functional L induces a topology on $\mathscr{S}_L$, since L is the integral of a positive definite quadratic form. The integrand of (2.3.381) being a positive definite quadratic form can be recognized as follows:

$$\bar{w}_{11}^2 + 2\bar{w}_{11}\bar{w}_{22} + \bar{w}_{22}^2 + 2(1-\sigma)(\bar{w}_{12}^2 - \bar{w}_{11}\bar{w}_{22})$$

$$= [\bar{w}_{11}^2 + 2\sigma\bar{w}_{11}\bar{w}_{22} + \bar{w}_{22}^2] + 2(1-\sigma)\bar{w}_{12}^2;$$

since $\sigma < 1$ and since

$$\bar{w}_{11}^2 + 2\sigma\bar{w}_{11}\bar{w}_{22} + \bar{w}_{22}^2$$

is a positive definite form, as its discriminant is equal to $1 - \sigma^2 > 0$, the whole integrand is a positive definite quadratic form. Hence, concerning L, the situation is as described earlier.

Now, let it be shown that for the assumed boundary conditions, a *topological equivalence* with the Soboloev norm

$$(\| w \|_2^2)^2 = \int_\Omega [\bar{w}^2 + \bar{w}_1^2 + \bar{w}_2^2 + \bar{w}_{11}^2 + \bar{w}_{12}^2 + \bar{w}_{22}^2] \, d\Omega \qquad (2.3.388)$$

can be established.

From (2.3.381) follows

$$L > \int_\Omega \left[\mathcal{E} - \frac{q}{2}(a - x_1)\bar{w}_1^2 \right] d\Omega, \qquad (2.3.389)$$

where

$$\mathcal{E} = \tfrac{1}{2} D[\bar{w}_{11}^2 + 2\bar{w}_{11}\bar{w}_{22} + \bar{w}_{22}^2 + 2(1 - \sigma)(\bar{w}_{11}^2 - \bar{w}_{11}\bar{w}_{22})]. \qquad (2.3.390)$$

Hence, also

$$L > \int_\Omega [\eta\mathcal{E} + (1 - \eta)\mathcal{E} - \tfrac{1}{2}q(a - x_1)\bar{w}_1^2] \, d\Omega, \qquad 0 < \eta < 1. \qquad (2.3.391)$$

But for the given boundary conditions, the elastic energy $\mathcal{E}$ of the plate reads

$$E = \int_\Omega \mathcal{E} \, d\Omega = \int_\Omega \tfrac{1}{2} D(\nabla^2 \bar{w})^2 \, d\Omega.$$

Therefore, according to (2.3.380),

$$\int_\Omega \mathcal{E} \, d\Omega \geqslant \tfrac{1}{2} q_c \int_\Omega (a - x_1) w_1^2 \, d\Omega. \qquad (2.3.392)$$

Consequently,

$$L > \int_\Omega \{\eta\mathcal{E} + \tfrac{1}{2} [(1 - \eta)q_c - q] (a - x_1)\bar{w}_1^2\} \, d\Omega \qquad (2.3.393)$$

for

$$q < (1 - \eta)q_c, \qquad (2.3.394)$$

which is a stronger condition than (2.3.386). Finally, assuming (2.3.394) to hold,

$$L > \int_\Omega \eta\mathcal{E} \, d\Omega = \eta E. \qquad (2.3.395)$$

By virtue of (2.3.390),

$$E = \int_\Omega \mathcal{E} \, \mathscr{D}\Omega = \int_\Omega \tfrac{1}{2} D[\bar{w}_{11}^2 + \bar{w}_{22}^2 + 2\sigma\bar{w}_{11}\bar{w}_{22} + 2(1 - \sigma)\bar{w}_{12}^2] \, d\Omega. \qquad (2.3.396)$$

Hence,

$$E < \int_\Omega \tfrac{1}{2} D [2\bar{w}_{11}^2 + 2\bar{w}_{22}^2 + 2\bar{w}_{12}^2] \, d\Omega$$

and

$$E < D \int_\Omega [\bar{w}^2 + \bar{w}_1^2 + \bar{w}_2^2 + \bar{w}_{11}^2 + \bar{w}_{12}^2 + \bar{w}_{22}^2] \, d\Omega. \qquad (2.3.397)$$

By virtue of (2.3.388), inequality (2.3.397) yields

$$E < D(\| w \|_2^2)^2. \qquad (2.3.398)$$

On the other side, (2.3.396) yields

$$E > \frac{(1 - \sigma)D}{2} \int_\Omega (\bar{w}_{11}^2 + 2\bar{w}_{12}^2 + \bar{w}_{22}^2) \, d\Omega, \qquad (2.3.399)$$

since inequality (2.3.338) can be applied. Moreover, for the assumed boundary conditions, inequality (2.3.341) holds true. Thus,

$$E > \frac{(1 - \sigma)D}{2k} (\| w \|_2^2)^2 \qquad (2.3.400)$$

can be derived. Combining inequalities (2.3.398) and (2.3.400) yields the topological equivalence of the elastic energy E and the Sobolev norm

$$\frac{(1 - \sigma)D}{2k} (\| w \|_2^2)^2 < E < D(\| w \|_2^2)^2. \qquad (2.3.401)$$

The equivalence relation (2.3.401) can be used for a stability evaluation as follows. Let stability condition (2.3.394) hold. Then, from (2.3.395)

$$L(t) > \eta E(t). \qquad (2.3.402)$$

For an appropriate initial perturbation of the plate, (2.3.402) yields

$$\eta E(t_0) < L(t_0) < \varepsilon. \qquad (2.3.403)$$

Since $\dot{L} = 0$, and by virtue of (2.3.402) and (2.3.403)

$$\eta E(t) < L(t_0) = L(t_0) < \varepsilon. \qquad (2.3.404)$$

From (2.3.401) follows

$$E(t_0) < D(\| w(t_0) \|^2_2)^2. \qquad (2.3.405)$$

Moreover, $E > 0$, by definition of E. Hence, from (2.3.404) and (2.3.405)

$$\eta E(t) < \varepsilon + \eta E(t_0) < \varepsilon + \eta D(\| w(t_0) \|_2^2)^2. \qquad (2.3.406)$$

Thus,

$$\mathrm{E}(t) < \frac{\varepsilon}{\eta} + D(\| w(t_0) \|_2^2)^2. \qquad (2.3.407)$$

320

Let the initial perturbation, which caused $L(t_0) < \varepsilon$, cause as well

$$D(\| w(t_0) \|_2^2)^2 < \delta \tag{2.3.408}$$

Then, finally,

$$E(t) < \frac{\varepsilon}{\eta} + \delta. \tag{2.3.409}$$

Combining inequalities (2.3.400) and (2.3.409) yields

$$(\| w \|_2^2)^2 < \frac{2k}{(1 - \sigma)D} \left(\frac{\varepsilon}{\eta} + \delta \right) = \varepsilon^* \qquad \text{for any } t > t_0. \tag{2.3.410}$$

Moreover, for a completely supported plate, inequality (2.3.342) holds true. This inequality makes it possible to introduce the *design norm* $\| w \|_D \equiv \sup | w |$. Combining (2.3.342) and (2.3.410) leads to

$$\sup |w| < C \left[\frac{2k}{(1 - \sigma)D} \left(\frac{\varepsilon}{\eta} + \delta \right) \right]^{1/2} < \varepsilon^{**} \tag{2.3.411}$$

$$\varepsilon^{**} = C \left[\frac{2k}{(1 - \sigma)D} \left(\frac{\varepsilon}{\eta} + \delta \right) \right]^{1/2}, \qquad \text{for any } t > t_0. \tag{2.3.411}$$

The conclusion is that for

$$q < (1 - \eta)q_c$$

there is stability in the sense of (2.3.411), i.e., the deflections of the plate remain small. It is noteworthy that this smallness depends on η^{-1}. One would therefore tend to let η come close to the value 'one'. However, this adversely affects, according to (2.3.394), adversely the stability domain: the closer η is to 'one' the smaller is the range of q-values for which one can expect stability. Thus, it is again a fact that one has only a result which is weaker than the one the classical energy approach to stability would provide.

Now, let it be shown, how much any inference on stability depends on the *boundary conditions* of the plate under investigation as well as on the *inequalities* being used for the assessment of stability. For this purpose, let it be assumed that the boundary conditions of the plate are as shown in Fig. 95: at least two edges ($x_1 = a$, $x_2 = b$) are supported, two edges ($x_1 = 0$, $x_2 = 0$) free or by choice supported as well. These boundary conditions are of course more general than those for the preceding example. Let again $L = H$, i.e. (2.3.381) be the choice for the Lyapunov functional.

For the boundary conditions in Fig. 95, Rayleigh's inequality reads

$$q_c \leqslant \frac{2E}{\displaystyle\int_{\Omega} (a - x_1)\bar{w}_1^2 \, d\Omega}, \tag{2.3.412}$$

a version which holds for any kind of the plate's edge conditions (including free edges).

Fig. 95. = Supported edge: clamped or freely supported. — Either free or supported edge (arbitrary)

According to (2.3.389) and (2.3.396), as before,

$$L > E - \int_\Omega \frac{q}{2}(a - x_1)\bar{w}_1^2 \, d\Omega. \qquad (2.3.413)$$

Using (2.3.412) in (2.3.413) yields

$$L > \int_\Omega [\tfrac{1}{2}(q_c - q)(a - x_1)\bar{w}_1^2] \, d\Omega.$$

Hence, $L \geqslant 0$ for $q \leqslant q_c$. This is the former condition (2.3.386).

On the other hand, in agreement with (2.3.391), (2.3.393) and (2.3.396), again,

$$L > \eta E + (1 - \eta)E - \int_\Omega \tfrac{1}{2}q(a - x_1)\bar{w}_1^2 \, d\Omega. \qquad (2.3.414)$$

Applying (2.3.412) to (2.3.414) yields

$$L > \eta E + \tfrac{1}{2}[(1 - \eta)q_c - q] \int_\Omega (a - x_1)\bar{w}_1^2 \, d\Omega. \qquad (2.3.415)$$

Hence,

$$L > \eta E,$$

i.e., (2.3.395), if $q < (1 - \eta)q_c$. This is the very condition (2.3.394).

From $\dot{L} = 0$, which holds true because $L = H$ and $\dot{H} = 0$, it follows that for an appropriate initial perturbation of the plate, and by virtue of (2.3.395),

$$\varepsilon < L(t_0) = L(t) > \eta E(t) \qquad \text{for } t > t_0. \qquad (2.3.416)$$

From (2.3.396), and the fact that $\bar{w}_{11}^2 + \bar{w}_{22}^2 + 2\sigma\bar{w}_{11}\bar{w}_{22}$ is a positive definite form, it follows that

$$E > D(1 - \sigma) \int_\Omega \bar{w}_{12}^2 \, d\Omega. \qquad (2.3.417)$$

Moreover,

$$\int_\Omega \bar{w}_{12}^2 \, d\Omega = \int_0^b \int_0^a \bar{w}_{12}^2 \, dx_1 \, dx_2, \qquad (2.3.418)$$

and by virtue of Schwarz' inequality,

$$\int_0^a \bar{w}_{12}^2 \, dx_1 > \int_{x_1}^a \bar{w}_{12}^2 \, dx_1 > \frac{1}{a} \left[\int_{x_1}^a \bar{w}_{12} \, dx_1 \right]^2$$

$$= \frac{1}{a} \left[\bar{w}_2(a, x_2, t) - \bar{w}_2(x_1, x_2, t) \right]^2.$$

Yet, $\bar{w}_2(a, x_2, t) = 0$ due to the boundary conditions. Hence,

$$\int_0^a \bar{w}_{12}^2 \, dx_1 > \frac{1}{a} (\bar{w}_2(x_1, x_2, t))^2. \tag{2.3.419}$$

Using (2.3.419) in (2.3.418) and again Schwarz' inequality yields

$$\int_\Omega \bar{w}_{12}^2 \, d\Omega > \frac{1}{a} \int_0^b \left[\bar{w}_2(x_1, x_2, t) \right]^2 \, dx_2 > \frac{1}{a} \int_{x_2}^b \bar{w}_2^2 \, dx_2 > \frac{1}{ab} \left[\int_{x_2}^b w_2 \, dx_2 \right]^2,$$

$$\int_\Omega \bar{w}_{12}^2 \, d\Omega > \frac{1}{ab} \left[w(x_1, b, t) - w(x_1, x_2, t) \right]^2 = \frac{1}{ab} w^2, \tag{2.3.420}$$

since $w = 0$ along $x_2 = b$.

Applying (2.3.420) to (2.3.417) leads to

$$E > \frac{D(1 - \sigma)}{\Omega} w^2, \tag{2.3.421}$$

since $\Omega = ab$.

Considering (2.3.421) together with (2.3.416) yields

$$w^2 < \frac{\Omega}{\eta D(1 - \sigma)} \varepsilon, \qquad \text{i.e., } |w| < \varepsilon^* \text{ for } t > t_0$$

$$\varepsilon^* = [\Omega \varepsilon / \eta D(1 - \sigma)]^{1/2}. \tag{2.3.422}$$

Relationships (2.3.422) express stability of the plate in the sense of permanently small deflections under stability condition (2.3.394).

It is interesting to compare the set of inequalities (2.3.394), (2.3.411) and the boundary conditions of the first example, Fig. 94, with the set of inequalities (2.3.394), (2.3.422) and the boundary conditions of the second example, Fig. 95. In both cases one has permanent smallness of deflections under the same stability condition for the load intensity. Yet, in the second case, the result holds under much more general boundary conditions (two adjacent edges supported, the other two edges either free or arbitrarily supported) and is therefore stronger than the result valid for the first case in which *all* edges are required to be supported. This comparison points out very clearly how much the statements on stability which one may be able to make depend with regard to their breadth on the choice of the appropriate inequalities.

Let the effort be continued to make the theory as flexible with respect to the plate's boundary conditions as possible. Let therefore, the plate shown in Fig. 96 be considered. For this plate, edge B_1 (i.e., $x_1 = 0$) is clamped. The

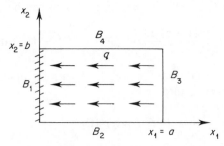

Fig. 96. Edge B_1 clamped, edges B_2, B_3, B_4
free or by choice arbitrarily supported

other edges, B_2, B_3, and B_4 may be free or by choice supported. If B_2, B_3, and B_4 were simultaneously supported, we had a completely supported plate whose case has already been treated. If B_2 (or B_4) or B_2 and B_4 were supported, we had the case shown in Fig. 95. This would also be the case if B_2 and B_3 (or B_4 and B_3) were supported. But the case shown in Fig. 96 has already been dealt with as well. Yet, a new situation arises if B_2, B_3, and B_4 are simultaneously *free*, or if at least B_2 and B_4 are simultaneously *free*.

For this new situation, a special version of Friedrich's inequality may be applied.[1] According to this version

$$\int_\Omega \bar{w}^2 \, d\Omega \leqslant \frac{4a^2}{\pi^2} \int_\Omega (\bar{w}_1^2 + \bar{w}_2^2) \, d\Omega, \qquad (2.3.423)$$

$$\int_\Omega (\bar{w}_1^2 + \bar{w}_2^2) \, d\Omega \leqslant \frac{4a^2}{\pi^2} \int_\Omega (\bar{w}_{11}^2 + 2\bar{w}_{12}^2 + \bar{w}_{22}^2) \, d\Omega, \qquad (2.3.424)$$

and thus,

$$\int_\Omega \bar{w}^2 \, d\Omega \leqslant \frac{16a^4}{\pi^4} \int_\Omega (\bar{w}_{11}^2 + 2\bar{w}_{12}^2 + \bar{w}_{22}^2) \, d\Omega \qquad (2.3.425)$$

holds true for the plate in Fig. 96.

Let the Hamiltonian H again be the Lyapunov functional L, i.e., (2.3.381). Hence, (2.3.413) holds true. But, also (2.3.412), being a general expression valid for any edge conditions, can be applied. Hence, rewriting (2.3.413) to yield (2.3.414), and applying (2.3.412) to (2.3.414) in order to yield (2.3.415), leads finally by means of (2.3.394) to (2.3.395), i.e., to the inequality

$$L > \eta E = \eta_2^{\frac{1}{2}} D \int_\Omega [\bar{w}_{11}^2 + \bar{w}_{22}^2 + 2\sigma\bar{w}_{11}\bar{w}_{22} + 2(1-\sigma)\bar{w}_{12}^2] \, d\Omega. \qquad (2.3.426)$$

[1]Leipholz, H., On a special case of Friedrich's inequality, *Mech. Res. Commun.* **7**, 329–333, 1980, and **8**, 259–261, 1981.

Yet, according to (2.3.338), inequality (2.3.426) can be changed into

$$L > \frac{(1-\sigma)\eta D}{2} \int_\Omega [\,\bar{w}_{11}^2 + 2\bar{w}_{12}^2 + \bar{w}_{22}^2\,]\,d\Omega. \qquad (2.3.427)$$

Let v satisfy the condition $0 < v < 1$. Then, inequality (2.3.427) can be changed by means of v into

$$L > \frac{v(1-\sigma)\eta D}{2} \int_\Omega (\bar{w}_{21}^2 + 2\bar{w}_{12}^2 + \bar{w}_{22}^2)\,d\Omega$$

$$+ \frac{(1-v)(1-\sigma)\eta D}{2} \int_\Omega (\bar{w}_{11}^2 + 2\bar{w}_{12}^2 + \bar{w}_{22}^2)\,d\Omega.$$

Applying to this ineqality the inequality (2.3.424) yields

$$L > \frac{v(1-\sigma)\eta D}{2} \frac{\pi^2}{4a^2} \int_\Omega (\bar{w}_1^2 + \bar{w}_2^2)\,d\Omega + \frac{(1-v)(1-\sigma)\eta D}{2}$$

$$\int_\Omega (\bar{w}_{11}^2 + 2\bar{w}_{12}^2 + \bar{w}_{22}^2 7\,d\Omega.$$

Let ξ satisfy the condition $0 < \xi < 1$. Then, the preceding inequality can be rewritten by means of ξ to yield

$$L > \frac{\xi v(1-\sigma)\eta D}{2} \frac{\pi^2}{4a^2} \int_\Omega (\bar{w}_1^2 + \bar{w}_2^2)\,d\Omega + \frac{(1-\xi)v(1-\sigma)\eta D}{2}$$

$$\times \frac{\pi^2}{4a^2} \int_\Omega (\bar{w}_1^2 + \bar{w}_2^2)\,d\Omega + \frac{(1-v)(1-\sigma)\eta D}{2} \int_\Omega (\bar{w}_{11}^2 + 2\bar{w}_{12}^2 + \bar{w}_{22}^2)\,d\Omega.$$

Applying to this inequality the inequality (2.3.423) yields

$$L > \frac{\xi v(1-\sigma)\eta D}{2} \frac{\pi^4}{16a^4} \int_\Omega \bar{w}^2\,d\Omega + \frac{(1-\xi)v(1-\sigma)\eta D}{2}$$

$$\times \frac{\pi^2}{4a^2} \int_\Omega (\bar{w}_1^2 + \bar{w}_2^2)\,d\Omega + \frac{(1-v)(1-\sigma)\eta D}{2} \int_\Omega (\bar{w}_{11}^2 + 2\bar{w}_{12}^2 + \bar{w}_{22}^2)\,d\Omega.$$

$$(2.3.428)$$

Set

$$x = \min\left[\frac{\pi^4 \xi v \eta (1-\sigma)D}{32a^4}, \; \frac{v\eta(1-\xi)(1-\sigma)D\pi^2}{8a^2}, \; \frac{(1-v)(1-\sigma)\eta D}{2}\right].$$

$$(2.3.429)$$

Then, using (2.3.388), (2.3.428) can be transformed into

$$L > x \int_\Omega (\bar{w}^2 + \bar{w}_1^2 + \bar{w}_2^2 + \bar{w}_{11}^2 + \bar{w}_{12}^2 + \bar{w}_{22}^2)\,d\Omega = x\,(\| w \|_2^2)^2. \qquad (2.3.430)$$

Moreover, (2.3.342) holds true for $\| w \|_2^2$. Therefore, (2.3.430) yields together

with (2.3.432) the inequalities

$$L > \varkappa (\| w \|_2^2)^2 > \frac{\varkappa}{C^2} \sup |w|^2.$$

But, for an appropriate initial perturbation, and due to $\dot{L} = 0$,

$$\varepsilon > L(t_0) = L(t) > \frac{\varkappa}{C^2} \sup |w|^2. \qquad (2.3.432)$$

Hence, from (2.3.432) it follows under condition $q < (1 - \eta)q_c$ that

$$\sup |w| < C \left(\frac{\varepsilon}{\varkappa}\right)^{1/2} \qquad \text{for any } t > t_0. \qquad (2.3.433)$$

This inequality explains again in which sense the plate in Fig. 96 will be stable for sufficiently small q, even if some of the edges B_2, B_3, and B_4 (or all of them) are free and not supported.

So far, the smallness of the plate's deflection could be warranted in the case of stability for fairly general boundary conditions, yet, only for a load intensity that *is smaller than the one of the classical buckling load*. Moreover, the design norm did allow us to infer solely on the behaviour of the *deflections*. Nothing could be said about *slopes and curvatures*. Hence, on the basis of the theory developed here, *local wrinkles of the plate cannot be ruled out even in the case of so-called stability*.

Consider now a non-conservative plate problem like the one in Fig. 97 involving distributed follower forces of intensity q. The differential equation describing the plate's small vibrations, taking viscous damping (coefficient β) into account, reads

$$\mu \ddot{\bar{w}} + D \nabla^2 \nabla^2 \bar{w} + q(a - x_1)\bar{w}_{11} + \beta \dot{\bar{w}} = 0. \qquad (2.3.434)$$

Since the problem is a divergence-type problem[1], inequality (2.3.376) holds

Fig. 97. Simply supported plate subjected to non-conservative follower loads

[1]Leipholz H. and Pfendt, F., On the Stability of rectangular, completely supported plates with uncoupling boundary conditions subjected to uniformly distributed follower forces, *Comp. Math. Appl. Mech. Eng.*, **32** 19–52, 1982.

true for this problem. Moreover, let the Lyapunov functional be chosen as

$$L = H + \int_\Omega \left(\frac{\beta^2}{4\mu} \bar{w}^2 + \frac{\beta}{2} \bar{w}\dot{\bar{w}} \right) d\Omega, \qquad (2.3.435)$$

where,

$$H = \int_\Omega \left\{ \frac{\mu}{2} \dot{\bar{w}}^2 + \frac{D}{2} \left[(\nabla^2 \bar{w})^2 + \frac{q}{D}(a - x_1)\bar{w}_1^2 \right] \right\} d\Omega. \qquad (2.3.436)$$

Then, as derived in [1],

$$\dot{L} = - \int_\Omega \left[\frac{\beta}{2} \dot{\bar{w}}^2 + q\bar{w}_1\dot{\bar{w}} + \frac{\beta D}{2\mu}(\nabla^2 \bar{w})^2 - \frac{\beta q}{2}(a - x_1)\bar{w}_1^2 \right] d\Omega. \qquad (2.3.437)$$

Let now the positive definiteness of L be established. Since

$$\int_\Omega \left(\frac{\mu}{2} \dot{\bar{w}}^2 + \frac{\beta^2}{4\mu} \bar{w}^2 + \frac{\beta}{2} \bar{w}\dot{\bar{w}} \right) d\Omega$$

is positive definite, it follows from (2.3.435) and (2.3.436) that

$$L \geqslant \int_\Omega \frac{D}{2} \left[(\nabla^2 \bar{w})^2 + \frac{q}{D}(a - x_1)\bar{w}_1^2 \right] d\Omega. \qquad (2.3.438)$$

Applying to (2.3.438) inequality (2.3.376) yields

$$L \geqslant \int_\Omega [\tfrac{1}{2}(q_c - q)(a - x_1)\bar{w}_1^2] d\Omega. \qquad (2.3.439)$$

Hence, L is positive definite in the subspace $\mathcal{S}_L$ spanned by $\bar{w}_1$ under the assumption that

$$q \leqslant q_c.$$

This is the already frequently encountered condition (2.3.386).

Due to the complete support of the plate,

$$\int_\Omega D(1 - \sigma)(\bar{w}_{12}^2 - \bar{w}_{11}\bar{w}_{22}) d\Omega = 0. \qquad (2.3.440)$$

Hence, (2.3.438) can be changed into

$$L > \int_\Omega \left[\mathcal{E} - \frac{q}{2}(a - x_1)\bar{w}_1^2 \right] d\Omega, \qquad (2.3.441)$$

which is (2.3.389), and where $\mathcal{E}$ is defined by (2.3.390). Subsequently, proceed exactly as described by relationships (2.3.391) to (2.3.403). The interesting fact is that the elastic energy E is again very relevant for the evaluation of stability, although the problem is a non-conservative one.

---

[1]Leipholz, H., *Stability of Elastic Systems*, Sijthoff and Noordhoff, Alphen aan den Rijn, The Netherlands, pp. 439–448, 1980.

Let now the requirement be pursued that $\dot{L}$ be negative definite. from (2.3.437) follows by means of (2.3.376) that

$$\dot{L} < - \int_\Omega \left[\frac{\beta}{2} \dot{\bar{w}}^2 + \frac{\beta}{2\mu} (q_c - q)(a - x_1)\bar{w}_1^2 + q\bar{w}_1\bar{w} \right] d\Omega. \quad (2.3.442)$$

Hence,

$$\dot{L} < 0 \quad (2.3.443)$$

holds true if

$$\frac{\beta}{2} \left[\frac{\beta}{2\mu} (q_c - q)M \right] - \frac{q^2}{4} > 0, \quad (2.3.444)$$

where

$$M = \int_\Omega (a - x_1)\bar{w}_1^2 \, d\Omega \Big/ \int_\Omega \bar{w}_1^2 \, d\Omega.$$

Inequality (2.3.444) yields a third stability condition,

$$\beta > q \left[\frac{\mu}{M(q_c - q)} \right]^{1/2}. \quad (2.3.445)$$

The result of this investigation holds not only for the specific plate in Fig. 97 but also for those plates which are 'rod-like'. Plates of that kind have been studied by Leipholz and Pfendt as mentioned earlier. They are divergence systems and involve simply supported, clamped plates, as well as plates for which two opposite edges are simply supported, the other two edges clamped.

In comparison with conservatively loaded plates there is little difference: The stability conditions (2.3.386) and (2.3.394) hold for conservatively loaded plates as well as for non-conservatively loaded rod-like plates. Yet, for the latter ones, condition (2.3.445) comes additionally into play. This condition requires that the damping intensity must be sufficiently large.

As far as the nature of stability is concerned, which is warranted by stability conditions (2.3.386), and (2.3.394), and (2.3.445), all the relations (2.3.404) to (2.3.411) hold again, except that now $L(t) < L(t_0)$, and not $L(t) = L(t_0)$, as for the conservatively loaded plate. Hence, (2.3.404) must be slightly changed, however, with no effect on the subsequent inequalities. Hence, (2.3.411) remains valid and the smallness of the plate's deflections is again secured for all time by conditions (2.3.386), (2.3.394), and (2.3.445) under the assumption of a sufficiently small initial perturbation.

Consider now a plate with more general edge conditions, see Fig. 98. In this figure it has been assumed that edges B_2, B_3, and B_4 are free. That is the extreme case. Yet, some or all of these three edges may be supported by choice.

For a plate of this kind it cannot be assumed in general that one is dealing with a divergence type system. Therefore, inequalities (2.3.376), (2.3.412), respectively, may not always be applicable. In order to circumvent this difficulty, let another inequality be derived as a substitute.

328

Fig. 98. Plate clamped along edge B_1 and subjected to non-conservative follower loads

Consider the expression

$$U = \int_\Omega [\mathcal{E} - \tfrac{1}{2}q(a - x_1)\bar{w}_1^2]\, d\Omega, \qquad (2.3.446)$$

which is the potential energy of the plate. By virtue of (2.3.330) and (2.3.338),

$$U > \int_\Omega [\tfrac{1}{2}D(1 - \sigma)(\bar{w}_{11}^2 + \bar{w}_{12}^2 + \bar{w}_{22}^2) - \tfrac{1}{2}qa\bar{w}_1^2]\, d\Omega. \qquad (2.3.447)$$

For plates of the kind shown in Fig. 98, inequalities (2.3.423), (2.3.424), and (2.3.425) hold true. Hence, by virtue of (2.3.424),

$$U > \int_\Omega \left[\frac{D(1 - \sigma)\pi^2}{8a^2}(\bar{w}_1^2 + \bar{w}_2^2) - \tfrac{1}{2}qa\bar{w}_1^2 \right] d\Omega. \qquad (2.3.448)$$

But, then also

$$U > \int_\Omega \left[\frac{D(1 - \sigma)\pi^2}{8a^2} - \tfrac{1}{2}qa \right] \bar{w}_1^2\, d\Omega. \qquad (2.3.449)$$

This is the desired inequality.

The differential equation of the plate in Fig. 98 is again (2.3.434). Hence, as before, the Lyapunov functional may be chosen equal to (2.3.435), i.e.

$$L = H + \int_\Omega \left(\frac{\beta^2}{4\mu}\bar{w}^2 + \frac{\beta}{2}\bar{w}\dot{w} \right) d\Omega, \qquad (2.3.450)$$

where

$$H = \int_\Omega \tfrac{1}{2}\mu\dot{\bar{w}}^2\, d\Omega + U.$$

Moreover, U is given by (2.3.446), and $\mathscr{E}$ is given by (2.3.330).
  Since

$$\int_\Omega \left(\frac{\mu}{2} \, \bar{w}^2 + \frac{\beta^2}{4\mu} \, \bar{w}^2 + \frac{\beta}{2} \, \bar{w}\dot{\bar{w}} \right) d\Omega$$

is positive definite, as already stated,

$$L > U.$$

Hence, by virtue of (2.3.449),

$$L > \int_\Omega \left[\frac{D(1-\sigma)\pi^2}{8a^2} - \tfrac{1}{2}qa \right] \bar{w}_1^2 \, d\Omega. \qquad (2.3.451)$$

thus, L is positive definite in the subspace S_L spanned by $\bar{w}_1$, if

$$q < \frac{D(1-\sigma)\pi^2}{4a^3}. \qquad (2.3.452)$$

This is a first-stability condition.
  For the boundary conditions of the plate in Fig. 98, $\dot{L}$ is given according to
H. Leipholz in *Stability of Elastic Systems*, p. 443, by

$$\dot{L} = - \int_\Omega \left[\frac{\beta}{2} \, \dot{\bar{w}}^2 + q\bar{w}_1\dot{\bar{w}} + \frac{\beta D}{2\mu} \, \bar{w} \nabla^2 \nabla^2 \bar{w} + \frac{\beta q}{2\mu} (a - x_1)\bar{w}\bar{w}_{11} \right] d\Omega.$$

This expression can be changed into

$$\dot{L} = - \int_\Omega \left[\frac{\beta}{2} \, \dot{\bar{w}}^2 + q\bar{w}_1\dot{\bar{w}} + \frac{\beta}{\mu} \, \mathscr{E} + \frac{\beta}{2\mu} \, q(a - x_1)\bar{w}\bar{w}_{11} \right] d\Omega, \qquad (2.3.453)$$

since

$$\int_\Omega (a - x_1)\bar{w}\bar{w}_{11} \, d\Omega = - \int_\Omega (a - x_1)\bar{w}_1^2 \, d\Omega + \int_\Omega \bar{w}\bar{w}_1 \, d\Omega$$

$$+ \int_B \bar{w}(a - x_1)\bar{w}_1 \cos(n, x_1) \, dB.$$

yet,

$$\int_B \bar{w}(a - x_1)\bar{w}_1 \cos(n, x_1) \, dB = 0, \qquad (2.3.455)$$

as $\bar{w} = 0$ along B_1, $\cos(n, x_1) = 0$ along B_2 and B_4, and $a - x_1 = 0$ along B_3.
  Furthermore,

$$\int_\Omega \bar{w}\bar{w}_1 \, d\Omega S = \tfrac{1}{2} \int \bar{w}^2 \, dB. \qquad (2.3.456)$$

Using (2.3.455) and (2.3.456) in (2.3.453) yields

$$\dot{L} = - \int_\Omega \left[\frac{\beta}{2} \, \dot{\bar{w}}^2 + q\bar{w}_1\dot{\bar{w}} + \frac{\beta}{\mu} \, \mathscr{E} - \frac{\beta}{2\mu} \, q(a - x_1)\bar{w}_1^2 \right] d\Omega - \tfrac{1}{2} \int \bar{w}^2 \, dB. \qquad (2.3.457)$$

330

Hence,

$$\dot{L} < - \int_\Omega \left[\frac{\beta}{2} \bar{w}^2 + q\bar{w}_1\dot{\bar{w}} + \frac{\beta}{\mu} \mathcal{E} - \frac{\beta}{2\mu} q(a - x_1)\bar{w}_1^2 \right] d\Omega,$$

which is

$$\dot{L} < - \int_\Omega \left\{ \frac{\beta}{2} \dot{\bar{w}}^2 + q\bar{w}_1\dot{\bar{w}} + \frac{\beta}{\mu} \left[\mathcal{E} - \tfrac{1}{2}q(a - x_1)\bar{w}_1^2 \right] \right\} d\Omega. \quad (2.3.458)$$

By virtue of (2.3.446), inequality (2.3.458) reads

$$\dot{L} < - \int_\Omega \left[\frac{\beta}{2} \dot{\bar{w}}^2 + q\bar{w}_1\dot{\bar{w}} \right] d\Omega - \frac{\beta}{\mu} U. \quad (2.3.459)$$

To (2.3.459) apply inequality (2.3.450) in order to obtain

$$\dot{L} < - \int_\Omega \left\{ \frac{\beta}{2} \dot{\bar{w}}^2 + q\bar{w}_1\dot{\bar{w}} + \frac{\beta}{\mu} \left[\frac{D(1 - \sigma)\pi^2}{8a^2} - \frac{1}{2} qa \right] \bar{w}_1^2 \right\} d\Omega. \quad (2.3.460)$$

For $\dot{L}$ to be never positive, the following condition must be satisfied:

$$\frac{\beta^2}{4\mu} \left[\frac{D(1 - \sigma)\pi^2}{4a^2} - qa \right] - \frac{q^2}{4} > 0. \quad (2.3.461)$$

This will be the case if (2.3.452) is being satisfied by q. Let this value of q following from (2.3.452) be called q^*. Then, the damping factor β must satisfy the inequality

$$\beta > \left[\frac{4a^2\mu(q^*)^2}{D(1 - \sigma)\pi^2 - 4a^3q^*} \right]^{1/2}. \quad (2.3.462)$$

The preceding deliberations and their results may now be summarized: Stability conditions (2.3.452) and (2.3.462) ensure stability for plates like those corresponding to Fig. 98. The nature of this stability can be discussed taking the fact into account that

$$L > U, \qquad \dot{L} \leqslant 0. \quad (2.3.463)$$

One has first from (2.3.435) and (2.3.450)

$$L = \int_\Omega \left[\frac{\mu}{2} \dot{\bar{w}}^2 + \frac{\beta^2}{4\mu} \bar{w}^2 + \frac{\beta}{2} \dot{\bar{w}}\bar{w} \right] d\Omega + U \quad (2.3.464)$$

where

$$U = \int_\Omega [\mathcal{E} - \tfrac{1}{2}q(a - x_1)\bar{w}_1^2] \, d\Omega. \quad (2.3.465)$$

But,

$$\frac{\beta^2}{\mu 4} \bar{w}^2 + \frac{\beta}{2} \bar{w}\dot{\bar{w}} + \frac{\mu}{2} \dot{\bar{w}}^2 < \mu\dot{\bar{w}}^2 + \frac{\beta^2}{2\mu} \bar{w}^2,$$

and

$$U < \int_\Omega \mathscr{E} \, d\Omega.$$

Hence, from (2.3.464) follows

$$L < \int_\Omega \left[\mu \dot{\overline{w}}^2 + \frac{\beta^2}{2\mu} \, \overline{w}^2 + \mathscr{E} \right] d\Omega.$$

Using for $\mathscr{E}$ relationship (2.3.338), the preceding inequality changes into

$$L < \int_\Omega \left\{ \mu \dot{\overline{w}}^2 + \frac{\beta}{2\mu} \, \overline{w}^2 + \frac{D}{2} \, [\, \overline{w}_{11}^2 + 2\sigma \overline{w}_{11} \overline{w}_{22} + \overline{w}_{22}^2 + 2(1 - \sigma) \overline{w}_{11}^2 \,] \right\} d\Omega.$$

Since $\sigma < 1$, also

$$L < \int_\Omega \left[\mu \dot{\overline{w}}^2 + \frac{\beta}{2\mu} \cdot \overline{w}^2 + \frac{D}{2} \, (\nabla^2 \overline{w})^2 + 2\overline{w}_{12}^2 \right] d\Omega. \qquad (2.3.466)$$

Assume an initial perturbation of the plate for which deflections, velocities, and curvatures are small throughout the plate. Then,

$$L(t_0) < \varepsilon \qquad \text{at } t = t_0 \qquad (2.3.467)$$

according to (2.3.466). Taking (2.3.463) into account yields

$$U(t) < L(t) \leqslant L(t_0) < \varepsilon \qquad \text{at } t > t_0. \qquad (2.3.468)$$

Also, by virtue of (2.3.330) and (2.3.338), inequality (2.3.468) yields

$$\int_\Omega [\tfrac{1}{2} D(1 - \sigma)(\overline{w}_{11}^2 + 2\overline{w}_{12}^2 + \overline{w}_{22}^2) - \tfrac{1}{2} qa\overline{w}_1^2] \, d\Omega < \varepsilon \qquad \text{at } t > t_0.$$
$$(2.3.469)$$

Now, let inequalities (2.3.423) to (2.3.425) be applied gradually to (2.3.469). One then arrives first at

$$\int_\Omega \frac{\eta D(1 - \sigma)}{2} \, (\overline{w}_{11}^2 + 2\overline{w}_{12}^2 + \overline{w}_{22}^2) \, d\Omega < \varepsilon \qquad \text{for } t > t_0 \quad (2.3.470)$$

if

$$q < \frac{(1 - \eta)(1 - \sigma)D\pi^2}{4a^3}, \qquad 0 < \eta < 1, \qquad (2.3.471)$$

and secondly at

$$\varkappa \int_\Omega (\overline{w}_{11}^2 + 2\overline{w}_{12}^2 + \overline{w}_{22}^2 + \overline{w}_1^2 + \overline{w}_2^2 + \overline{w}^2) \, d\Omega < \varepsilon \qquad \text{at } t > t_0 \quad (2.3.472)$$

where

$$\varkappa = \min \left[\frac{\pi^4 \upsilon \eta (1 - \sigma)D}{32a^2}, \quad \frac{\upsilon \eta (1 - \xi)(1 - \sigma)D\pi^2}{8a^2}, \quad \frac{(1 - \upsilon)(1 - \sigma)\eta D}{2} \right],$$

and

$$0 < \xi < 1, \qquad 0 < v < 1.$$

Inequality (2.3.471) is an additional stability condition to be added to conditions (2.3.452) and (2.3.462).

Since

$$(\| w \|_2^2)^2 = \int_\Omega [\bar{w}^2 + \bar{w}_1^2 + \bar{w}_2^2 + \bar{w}_{11}^2 + \bar{w}_{12}^2 + \bar{w}_{22}^2] \, d\Omega,$$

the inequality

$$\varkappa(\| w \|_2^2)^2 < \varepsilon \qquad \text{at } t > t_0 \tag{2.3.473}$$

follows from (2.3.472). Yet, (2.3.342) holds. Hence, one has finally

$$\sup |w|^2 < C^2 \frac{\varepsilon}{\varkappa}, \qquad \sup |w| < C \left(\frac{\varepsilon}{\varkappa}\right)^{1/2}. \tag{2.3.474}$$

This is formally the same result as (2.3.433) for the conservatively loaded plate in Fig. 96.

The final conclusion for the plate in Fig. 98 subjected to follower forces is the following: Again, the potential energy U and the strain energy E play a substantial role in warranting stability of the plate in terms of permanently small deflections in spite of the presence of follower forces. However, while the conservatively loaded plate has to satisfy the stability conditions $q < q_c$, $q < (1 - \eta)q_c$, the non-conservatively loaded plate must satisfy the more stringent conditions

$$q < \frac{D(1 - \sigma)\pi^2}{4a^2}, \qquad q < \frac{(1 - \eta)(1 - \sigma)D\pi^2}{4a^2}, \qquad 0 < \eta < 1.$$

In addition, the presence of sufficiently large damping is required by conditions (2.3.462).

Specifically, this part of section 2.3.7 has shown how versatile one must be with inequalities for the various boundary conditions in order to achieve a satisfactory discussion of stability of a continuous elastic body.

Attention will now be paid to one of the questions raised earlier: when is one allowed to work with linearized systems?

Real systems are non-linear. Mathematical difficulties frequently enforce, however, the introduction of *linearized models* for the description of the systems under investigation. One of the most important questions, to which Lyapunov has devoted himself, is therefore the one concerning the validity of stability criteria obtained by means of linearized models. Lyapunov has solved this question for discrete systems. Here, the same question will be investigated for continuous systems.

Let H be the Hamiltonian of the non-linear, continuous system. It is

$$H = H_L + H_N, \tag{2.3.475}$$

where H_L is that part whose density involves quadratic terms in w, and H_N is that part which includes terms of higher order. w is the characteristic of the system under consideration. The system is assumed to be conservative.

First, let *stability in the first approximation* be considered using the linearized model. The linear canonical equations of the linearized model are obtained using H_L. Therefore,

$$\dot{H}_L = 0, \qquad H_L = \text{const}$$

is obtained, applying the *linear* canonical equations to $\dot{H}_L$. Hence, H_L can be used as a Lyapunov functional for the linearized model.

It may be possible to obtain the estimate

$$H_L < \varkappa_{(2)} \| w \|_{(2, L)}^2. \tag{2.3.477}$$

Choosing proper initial conditions, and because H_L is constant, one can always arrive at

$$H_L < \varkappa_{(2)} \| w \|_{(2, L)}^2 < \varepsilon \qquad \text{for } t \geqslant t_0. \tag{2.3.478}$$

It may in addition be possible to find the estimate

$$H_L > \varkappa_{(1)} \| w \|_{(1)}^2 > 0 \qquad \text{for } t \geqslant t_0, \tag{2.3.479}$$

if a certain 'stability condition C' is being satisfied.

Combining (2.3.447) and (2.3.479) yields

$$\varepsilon > \varkappa_{(1)} \| w \|_{(1)}^2, \tag{2.3.480}$$

$$\| w \|_{(1)} < \sqrt{\frac{\varepsilon}{\varkappa_{(1)}}} \qquad \text{for } t \geqslant t_0.$$

This means that there is *stability in the first approximation in the sense of norm* $\| w \|_{(1)}$ *under* 'stability condition C'.

Such a result in the first approximation is sufficient and dependable, if a calculation using the *non-linear* model of the system yields the same stability criterion (2.3.480) as in the linear case, and if 'stability condition C' remains valid.

Let a sufficient condition for the stability in the first approximation to be valid be derived. The *non-linear* canonical equations may be obtained by means of H. Since the system is supposed to be conservative, these equations will lead to

$$\dot{H} = 0, \qquad H = \text{const.} \tag{2.3.481}$$

It may be possible to find the estimate

$$H < \bar{\varkappa}_{(2)} \| w \|_{(2)}^2. \tag{2.3.482}$$

As before, one is able to enforce the condition

$$H < \bar{\varkappa}_{(2)} \| w \|_{(2)}^2 < \bar{\varepsilon} \qquad \text{for } t \geqslant t_0 \tag{2.3.483}$$

since H is constant, and appropriate initial conditions can always be chosen to that effect.

334

Moreover, it may be possible under the 'stability condition C' not only to obtain the inequality

$$H_L > x_{(1)} \| w \|_{(1)}^2 \qquad (2.3.483)$$

but also the inequality

$$H_N > p, \qquad (2.3.484)$$

where p is a sufficiently small, *positive* quantity. Since H is constant according to (2.3.481), inequalities (2.3.483) and (2.3.484) combined yield

$$H = H_L + H_N > x_{(1)} \| w \|_{(1)}^2 + p > 0 \qquad \text{for } t \geqslant t_0. \qquad (2.3.485)$$

Inequalities (2.3.483) and (2.3.485) lead first to

$$\bar{\varepsilon} > x_{(1)} \| w \|_{(1)}^2 + p.$$

But since $p > 0$, also to

$$\bar{\varepsilon} > x_{(1)} \| w \|_{(1)}^2$$

or

$$\| w \|_{(1)} < \sqrt{\frac{\bar{\varepsilon}}{x_{(1)}}} \qquad \text{for } t \geqslant t_0. \qquad (2.3.486)$$

(2.3.486) is a stability criterion corresponding totally to the criterion (2.3.480) obtained in the first approximation. In addition, the 'stability condition C' used in the linear case remained valid in the non-linear case. Hence, the non-linear calculation yields nothing new with respect to stability as compared with the linear calculation. Therefore, it is sufficient to investigate stability using the linear model, i.e., it is sufficient to investigate stability in the first approximation. Summarizing, 'theorem L' can be formulated:

If inequality (2.3.483) holds true in the first approximation under 'stability condition C', it is sufficient for the dependability of a stability investigation in the first approximation that under 'stability condition C', also inequality (2.3.484) holds true.

As an example, choose the elastic rod subjected to a unidirectional force P. This problem is conservative. Following *C. L. Dym*[1],

$$H = \frac{1}{2} \int_0^l \left\{ \mu \dot{v}^2 + \alpha \frac{(v'')^2}{1 - (v')^2} + 2P[(1 - (v')^2)^{1/2} - 1] \right\} dx. \qquad (2.3.487)$$

Because of

$$\frac{1}{1 - (v')^2} \approx 1 + (v')^2, \qquad (1 - (v')^2)^{1/2} \approx 1 - \frac{(v')^2}{2} + \frac{1}{8}(v')^4,$$

[1]Dym, C. L., *Stability Theory and Its Application to Structural Mechanics*, Noordhoff Int. Publ., Leyden, pp. 70–74, 1974.

also

$$H = \frac{1}{2} \int_0^l \{\mu\dot{v}^2 + \alpha(v'')^2 - P(v')^2\} \, dx + \frac{1}{2} \int_0^l \left\{ \alpha(v'')^2(v')^2 - \frac{P}{4}(v')^4 \right\} \, dx.$$

$$(2.3.488)$$

In (2.3.488), μ is the linear mass density, α the bending stiffness, and l the length of the rod. Let

$$\dot{v} = \frac{\partial v}{\partial t}, \qquad v' = \frac{\partial v}{\partial x} \text{ etc.}$$

where $v(x, t)$ is the lateral deflection of the rod.

It is obvious that in (2.3.488)

$$H_L = \frac{1}{2} \int_0^l \{\mu\dot{v}^2 + \alpha(v'')^2 - P(v')^2\} \, dx, \qquad (2.3.489)$$

$$H_N = \frac{1}{2} \int_0^l \{\alpha(v'')^2(v')^2 - \frac{1}{4}P(v')^4\} \, dx. \qquad (2.3.490)$$

The linear calculation yields

$$\left. \begin{array}{l} H_l < \varkappa_{(2)} \|v\|_{(2)} \qquad \text{for } t \geqslant t_0, \\[2mm] \varkappa_{(2)} = \max \left(\dfrac{l\mu}{2}, \dfrac{l\alpha}{2} \right), \\[2mm] \|v\|_{(2)} = \sup |\dot{v}|^2 + \sup |v''|^2. \end{array} \right\} \qquad (2.3.491)$$

Choosing at $t = t_0$ the quantities $|\dot{v}|$ and $|v''|$ small enough, $H_L < \varepsilon$.

Rayleigh's inequality yields

$$\int_0^l \alpha(v'')^2 \, dx \geqslant \int_0^l P_{\text{crit}}(v')^2 \, dx, \qquad (2.3.492)$$

where P_{crit} is Euler's buckling load. By means of (2.3.492), (2.3.489) can be changed into

$$H_L > \frac{1}{2} \int_0^l (P_{\text{crit}} - P)(v')^2 \, dx,$$

and by means of Schwarz' inequality into

$$H_L > \frac{P_{\text{crit}} - P}{2l} v^2. \qquad (2.3.493)$$

The 'stability condition C' reads

$$P < P_{\text{crit}}. \qquad (2.3.494)$$

Under this condition,

$$H_L > \varkappa_{(1)} \|v\|_{(1)}^2 > 0 \qquad \text{for } t \geqslant t_0, \qquad (2.3.495)$$

336

where

$$x_{(1)} = \frac{P_{\text{crit}} - P}{2l} > 0$$

and

$$\| v \|_{(1)}^2 = | v |^2.$$ (2.3.496)

Since the problem is conservative, and therefore $H_L = \text{const}$, (2.3.491) and (2.3.495) yield

$$x_{(1)} \| v \|_{(1)}^2 < \varepsilon$$

or

$$| v | < \sqrt{\frac{\varepsilon}{x_{(1)}}} \qquad \text{for } t \geqslant t_0.$$ (2.3.497)

This is stability in the sense of pointwise smallness of the rod's deflections v under stability condition (2.3.494).

Now, let H_N be investigated. Using the mean value theorem of integration, (2.3.490) yields

$$H_N = \tfrac{1}{2}[v'(\xi)]^2 \int_0^l \{\alpha(v'')^2 - \tfrac{1}{4}P(v')^2\} \, dx, \qquad 0 < \xi < l.$$ (2.3.498)

By means of (2.3.492), this relationship can be rewritten to yield

$$H_N > \tfrac{1}{2}[v'(\xi)]^2 \int_0^l (P_{\text{crit}} - \tfrac{1}{4}P)(v')^2 \, dx$$

and using (2.3.493),

$$H_N > [v'(\xi)]^2 \frac{4P_{\text{crit}} - P}{4l} v^2.$$ (2.3.499)

Hence, under the stability condition (2.3.494) of the linear case,

$$H_N > p,$$ (2.3.500)

where

$$p = \tfrac{1}{2}[v'(\xi)]^2 \frac{4P_{\text{crit}} - P}{4l} v^2 > 0,$$

since $4P_{\text{crit}} - P > 0$ according to (2.3.494). But then, 'thereom L' asserts that stability can be investigated linearly, i.e., in the first approximation.

The correctness of this assertion can be shown in the following way:

$$H < \tfrac{1}{2} \int_0^l \{\mu(\dot{v})^2 + \alpha(v'')^2 + \alpha(v'')^2(v')^2\} \, dx < \bar{\varepsilon}.$$ (2.3.501)

Because of (2.3.493), (2.3.500), and the fact hat $H = H_L + H_N$, also

$$H < \frac{P_{\text{crit}} - P}{2l} v^2 + p > 0.$$ (2.3.502)

From (2.3.501) and (2.3.502), the inequality

$$0 < \frac{P_{\mathrm{crit}} - P}{2l} \, v^2 + p < \bar{\varepsilon} \qquad (2.3.503)$$

can be derived, which holds true at any time, since H is constant. Therefore, it follows from the non-linear calculation, i.e., from (2.3.503), that

$$\left.\begin{array}{c} |\,v\,| < \sqrt{\dfrac{\bar{\varepsilon}}{\varkappa_{(1)}}} \qquad \text{for } t \geqslant t_0 \\[12pt] \varkappa_{(1)} = \dfrac{P_{\mathrm{crit}} - P}{2l} > 0 \end{array}\right\} \qquad (2.3.504)$$

holds true, if the fulfillment of (2.3.494) is presupposed. This result is completely analogous to the result obtained by the linear calculation, that is to say (2.3.497). The non-linear calculation has not brought anything new into the picture as compared to the linear calculation. Therefore, it is sufficient to investigate the stability of the rod by means of a linear calculation.

As a conclusion, let these remarks be made.

(i) If the problem is non-linear and non-conservative, everything runs as before, except that a more general Lyapunov functional has to be used instead of the Hamiltonian H.

(ii) It may frequently be sufficient to determine the stability boundaries of a system under certain circumstances by using its linearized model. Yet, if one is interested in the *postbuckling behaviour* of the system, one must definitely use a non-linear model. A good example for the postbuckling analysis of a rod has been given by E. L. Reis[1].

2.3.8. Problems of Elastokinetics

The previously discussed elastostatic problems were obtained by considering the perturbed motion around an equilibrium position. We refer to 'stability problems of elastostatics' because the unperturbed original state was an equilibrium position and because the loading was assumed to be time-independent.

Let us generalize and consider a certain (if possible stationary) *motion* as the original state of the elastic body. We then investigate motions that are adjacent to the undisturbed motion, which leads to the concept of stability problems of elastokinetics. This is a restricted definition, because any stability problem is kinetic in the broadest sense. An equilibrium position can always be considered as a special case of an undisturbed original motion. There will always be disturbed motions around the original state. This 'kinetic' content of any stability theory was already apparent in stability problems of elastostatics,

---

[1]Reis, E. L. *Column Buckling, Bifurcation Theory and Nonlinear Eigenvalue Problems*, ed. J. B. Keller and S. T. Antmann, W. A. Benjamin Inc., N.Y., pp. 1–16, 1969.

because there we considered small oscillations and used the kinetic stability criterion.

Therefore, we will now use the motion of the elastic body as the undisturbed state. It is assumed that the points of the body have the displacements $u_i^{(0)}$, $i = 1, 2, 3$, for this original motion. The disturbance of the original state results in adjacent motions with the displacements $u_i^{(0)} + u_i$. The u_i are the 'variations'. The most important task is to establish the differential equations for the u_i within the framework of the general theory — that is, the variational equations. The solution of these equations determines the nature of the motions adjacent to the original state. If the u_i, which are obtained from the variational equations, can be made arbitrarily small at all times for sufficiently small disturbances, the undisturbed original motion is called 'stable' according to Lyapunov's definition. Strictly speaking, it would again be necessary to determine that the strains remain sufficiently small, so that we are certain that none of the foundations of the theory of elasticity have been violated. However, for purposes of simplicity, we will not do this and shall simply assume that the strains are small enough.

The variational equations can again be obtained from the principle of virtual work. There are two ways in which this can be done, just as was the case for elastostatic problems, which result from the distinction between conservative or monogenic problems and non-conservative or polygenic problems.

Let us first consider *conservative* or *monogenic* problems. The theory was advanced by Mettler.[1]

The principle of virtual work results in

$$\int_v \mu \ddot{u}\, \delta u\, \mathrm{d}V + \delta(\Pi_i + \Pi_a) = 0. \tag{2.3.505}$$

Let

$$T = \frac{1}{2} \int_V \mu \dot{u}^2\, \mathrm{d}V$$

be the kinetic energy. Then $\delta T = \int_V \mu \dot{u}\, \delta \dot{u}\, \mathrm{d}V$. If δT is added and subtracted from the left-hand side of (2.3.505), the following result is obtained:

$$\delta(T - \Pi_i - \Pi_a) = (\mu \ddot{u}\, \delta u + \mu \dot{u}\, \delta \dot{u})\, \mathrm{d}V.$$

Making use of the Lagrangian $L = T - (\Pi_i + \Pi_a)$, the following relationship

$$\delta L = \frac{\mathrm{d}}{\mathrm{d}t} \int_V \mu \dot{u}\, \delta u\, \mathrm{d}V \tag{2.3.506}$$

is obtained. If both sides of (2.3.506) are integrated with respect to time t between the limits t_1, t_2, the result is

$$\int_{t_1}^{t_2} \delta L\, \mathrm{d}t = \int_V \left[\mu \dot{u}\, \delta u\right]_{t_1}^{t_2} \mathrm{d}V,$$

[1]Mettler, E., A theory of stability of elastic motion, *Ing. Arch.* 16, 135–146, 1947.

and the final result is

$$\delta \int_{t_1}^{t_2} L\, dt = 0, \qquad (2.3.507)$$

which is the Hamilton principle if it is assumed that the variations satisfy $\delta u(t_1) = \delta u(t_2) = 0$.

The Lagrangian L of the adjacent motions can be expanded with respect to the Lagrangian L_g of the undisturbed original motion, so that

$$L = L_g + \delta L_g + \tfrac{1}{2}\delta^2 L_g. \qquad (2.3.508)$$

When (2.3.508) is substituted into (2.3.507), we obtain

$$\delta \int_{t_1}^{t_2} L_g\, dt + \delta\left[\delta \int_{t_1}^{t_2} L_g\, dt\right] + \tfrac{1}{2}\delta \int_{t_1}^{t_2} \delta^2 L_g\, dt = 0. \qquad (2.3.509)$$

The first two terms in (2.3.509) are zero, because the condition (2.3.507) also holds for the undisturbed original motion. Therefore

$$\delta \int_{t_1}^{t_2} L_g\, dt = 0.$$

The third term from (2.3.509) remains for the derivation of the variational equations for the u_i:

$$\delta \int_{t_1}^{t_2} \delta^2 L_g\, dt = 0. \qquad (2.3.510)$$

It is analogous to the equation

$$\delta(\delta^2 \Pi_g) = 0,$$

which holds in elastostatics and is given in Sections 2.3.1, 2.3.4, and 2.3.5.

If it is assumed that only small strains are present, we have

$$\delta \int_{t_1}^{t_2} \delta^2 L_g\, dt = -\int_V \left[\frac{\partial \Delta\sigma_{ik}}{\partial x_k} + \frac{\partial}{\partial x_k}\left(\sigma_{jk}\frac{\partial u_i}{\partial x_j}\right)\right]\delta u_i\, dV$$

$$-\int_O \left[\Delta\sigma_{ik} + \frac{\partial u_i}{\partial x_j}\sigma_{jk}\right]n_k\, \delta u_i\, dO + \int_V \mu \frac{\partial^2 u_i}{\partial t^2}\delta u_i\, dV = 0,$$

$$(2.3.511)$$

which can be found in Mettler's paper. V is the volume, O the surface area, μ the density of the elastic body. σ_{ij} is the stress tensor, $\Delta\sigma_{ij}$ the variation of the stress tensor for the disturbed motion due to the perturbation, u_i the vector of displacement variations, and n_i the unit vector in the direction of the surface normal.

Since the variations δu_i are independent, Eq. (2.3.511) results in the system of differential equations

$$\frac{\partial \Delta\sigma_{ik}}{\partial x_i} + \frac{\partial}{\partial x_k}\left(\sigma_{jk}\frac{\partial u_i}{\partial x_j}\right) = \mu \frac{\partial^2 u_i}{\partial t^2}, \qquad (2.3.512)$$

340

as well as the boundary conditions

$$[\Delta\sigma_{ik} + (\partial u_i/\partial x_j)\sigma_{jk}]\,n_k = 0. \tag{2.3.513}$$

It should be noted that $\sigma_{jk}n_k = p_j$ is the prescribed surface load. In addition, the relationship

$$\Delta\sigma_{ik} = \lambda^*\,\delta_{ik}\,\frac{\partial u_j}{\partial x_j} + \mu^*\left(\frac{\partial u_i}{\partial x_k} + \frac{\partial u_k}{\partial x_i}\right) \tag{2.3.514}$$

must be used which follows from Hooke's law (in which λ^*, μ^* are the Lamé constants and δ_{ik} is the Kronecker delta). The final differential equations of the displacement 'variations' u_i are given by

$$\mu^*\,\Delta^2 u_i + (\lambda^* + \mu^*)\,\frac{\partial^2 u_j}{\partial x_i \partial x_j} + \frac{\partial}{\partial x_k}\left(\sigma_{jk}\,\frac{\partial u_i}{\partial x_j}\right) = \mu\,\frac{\partial^2 u_i}{\partial t^2}, \tag{2.3.515}$$

which follows from (2.3.512) and holds for small displacements of the undisturbed state.

As already mentioned, the solutions of (2.3.515) determine the stability of the original state. This must be investigated in each individual case. The application of this theory to a practical example can be found in the paper by Mettler, which also contains additional references.

Additional terms must be added for *non-conservative, polygenic* problems. For example, it must be remembered that when the original motion is disturbed, there can also be changes in the surface and body forces. This is the case for 'follower' forces, i.e., when they change dependent on the variations in the displacements.

This time, let us use the equations of motion

$$\partial S_{ik}/\partial \xi_k + X_i = \mu\,\mathrm{d}^2 v_i/\mathrm{d}t^2. \tag{2.3.516}$$

S_{ik} is the stress tensor and ξ_k are the Euler coordinates, X_i the specific volume forces, and v_i the displacements. Equation (2.3.516) is transformed into Lagrangian coordinates as follows, assuming small strains:

$$\frac{\partial}{\partial x_k}\left[\left(\delta_{ij} + \frac{\partial v_i}{\partial x_j}\right)S_{jk}\right] + X_i = \mu\,\frac{\partial^2 v_i}{\partial t^2}. \tag{2.3.517}$$

The following equilibrium conditions also hold for any point of the loaded surface of the elastic body:

$$(\delta_{ij} + \partial v_i/\partial x_j)S_{jk}n_k = p_i. \tag{2.3.518}$$

It is assumed that the parameters appearing in (2.3.517), (2.3.518) are obtained as the result of a disturbance to the original state, so that

$$S_{ij} = \sigma_{ij} + \Delta\sigma_{ij}, \qquad p_i = p_{i0} + \Delta p_i,$$

$$X_i = X_{i0} + \Delta X_i, \qquad v_i = u_{i0} + u_i. \tag{2.3.519}$$

In (2.3.519), σ_{ij}, p_{i0}, X_{i0}, u_{i0} are parameters for the undisturbed state and

$\Delta\sigma_{ij}$, Δp_i, ΔX_i, u_i are the variations that arise because of a disturbance of the original state.

When (2.3.519) is substituted into (2.3.517), the result is

$$\frac{\partial}{\partial x_k}\left[\left(\delta_{ij} + \frac{\partial u_{i0}}{\partial x_j} + \frac{\partial u_i}{\partial x_j}\right)(\sigma_{jk} + \Delta\sigma_{jk})\right] + X_{i0} + \Delta X_i$$

$$= \mu\frac{\partial^2 u_{i0}}{\partial t^2} + \mu\frac{\partial^2 u_i}{\partial t^2}. \qquad (2.3.520)$$

Since the relationship (2.3.517) also holds for the original state, we have

$$\frac{\partial}{\partial x_k}\left[\left(\delta_{ij} + \frac{\partial u_{i0}}{\partial x_j}\right)\sigma_{jk}\right] + X_{i0} = \mu\frac{\partial^2 u_{i0}}{\partial t^2}.$$

Lete us also assume that the products

$$(\partial u_i/\partial x_j)\Delta\sigma_{jk} \qquad \text{and} \qquad (\partial u_{i0}/\partial x_j)\Delta\sigma_{jk} \qquad (2.3.521)$$

can be ignored because they are small. Then Eq. (2.3.520) becomes

$$\frac{\partial}{\partial x_k}\Delta\sigma_{ik} + \frac{\partial}{\partial x_k}\left(\sigma_{jk}\frac{\partial u_i}{\partial x_j}\right) + \Delta X_i = \mu\frac{\partial^2 u_i}{\partial t^2}, \qquad (2.3.522)$$

which coincides with Eq. (2.3.512) except for an additional term.

Again assuming that the products (2.3.521) are small and can be ignored, the following boundary conditions are obtained from (2.3.518) and (2.3.519):

$$[\Delta\sigma_{ik} + (\partial u_i/\partial x_j)\sigma_{jk}]n_k = \Delta p_i, \qquad (2.3.523)$$

which corresponds to the conditions (2.3.513).

The difference between (2.3.522), (2.3.523), and (2.3.512), (2.3.513) is that 'follower' loads are assumed to be present in the first case, so that $\Delta X_i \neq 0$, $\Delta p_i \neq 0$. The loads are 'unidirectional' in the conservative case and $\Delta X_i = \Delta p_i = 0$. In this case, the more general equations (2.3.522), (2.3.523) immediately become the special equations (2.3.512), (2.3.513).

Finally, (2.3.522) can be rewritten as follows using the Hooke's law:

$$\mu^*\Delta^2 u_i + (\lambda^* + \mu^*)\frac{\partial^2 u_j}{\partial x_i\partial x_j} + \frac{\partial}{\partial x_k}\left(\sigma_{ij}\frac{\partial u_i}{\partial x_j}\right) + \Delta X_i = \mu\frac{\partial^2 u_i}{\partial t^2}, \qquad (2.3.524)$$

which corresponds to (2.3.515).

In the mathematical discussion of this problem, we will introduce the following simplifications:

(1) $\Delta X_j = \Delta p_i = 0$, i.e., there are no 'follower' loads, and Eqs. (2.3.512), (2.3.513) are used.

(2) The undisturbed state is identified with the equilibrium state. This means that the quantities $s_{ik}^{(0)}$ and $s_{ik}^{(t)}$ which determine the original state of stress

$$\sigma_{ik} = -\alpha s_{ik}^{(0)} - \beta\Phi(t)s_{ik}^{(t)}, \qquad (2.3.525)$$

can be determined in the quasi-static way, i.e., the $s_{ik}^{(0)}$ can be determined from the static loads $p_i^{(0)}$, $X_i^{(0)}$, which depend on a parameter α, and the $s_{ik}^{(t)}$ can be obtained from the amplitude values of the variable loads $p_i^{(t)}$, $X_i^{(t)}$, which depend on a parameter β.

In addition, Hooke's law is written in the form $\Delta\sigma_{ij} = C_{ijkl}\varepsilon_{ki}$, where C_{ijkl} is the elasticity tensor and ε_{kl} the strain tensor. Taking the symmetry properties of the elasticity tensor into account, the following equations are obtained from (2.3.512), (2.3.513):

$$\frac{\partial}{\partial x_k}\left(C_{ikmn}\frac{\partial u_m}{\partial x_n}\right) - \alpha\frac{\partial}{\partial x_k}\left(s_{jk}^{(0)}\frac{\partial u_i}{\partial x_j}\right) - \beta\Phi(t)\frac{\partial}{\partial x_k}\left(s_{jk}^{(t)}\frac{\partial u_i}{\partial x_j}\right) - \mu\frac{\partial^2 u_i}{\partial t^2} = 0,$$
$$(2.3.526)$$

$$\Delta\sigma_{ik}n_k = \left(\alpha\frac{\partial u_i}{\partial x_j}s_{jk}^{(0)} + \beta\Phi(t)\frac{\partial u_i}{\partial x_j}s_{jk}^{(0)}\right)n_k. \qquad (2.3.527)$$

The following 'abbreviated' problem corresponds to (2.3.526), (2.3.527):

$$\frac{\partial}{\partial x_k}\left(C_{ikmn}\frac{\partial u_m}{\partial x_n}\right) - \mu\frac{\partial^2 u_i}{\partial t^2} = 0, \qquad \Delta\sigma_{ik}n_k = 0 \qquad (2.3.529)$$

which corresponds to the free eigen-oscillations of the elastic body. The solution of (2.3.529) can be written in the form of an integral equation. The classical problem of the theory of elasticity

$$\frac{\partial}{\partial x_k}\left(C_{ikmn}\frac{\partial u_m}{\partial x_n}\right) + X_i = 0, \qquad \Delta\sigma_{ik}n_k = p_i \qquad (2.3.530)$$

allows the solution

$$u_i(P) = \int_V G_{ik}(P, Q)X_k(Q)\,dV_Q + \int_0 G_{ik}(P, Q)p_k(Q)\,dO_Q, \qquad (2.3.531)$$

where G_{ik} is the Green tensor, V the volume, O the surface area, and P, Q are two points of the elastic body. A description of the Green tensor is given in the books by Bolotin (1.7, pp. 219–221).

A comparison of (2.3.529) and (2.3.530) shows that the solution (2.3.531) can be taken for (2.3.529), if the following substitution is made in (2.3.531):

$$X_i = -\mu\partial^2 u_i/\partial t^2, \qquad p_i = 0.$$

When $u_i(P, t) = \phi_i(P)e^{i\omega t}$ is introduced and if $\omega^2 = \lambda$, the result is

$$\phi_i(P) - \lambda\int_V \mu(Q)G_{ik}(P, Q)\phi_k(Q)\,dV_Q = 0. \qquad (2.3.532)$$

The system of integral equations (2.3.532) has the eigenvalues λ_k, $k = 1, 2, 3, \ldots$, and the eigenvectors $\phi_i^{(k)}(P)$, $i = 1, 2, 3$, correspond to each eigenvalue λ_k. The components of the eigenvector, the eigenfunctions $\phi_r^{(k)}$ (r fixed), are a complete and orthonormal function system. Each vector $u_i(P)$ generated by the Green tensor

$$u_i^{(P)} = \int_V G_{ik}(P, Q)X_k(Q)\,dV_Q$$

can be expanded in the absolutely and uniformly convergent series

$$u_i(P) = \sum_{m=1}^{\infty} \rho_m \phi_i{}^{(m)}(P), \qquad \rho_m = \int_V u_i(P)\phi_i{}^{(m)}\,dV. \qquad (2.3.533)$$

The components of the Green tensor can be written in the bilinear series

$$G_{ik}(P,Q) = \sum_{m=1}^{\infty} \frac{\phi_i{}^{(m)}(P)\phi_k{}^{(m)}(Q)}{\lambda_m}. \qquad (2.3.534)$$

Just as was done in the abbreviated problem (2.3.529), the solution (2.3.531) of (2.3.530) can be applied to our problem (2.3.526), (2.3.527). For this it is necessary to replace X_i in (2.3.531) by

$$-\alpha \frac{\partial}{\partial x_k}\left(s_{jk}^{(0)}\frac{\partial u_i}{\partial x_j}\right) - \beta\Phi(t)\frac{\partial}{\partial x_k}\left(s_{jk}^{(t)}\frac{\partial u_i}{\partial x_j}\right) - \mu\frac{\partial^2 u_i}{\partial t^2}$$

and to replace p_i by

$$[\alpha(\partial u_i/\partial x_j)s_{jk}^{(0)} + \beta\Phi(t)(\partial u_i/\partial x_j)s_{jk}^{(t)}]n_k.$$

The following solution is the solution of (2.3.526), (2.3.527):

$$u_i(P) = -\int_V \mu(Q)G_{ik}(P,Q)\frac{\partial^2 u_k(Q,t)}{\partial t^2}\,dV_Q$$

$$-\alpha\int_V G_{ik}(P,Q)\frac{\partial}{\partial \xi_l}\left[s_{jl}^{(0)}\frac{\partial u_k Q}{\partial \xi_j}\right]dV_Q$$

$$-\beta\Phi(t)\int_V G_{ik}(P,Q)\frac{\partial}{\partial \xi_l}\left[s_{jl}^{(t)}\frac{\partial u_k Q}{\partial \xi_j}\right]dV_Q$$

$$+\alpha\int_0 G_{ik}(P,Q)s_{jl}^{(0)}(Q)\frac{\partial u_k(Q)}{\partial \xi_j}n_l\,dO_Q$$

$$+\beta\Phi(t)\int_0 G_{ik}(P,Q)s_{jl}^{(t)}(Q)\frac{\partial u_k(Q)}{\partial \xi_j}n_l\,dO_Q. \qquad (2.3.535)$$

For example, we have

$$G_{ik}(P,Q)\frac{\partial}{\partial \xi_l}\left[s_{jl}^{(0)}\frac{\partial u_k(Q)}{\partial \xi_j}\right] \equiv \frac{\partial}{\partial \xi_l}\left[G_{ik}(P,Q)s_{jl}^{(0)}(Q)\frac{\partial u_k(Q)}{\partial \xi_j}\right]$$

$$-\frac{\partial G_{ik}(P,Q)}{\partial \xi_l}s_{jl}^{(0)}(Q)\frac{\partial u_k(Q)}{\partial \xi_j}.$$

Applying the Gaussian formula

$$\int_V \partial F_k/\partial x_k\,dV = \int_0 F_k n_k\,dO,$$

344

we obtain

$$
-\alpha \int_V G_{ik}(P,Q) \frac{\partial}{\partial \xi_l} \left[s_{jl}^{(0)} \frac{\partial u_k Q}{\partial \xi_j} \right] dV_Q
$$

$$
= \alpha \int_V \frac{\partial G_{ik}(P,Q)}{\partial \xi_l} s_{jl}^{(0)}(Q) \frac{\partial u_k(Q)}{\partial \xi_j} dV_Q
$$

$$
- \alpha \int_0 G_{ik}(P,Q) s_{jl}^{(0)}(Q) \frac{\partial u_i(Q)}{\partial \xi_j} n_l \, dO_Q. \tag{2.3.536}
$$

The third term on the right-hand side of (2.3.535) can be transformed in a similar way.

When (2.3.536) and the transformation of the third term in (2.3.535) is substituted, the following new relationship is obtained:

$$
u_i(P) = - \int_V \mu(Q) G_{ik}(P,Q) \frac{\partial^2 u_k(Q,t)}{\partial t^2} dV_Q
$$

$$
+ \alpha \int_V \frac{\partial G_{ik}(P,Q)}{\partial \xi_l} s_{jl}^{(0)}(Q) \frac{\partial u_k(Q)}{\partial \xi_j} dV_Q
$$

$$
+ \beta \Phi(t) \int_V \frac{\partial G_{ik}(P,Q)}{\partial \xi_l} s_{jl}^{(t)}(Q) \frac{\partial u_k(Q)}{\partial \xi_j} dV_Q, \tag{2.3.537}
$$

which will now be used as the basis of an approximate calculation. This is again justified by the certain convergence of the Galerkin method.

Let us see the trial solution

$$
u_i(P,t) = \sum_{m=1}^{\infty} f_m(t) \phi_i^{(m)}(P) \tag{2.3.538}
$$

of the Galerkin method for (2.3.537), where $\phi_i^{(m)}$ are the eigenvectors of the eigen-oscillation problem. The bilinear form (2.3.534) is used for the components of the Green tensor. Because of (2.3.532) we assume

$$
\int_V \mu(Q) G_{ik}(P,Q) \phi_k^{(m)}(Q) \frac{d^2 f_m}{dt^2} dV_Q = \frac{\phi_i^{(m)}(P)}{\lambda_m} \frac{d^2 f_m}{dt^2}.
$$

The comparison of the coefficients of the $\phi_i^{(m)}(P)$ results in the following system of differential equations:

$$
(1/\lambda_{(m)}\ddot{f}_m + f_m - \alpha a_{mn} f_n - \beta \Phi(t) b_{mn} f_n = 0, \qquad m = 1, 2, 3, \ldots, \tag{2.3.539}
$$

with

$$
a_{mn} = \frac{1}{\lambda_m} \int_V s_{jl}^{(0)} \frac{\partial \phi_k^{(m)}}{\partial \xi_j} \frac{\partial \phi_k^{(n)}}{\partial \xi_l} dV,
$$

$$
\tag{2.3.540}
$$

$$
b_{mn} = \frac{1}{\lambda_m} \int_V s_{jl}^{(0)} \frac{\partial \phi_k^{(m)}}{\partial \xi_j} \frac{\partial \phi_k^{(n)}}{\partial \xi_l} dV.
$$

Equations (2.3.539) are completely analogous to the system of differential equations (2.3.40) found for the theory of elastostatic stability problems. However, there is an important difference between (2.3.539) and (2.3.40). The coefficient in the last term of (2.3.539) is *variable* and usually periodic. This means that a different and more complicated method of solution is necessary than the one we found in elastostatics. Let us outline the case of periodic coefficients.

Equation (2.3.539) can also be written in the form

$$\ddot{f}_i + \Phi_{ik}(t)f_k = 0, \qquad i, k = 1, 2, 3, \ldots, \qquad (2.3.541)$$

where the abbreviation

$$\Phi_{ik}(t) = \lambda_{(i)}[\delta_{ik} - \alpha a_{ik} - \beta\Phi(t)b_{ik}]$$

has been used. It is also assumed that the Φ_{ik} are *periodic*, i.e., $\phi(t + T) = \Phi(t)$. If the new variables

$$\xi_i = f_i \qquad (i = 1, 2, \ldots, n), \qquad \xi_i = \dot{f}_{i-n} \qquad (i = n + 1, n + 2, \ldots, 2n)$$

are introduced, Eq. (2.3.541) is transformed into the linear system of differential equations

$$\dot{\xi}_i = \xi_{n+i}, \qquad i = 1, 2, \ldots, n,$$

$$\dot{\xi}_i = -\Phi_{i-n,k}\xi_k \qquad i = n + 1, n + 2, \ldots, 2n$$

which has periodic coefficients. This system may now be discussed as in Section 1.3.3 for linear variational equations with periodic coefficients. This means that the *characteristic exponents* or multipliers must be analyzed in order to investigate stability.

Another method will be used in practice to determine regions of instability. Use will be made of the fact that the system (2.3.541) has periodic solutions with the period T or $2T$ at the boundaries of the instability regions. Instability regions are bounded by two solutions with the same period and stability regions are bounded by two solutions with different periods. The critical value of the parameter contained in (2.3.541) is therefore easily obtained from the condition that (2.3.541) have periodic solutions.

In order to outline the calculation, let us set $\Phi(t) \equiv \cos \tau t$ in (2.3.539). When the trial solution

$$f(t) = \sum_{k=1,3,5} \left(A_k \sin \frac{k\tau t}{2} + B_k \cos \frac{k\tau t}{2} \right),$$

for which $f(t) = [f_i(t)]$, A_k, B_k are vectors, is substituted into (2.3.539), the following system of equations is obtained by a comparison of the coefficients of $\sin(k\tau t/2)$, $\cos(k\tau t/2)$:

$$(\mathbf{E} - \alpha\mathbf{A} + \tfrac{1}{2}\beta\mathbf{B} - \tfrac{1}{4}\tau^2\mathbf{C})\mathbf{A}_1 - \tfrac{1}{2}\beta\mathbf{B}\mathbf{A}_3 = 0$$

$$(\mathbf{E} - \alpha\mathbf{A} - \tfrac{1}{4}k^2\tau^2\mathbf{C})\mathbf{A}_k - \tfrac{1}{2}\beta\mathbf{B}(\mathbf{A}_{k-2} + \mathbf{A}_{k+2}) = 0, \qquad k = 3, 5, \ldots,$$

$$(\mathbf{E} - \alpha\mathbf{A} - \tfrac{1}{2}\beta\mathbf{B} - \tfrac{1}{4}\tau^2\mathbf{C})\mathbf{B}_1 - \tfrac{1}{2}\beta\mathbf{B}\mathbf{B}_3 = 0,$$

$$(\mathbf{E} - \alpha\mathbf{A} - \tfrac{1}{4}k^2\tau^2\mathbf{C})\mathbf{B}_k - \tfrac{1}{2}\beta\mathbf{B}(\mathbf{B}_{k-2}(\mathbf{B}_{k-2} + \mathbf{B}_{k+2}) = 0, \qquad k = 3, 5, \ldots,$$

with the matrices

$$\mathbf{E}, \mathbf{A} = (a_{ik}), \qquad \mathbf{B} = (b_{ik}), \qquad \mathbf{C} = (1/\lambda_i), \text{ diagonal}$$

and the vectors A_k, B_k. Solutions having the period $4\pi/\tau$ can occur if the conditions

$$\begin{vmatrix} \mathbf{E} - \alpha\mathbf{A} \pm \frac{1}{2}\beta\mathbf{B} - \frac{1}{4}\tau^2\mathbf{C} & -\frac{1}{2}\beta\mathbf{B} & 0 & \cdots \\ -\frac{1}{2}\beta\mathbf{B} & \mathbf{E} - \alpha\mathbf{A} - \frac{9}{4}\tau^2\mathbf{C} & -\frac{1}{2}\beta\mathbf{B} & \cdots \\ 0 & -\frac{1}{2}\beta\mathbf{B} & \mathbf{E} - \alpha\mathbf{A} - \frac{25}{4}\tau^2\mathbf{C} & \cdots \\ \cdots & \cdots & \cdots & \cdots \end{vmatrix} = 0$$

$$(2.3.542)$$

are satisfied. If the trial solution

$$f(t) = \frac{1}{2} B_0 + \sum_{k=2,4,6} A_k \sin(k_\tau t/2) + B_k \cos(k_\tau t/2)$$

is substituted into (2.3.539), the following conditions for the occurrence of solutions having the period $2\pi/\tau$ are obtained for $\Phi(t) = \cos \tau t$:

$$\begin{vmatrix} \mathbf{E} - \alpha\mathbf{A} - \tau^2\mathbf{C} & -\frac{1}{2}\beta\mathbf{B} & 0 & \cdots \\ -\frac{1}{2}\beta\mathbf{B} & \mathbf{E} - \alpha\mathbf{A} - 4\tau^2\mathbf{C} & -\frac{1}{2}\beta\mathbf{B} & \cdots \\ 0 & -\frac{1}{2}\beta\mathbf{B} & \mathbf{E} - \alpha\mathbf{A} - 16\tau^2\mathbf{C} & \cdots \\ \cdots & \cdots & \cdots & \cdots \end{vmatrix} = 0 \qquad (2.3.543)$$

and

$$\begin{vmatrix} \mathbf{E} - \alpha\mathbf{A} & -\beta\mathbf{B} & 0 & 0 & \cdots \\ -\frac{1}{2}\beta\mathbf{B} & \mathbf{E} - \alpha\mathbf{A} - \tau^2\mathbf{C} & -\frac{1}{2}\beta\mathbf{B} & 0 & \cdots \\ 0 & -\frac{1}{2}\beta\mathbf{B} & \mathbf{E} - \alpha\mathbf{A} - 4\tau^2\mathbf{C} & -\frac{1}{2}\beta\mathbf{B} & \cdots \\ 0 & 0 & -\frac{1}{2}\beta\mathbf{B} & \mathbf{E} - \alpha\mathbf{A} - 16\tau^2\mathbf{C} & \cdots \\ \cdots & \cdots & \cdots & \cdots & \cdots \end{vmatrix} = 0.$$

$$(2.3.544)$$

The *critical parameter values* are obtained from (2.3.542), (2.3.543), and (2.3.544). In the actual calculation these infinite determinants are truncated; and finite, reduced determinants are used. This is allowable, because the infinite determinants which occur are normal and therefore *convergent*, which was proven by Bolotin (1.7). The reduction method is allowed for normal determinants. It is often sufficient to know the *principal instability regions* that result for $k = 1$. An approximation for them is obtained if the trial solution

$$f(t) = \mathbf{A} \sin(\tau t/2) + \mathbf{B} \cos(\tau t/2)$$

of the harmonic balance method is substituted into (2.3.539). The approximate critical parameter values then follow from the condition

$$\det(\mathbf{E} - \alpha\mathbf{A} \pm \tfrac{1}{2}\beta\mathbf{B} - \tfrac{1}{4}\tau^2\mathbf{C}) = 0. \qquad (2.3.545)$$

This condition is expanded as follows and written as two lines and two columns

$$\begin{vmatrix} 1 - \alpha a_{11} \pm \tfrac{1}{2}\beta b_{11} - (\tau^2/4\lambda_1) & -\alpha a_{12} \pm \tfrac{1}{2}\beta b_{12} \\ -\alpha a_{21} \pm \tfrac{1}{2}\beta b_{21} & 1 - \alpha a_{22} \pm \tfrac{1}{2}\beta b_{22} - (\tau^2/4\lambda_2) \end{vmatrix} = 0.$$

The book by Bolotin (1.7) contains an excellent and very detailed discussion of stability problems of elastokinetics as well as a large number of examples. The interested reader is referred to this book.

We will describe the previous theory by means of the following classical problem.

Example: Let us consider a straight column with a pulsating longitudinal load (Fig. 99). The differential equation for the transverse vibrations is

$$EJ\frac{\partial^4 w}{\partial x^4} + (P_0 + P_t \cos \tau t)\frac{\partial^2 w}{\partial x^2} + \mu\frac{\partial^2 w}{\partial t^2} = 0. \qquad (2.3.546)$$

The trial solution $w(x,t) = f_m(t)\phi_m(x)$ is used, where

$$\phi_k = \sin(k\pi x/l)$$

are the eigenfunctions of the free column oscillations:

$$\phi_k'' = -(k^2\pi^2/l^2)\phi_k, \qquad \phi_k^{IV} = (k^4\pi^4/l^4)\phi_k. \qquad (2.3.547)$$

Equations (2.3.547) and (2.3.546) result in

$$EJ(k^4\pi^4/l^4)f_k(t)\phi_k(x) - (P_0 + P_t \cos \tau t)(k^2\pi^2/l^2)f_k(t)\phi_k(x)$$
$$+ \mu\ddot{f}_k(t)\phi_k(x) = 0,$$

Fig. 99. Buckling column with a pulsating compressive load

so that the following system of differential equations is obtained for the $f_k(t)$:

$$\mu \ddot{f}_k(t) + \left[\frac{k^4 \pi^4 EJ}{l^4} - (P_0 + P_t \cos \tau t) \frac{k^2 \pi^2}{l^2} \right] f_k(t) = 0, \qquad (2.3.548)$$

which corresponds to Eq. (2.3.539). A simple transformation of (2.3.548) results in

$$\ddot{f}_k(t) + \frac{k^2 \pi^2 (EJk^2 \pi^2 - P_0 l^2)}{\mu l^4} \left(1 - \frac{l^2 P_t}{EJk^2 \pi^2 - P_0 l^2} \cos \tau t \right) f_k(t) = 0, \quad (2.3.549)$$

which is now equivalent to the system (2.3.541). A comparison of (2.3.549) and (2.3.541) shows that we must have

$$\alpha = 0, \qquad \beta = 1, \qquad \mathbf{B} = (b_m), \text{ diagonal}, \quad b_m = \frac{l^2 P_t}{EJm^2 \pi^2 - l^2 P_0}$$

$$\mathbf{C} = \left(\frac{1}{\lambda_m} \right), \text{ diagonal}, \quad \lambda_m = \frac{m^2 \pi^2}{\mu l^4} (EJm^2 \pi^2 - P_0 l^2) \qquad (2.3.550)$$

for the rest of the calculation. The principal regions of instability are obtained from (2.3.545). When this condition is formulated for the assumptions (2.3.550) of the preceding problem, we have

$$\det(\mathbf{E} \pm \tfrac{1}{2} \mathbf{B} - (\tau^2/4)\mathbf{C}) = 0.$$

Since the matrices are all diagonal, we simply obtain

$$1 \pm \tfrac{1}{2} b_m - \tau^2/4 \lambda_m = 0.$$

The following approximate equation for the boundaries of the principal region is obtained for $m = 1$:

$$1 \pm \frac{l^2 P_t}{+ (EJ\pi^2 - l^2 P_0)} - \frac{\tau^2 \mu l^4}{4\pi^2 (EJ\pi^2 - P_0 l^2)} = 0,$$

which finally results in the well-known equation

$$\tau_{\text{crit}} = \frac{2\pi}{l^2} \left(\frac{EJ\pi^2 - P_0 l^2}{\mu} \right)^{1/2} \left(1 \pm \frac{P_t l^2}{+ (EJ\pi^2 - P_0 l^2)} \right)^{1/2}.$$

In this simple case, it is possible to give the exact solution, because equation (2.3.549) is the Mathieu differential equation, the solution of which is known. Usually examples that occur in practice are so complicated that it is not possible to make an exact calculation, and the above approximate calculation must be carried out.

Exercises 2.3

1) Using the classical energy method, derive a bound for the buckling load of the clamped–free rod subjected to a conservative (unidirectional) load

q that is uniformly distributed over the length of the rod. Prove that for a load q with a magnitude below that above mentioned bound, deflections and slopes of the rod remain sufficiently small.

2) Consider a membrane in an airstream which is infinitely long in the y-direction and has the length l in the x-direction. The membrane is supported along the edges in the y-direction and is subjected to compressive forces q which act in the x-direction and are uniformly distributed over the supported edges.

This boundary-value problem is mathematically defined as

$$\mu\ddot{w} + \alpha w^{IV} + qw'' + M^*w' = 0,$$

$$w = w(x, t),$$

$$w(0, t) = w(l, t) = w''(0, t) = w''(l, t) = 0,$$

μ, α are structural constants and M^* is another load parameter stemming from the effect of the airstream. Using Galerkin's method, determine the stability domain in the q, M^*-plane.

3) The boundary eigenvalue problem of an elastically supported rod with varying flexural rigidity and subject to a compressive longitudinal force is given by

$$[(4 - x^2)y'']'' + 50y = -\lambda y''$$

$$y(-1) = y''(-1) = y(+1) = y''(+1) = 0.$$

Using Galerkin's method, find the first critical value of λ.

4) Consider a hinged rod of length l, mass per unit length μ, flexural rigidity α, subjected to an axial torque T. Setting $\lambda_1 = T/\alpha$, $\lambda_0 = \mu\omega^2/\alpha$, the boundary value problem describing the small vibrations of the rod about its trivial equilibrium reads

$$y^{IV} + \lambda_1 i y''' - \lambda_0 y = 0,$$

$$y_0 = y_l = y_0'' = y_l'' = 0,$$

where i is the imaginary unit.

Using Galerkin's method calculate the lowest buckling load. Show also that all matrices involved are either symmetric or symmetrizable so that the static stability criterion is applicable to this problem.

5) Consider an elastic beam of length l, bending stiffness α, mass per unit length μ, subjected to uniformly distributed follower forces q. Let the lateral deflections of the beam be $w(x, t)$. Then, for the clamped–clamped beam, the eigenvalue-boundary value problem of the small vibrations of the beam about its trivial equilibrium position is given by

$$\mu\ddot{w} + \alpha w^{IV} + q(l - x)w'' + \beta_1, \dot{w}^{IV} + \beta_2\dot{w} = 0,$$

$$w(0, t) = w(l, t) = w'(0, t) = w'(l, t) = 0,$$

where β_1 and β_2 are, respectively, coefficients of internal and external damping.

Using a two-term Galerkin approach, show that for small damping, condition $a_{12}a_{21} > 0$, with

$$a_{ik} = \frac{1}{\mu} \left\{ \alpha \varkappa_i \, \delta_{ik} + q \int_0^l [(l - x)\phi_i'' \phi_k] \, dx \right\},$$

$$\varkappa_i = \phi_i^{IV}/\phi_i, \qquad \phi_i = \text{coordinate functions},$$

is sufficient for the stability condition of the undamped beam to already guarantee the stability of the damped beam. In other words: damping can be disregarded when calculating the stability range of this beam. By the way, for the hinged–hinged beam this is true for any amount of damping. Why?

6) For the clamped–free rod with given gravitational acceleration, total mass and follower force distribution, find the mass distribution such that the critical value for the follower force is maximized.

Let the mass distribution be of the form

$$m(x) = 0.25 + 0.75 \, \delta(x - \beta l),$$

where $0 \leqslant \beta \leqslant 1$. Assume a follower force distribution

$$F(x) = 2xf/l,$$

and neglect the weight.

Show that the governing mode of instability is flutter for all β, that the optimal flutter load is $f_{\text{flut}} = 57.5a/l^3$, and that this optimal value occurs for $\beta = 0.54$.

7) Using the extended version of the energy method, calculate the first flutter load of the clamped–hinged beam subjected to uniformly distributed follower-forces.

8) Using Lyapunov's direct method in its version appropriate for continuous bodies, find the stability conditions for a cylindrical shell with fixed ends, subjected to follower forces tangential to the shell's axis and uniformly distributed over the shell's wall.

Use as a Lyapunov functional

$$V = H + \int_S \left(\frac{\beta^2}{4\mu} w^2 + \frac{\beta}{2} w\dot{w} \right) \, dS,$$

where H is the Hamiltonian, β is the damping coefficient, μ is the mass per unit surface, S is the shell's surface, and w is the displacement vector $w = (u, v, w)$.

9) Consider a clamped–free elastic rod subjected to a conservative, undirectional compressive force P at its tip.

Let v be the lateral deflection, μ the mass per unit length, α the bending stiffness, l the length, then the Hamiltonian H of the system reads

$$H = \frac{1}{2} \int_0^l \left\{ \mu \dot{v}^2 + \alpha \, \frac{(v'')^2}{1 - (v')^2} + 2P[(1 - (v')^2)^{1/2} - 1] \right\} dx$$

$$\dot{v} = \frac{\partial v}{\partial t}, \qquad v' = \frac{\partial v}{\partial x_1}, \text{ etc.,}$$

taking geometric non-linearities into account.

Using Lyapunov's technique, show that in the case of this rod linearization is permissible as the linearized problem yields the same stability limit as the non-linear problem.

10) A clamped–hinged rod of length l, mass per unit length μ, bending stiffness α, is subjected to a compressive, longitudinal force $P(t) = P_0 + P_t \cos \theta t$. Calculate the first three critical frequencies θ if $P_0 = q\alpha/l^2$, $P_t = 0.3 P_0$.

References

(1) [1] Malkin, J. G., *Stability Theory of Motion*, Oldenbourg, Munich, 1959.

[2] Hahn, W., *Stability of Motion*, Springer, New York, 1967.

[3] LaSalle, J., and Lefschetz, S., *Stability by Lyapunov's Direct Method with Applications*, Academic Press, New York, 1961.

[4] Lefschetz, S., *Stability of Nonlinear Control Systems*, Academic Press, New York, 1965.

[5] Burgermeister, G., Steup, H., and Kretzschmar, H., *Stability Theory*, Vols. I, II, Deutscher Verlag der Wissenshaften, Berlin, 1957, 1963.

[6] Timoshenko, ST. P., and Gere, J. M., *Theory of Elastic Stability*, McGraw-Hill, New York, 1961.

[7] Bolotin, V. V., *The Dynamic Stability of Elastic Systems*, Holden-Day, Inc., San Francisco, London, Amsterdam, 1964.

[8] Bolotin, V. V., *Nonconservative Problems of the Theory of Elastic Stability*, Pergamon Press, Oxford, 1963.

[9] Kollbruner, C. F., and Meister, M., *Buckling*, Springer, Berlin, 1955.

[10] Letov, A. M., *Stability in Nonlinear Control Systems*, Princeton University Press, Princeton, New Jersey, 1961.

(2) [1] Minorsky, N., *Nonlinear Oscillations*, Van Nostrand, Princeton, New Jersey, 1962.

[2] Whittaker, E. T., *Analytic Dynamics of Particles and Rigid Bodies*, Cambridge University Press, London and New York, 1959.

[3] Klein, F., and Sommerfeld, A., *The Theory of the Gyroscope*, Johnson Reprint Corp. (Teubner), New York, Stuttgart 1965.

[4] Klotter, K., *Technical Vibration Theory*, Vol. 2. Springer, Berlin, 1960.

[5] Solodovnikov, V. V., *Foundations of Autonomous Control*, Oldenbourg, Munich, 1959.

[6] Wang, C. -T., *Applied Elasticity*, McGraw-Hill, New York, 1953.

[7] Budó, A., *Theoretical Mechanics*, Springer, Berlin, 1963.

[8] Szabó, I., *Advanced Technical Mechanics*, Springer, Berlin, 1960, and Goldstein, H., *Classical Mechanics* 2nd edn., Addison-Wesley, Reading, Mass., Menlo Park, Ca., London, Amsterdam, Don Mills, Ontario, Sydner, 1981.

[9] Cunningham, W. J., *Introduction to Nonlinear Analysis*, McGraw-Hill, New York, 1958.

[10] Kauderer, H., *Non-Linear Mechanics*, Springer, Berlin, 1958.

[11] Oppelt, W., *Small Handbook of Technical Control Processes*, Verlag Chemie, Weinheim, 1964.

[12] Popov, E. P., *Dynamics of Automatic Control Systems*, Pergamon Press, Oxford, 1958.

[13] Popov, E. P., and Paltov, J. P., *Approximate Methods for the Investigation of Non-Linear Control Systems*, Akademische Verlag Gesellschaft, Leipzig, 1963.

[14] Magnus, K., *Vibrations.*, Teubner, Stuttgart, 1961.

(3) Zurmühl, R., *Applied Mathematics*, p. 782. Springer, Berlin, 1953, and Gantmacher, F. R., *The Theory of Matrices*, Vol. 2, Chelsea Publishing Company, New York, 1960.

(4) Kamke, E., *Differential Equations of Real Functions*, pp. 120–193, Akademische Verlag Gesellschaft Geest and Portig, K. -G., Leipzig, 1952.

(5) Duschek, A., *Advanced Mathematics*, Vol. IV, p. 217, Springer, Vienna, 1961.

(6) Lanczos, C., *The Variational Principles of Mechanics*, University of Toronto Press, Toronto, 1964.

(7) Frank, P., and Von Mises, R., *Differential and Integral Equations of Mechanics and Physics*, Vol. II, pp. 44–190, Vieweg, Braunschweig, New York, 1961.

(8) [1] Pflüger, A., *Stability Problems of Elastostatics*, Springer, Berlin, 1964.

[2] Bürgermeister, G., Steup, H. and Kretzschmar, H., *Stability Theory*, Vols. I, II Deutscher Verlag der Wissenschaften, Berlin, 1957, 1963.

[3] Gerard, G., *Introduction to Structural Stability Theory*, McGraw-Hill, New York, 1962.

[4] Marguerre, K., *Modern Problems for Engineers*, pp. 189–249, Springer, Berlin, 1950.

[5] Langhaar, H. L., *Energy Methods in Applied Mechanics*, pp. 201–232, Wiley, New York, London, 1962.

(9) Tricomi, R., *Elliptical Functions*, pp. 263–271, Akademische Verlag Gesellschaft Geest und Portig, K. -G., Leipzig, 1948.

(10) Wang, C. -T., *Applied Elasticity*, pp. 217–222, McGraw-Hill, New York, 1953.

(11) Koiter, W. T., Elastic stability and post-buckling behaviour, in *Nonlinear Problems* (R. E. Langer, ed.), pp. 257–275, University of Wisconsin Press, Madison, Wisconsin, 1963.

(12) Michlin, S. G., *Variational Methods of Mathematical Physics*, pp. 365–489, Akademie Verlag, Berlin, 1962.

(13) Bisplinghoff, R. L., and Ashley, H., *Principles of Aeroelasticity*, pp. 418–424, Wiley, New York, 1962.

Author Index

Subject Index